3D Concrete Printing Technology

Construction and Building Applications

3D Concrete Printing Technology

Construction and Building Applications

Edited by

JAY G. SANJAYAN
Centre for Sustainable Infrastructure, Faculty of Science, Engineering and Technology, Swinburne University of Technology, Hawthorn, VIC, Australia

ALI NAZARI
Centre for Sustainable Infrastructure, Faculty of Science, Engineering and Technology, Swinburne University of Technology, Hawthorn, VIC, Australia

BEHZAD NEMATOLLAHI
Centre for Sustainable Infrastructure, Faculty of Science, Engineering and Technology, Swinburne University of Technology, Hawthorn, VIC, Australia

ELSEVIER

Butterworth-Heinemann
An imprint of Elsevier

Butterworth-Heinemann is an imprint of Elsevier
The Boulevard, Langford Lane, Kidlington, Oxford OX5 1GB, United Kingdom
50 Hampshire Street, 5th Floor, Cambridge, MA 02139, United States

British Library Cataloguing-in-Publication Data
A catalogue record for this book is available from the British Library

Library of Congress Cataloging-in-Publication Data
A catalog record for this book is available from the Library of Congress

ISBN: 978-0-12-815481-6

For Information on all Butterworth-Heinemann publications
visit our website at https://www.elsevier.com/books-and-journals

Publisher: Matthew Deans
Acquisition Editor: Ken McCombs
Editorial Project Manager: Peter Adamson
Production Project Manager: Omer Mukthar
Cover Designer: Greg Harris

Typeset by MPS Limited, Chennai, India

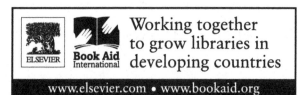

DEDICATION

3D concrete printing is an entirely new field and putting it into practice in real life requires years of efforts. Not only equipment and facilities are necessary, but trained researchers, industry professionals, and investors are essential to develop this exciting technology. People who are currently developing 3D concrete printing are the pioneers in this field whose perseverance and belief in the technology will definitely be appreciated by the wider community in the not-so-distant future. This book is dedicated to all those who believe in the technology and make serious efforts to make 3D concrete printing a commercially successful technology.

CONTENTS

LIST OF CONTRIBUTORS

Ghassan K. Al-Chaar
Construction Engineering Research Laboratory, US Army Engineer Research and Development Center, Champaign, IL, United States

Fatima AlSakka
Civil and Environmental Engineering, American University of Beirut, Beirut, Lebanon

Daniel Avrutis
Centre for Sustainable Infrastructure, Faculty of Science, Engineering and Technology, Swinburne University of Technology, Hawthorn, VIC, Australia

Lynette A. Barna
Cold Regions Research and Engineering Laboratory, US Army Engineer Research and Development Center, Hanover, NH, United States

Dale P. Bentz
Materials and Structural Systems Division, Engineering Laboratory, National Institute of Standards and Technology, Gaithersburg, MD, United States

Isaiah R. Bentz
Materials and Structural Systems Division, Engineering Laboratory, National Institute of Standards and Technology, Gaithersburg, MD, United States

C. Bouyssou
XtreeE, Rungis, France

Jedadiah F. Burroughs
Geotechnical and Structures Laboratory, US Army Engineer Research and Development Center, Vicksburg, MS, United States

Xiangpeng Cao
Shenzhen Mingyuan Building Technology Co., Ltd., Shenzhen, P.R. China

Michael P. Case
Construction Engineering Research Laboratory, US Army Engineer Research and Development Center, Champaign, IL, United States

Jian-Fei Chen
School of Planning, Architecture and Civil Engineering, Queen's University Belfast, Northern Ireland, United Kingdom

J. Dirrenberger
XtreeE, Rungis, France; Laboratoire PIMM, Arts et Métiers-ParisTech, Cnam, CNRS UMR 8006, Paris, France

R. Duballet
XtreeE, Rungis, France; Laboratoire Navier, UMR 8205, Ecole des Ponts, IFSTTAR, CNRS, UPE, Paris, France

Laurie Edwards
Boral Innovation Factory, Australia

Peng Feng
Department of Civil Engineering, Tsinghua University, Beijing, P.R. China

N. Gaudillière
XtreeE, Rungis, France; Laboratoire GSA, Ecole Nationale Supérieure d'Architecture Paris-Malaquais, Paris, France

Manuel Hambach
Chair of Solid State and Materials Chemistry, University of Augsburg, Augsburg, Germany

Farook Hamzeh
Civil and Environmental Engineering, American University of Beirut, Beirut, Lebanon

Camille Holt
Boral Innovation Factory, Australia

Young Kwang Hwang
School of Civil and Environmental Engineering, College of Engineering, Yonsei University, Seoul, Republic of Korea

Scott Z. Jones
Materials and Structural Systems Division, Engineering Laboratory, National Institute of Standards and Technology, Gaithersburg, MD, United States

Ali Kazemian
The Sonny Astani Department of Civil and Environmental Engineering, University of Southern California, Los Angeles, CA, United States; Department of Computer Science, University of Southern California, Los Angeles, CA, United States; Contour Crafting Corporation, El Segundo, CA, United States

Louise Keyte
Boral Innovation Factory, Australia

Behrokh Khoshnevis
The Sonny Astani Department of Civil and Environmental Engineering, University of Southern California, Los Angeles, CA, United States; Contour Crafting Corporation, El Segundo, CA, United States; Department of Industrial and Systems Engineering, University of Southern California, Los Angeles, CA, United States

Megan A. Kreiger
Construction Engineering Research Laboratory, US Army Engineer Research and Development Center, Champaign, IL, United States

Mingyang Li
Singapore Centre for 3D Printing, School of Mechanical and Aerospace Engineering, Nanyang Technological University, Singapore, Singapore

Zhijian Li
College of Architecture and Civil Engineering, Beijing University of Technology, Beijing, P.R. China; School of Civil Engineering and Transportation, Hebei University of Technology, Tianjin, P.R. China

Zongjin Li
Institute of Applied Physics and Materials Engineering, University of Macau, Macau, P.R. China

Yun Mook Lim
School of Civil and Environmental Engineering, College of Engineering, Yonsei University, Seoul, Republic of Korea

Guowei Ma
School of Civil Engineering and Transportation, Hebei University of Technology, Tianjin, P.R. China; School of Civil, Environmental and Mining Engineering, The University of Western Australia, Crawley, WA, Australia

Zeina Malaeb
Civil and Environmental Engineering, American University of Beirut, Beirut, Lebanon

A. Mallet
XtreeE, Rungis, France

Taylor Marchment
Centre for Sustainable Infrastructure, Faculty of Science, Engineering and Technology, Swinburne University of Technology, Hawthorn, VIC, Australia

Viktor Mechtcherine
Technische Universität Dresden, Institute of Construction Materials, Dresden, Germany

Ryan Meier
The Sonny Astani Department of Civil and Environmental Engineering, University of Southern California, Los Angeles, CA, United States

Xinmiao Meng
Department of Civil Engineering, Beijing Forestry University, Beijing, P.R. China

Farzad Moghaddam
Boral Innovation Factory, Australia

Young Jun Nam
School of Civil and Environmental Engineering, College of Engineering, Yonsei University, Seoul, Republic of Korea

Ali Nazari
Centre for Sustainable Infrastructure, Faculty of Science, Engineering and Technology, Swinburne University of Technology, Hawthorn, VIC, Australia

Behzad Nematollahi
Centre for Sustainable Infrastructure, Faculty of Science, Engineering and Technology, Swinburne University of Technology, Hawthorn, VIC, Australia

Venkatesh Naidu Nerella
Technische Universität Dresden, Institute of Construction Materials, Dresden, Germany

Ji Woon Park
School of Civil and Environmental Engineering, College of Engineering, Yonsei University, Seoul, Republic of Korea

Max A. Peltz
Materials and Structural Systems Division, Engineering Laboratory, National Institute of Standards and Technology, Gaithersburg, MD, United States

Shunzhi Qian
Singapore Centre for 3D Printing, School of Mechanical and Aerospace Engineering, Nanyang Technological University, Singapore, Singapore; School of Civil and Environment Engineering, Nanyang Technological University, Singapore, Singapore

Ph. Roux
XtreeE, Rungis, France

Todd S. Rushing
Geotechnical and Structures Laboratory, US Army Engineer Research and Development Center, Vicksburg, MS, United States

Matthias Rutzen
Chair of Solid State and Materials Chemistry, University of Augsburg, Augsburg, Germany

Jay G. Sanjayan
Centre for Sustainable Infrastructure, Faculty of Science, Engineering and Technology, Swinburne University of Technology, Hawthorn, VIC, Australia

Jameson D. Shannon
Geotechnical and Structures Laboratory, US Army Engineer Research and Development Center, Vicksburg, MS, United States

Peter B. Stynoski
Construction Engineering Research Laboratory, US Army Engineer Research and Development Center, Champaign, IL, United States

Ming Jen Tan
Singapore Centre for 3D Printing, School of Mechanical and Aerospace Engineering, Nanyang Technological University, Singapore, Singapore

Belinda Townsend
Boral Innovation Factory, Australia

Praful Vijay
Institute of Construction Materials, Faculty of Civil Engineering, TU Dresden, Dresden, Germany

Dirk Volkmer
Chair of Solid State and Materials Chemistry, University of Augsburg, Augsburg, Germany

Li Wang
School of Civil Engineering and Transportation, Hebei University of Technology, Tianjin, P.R. China

Yiwei Weng
Singapore Centre for 3D Printing, School of Mechanical and Aerospace Engineering, Nanyang Technological University, Singapore, Singapore; School of Civil and Environment Engineering, Nanyang Technological University, Singapore, Singapore

Ming Xia
Centre for Sustainable Infrastructure, Faculty of Science, Engineering and Technology, Swinburne University of Technology, Hawthorn, VIC, Australia

Lieping Ye
Department of Civil Engineering, Tsinghua University, Beijing, P.R. China

Xiao Yuan
Contour Crafting Corporation, El Segundo, CA, United States

M. Zakeri
XtreeE, Rungis, France

PREFACE

Construction-related spending is $10 trillion globally, equivalent to 13% of GDP. This makes construction one of the largest sectors of the world economy. However, construction has suffered for decades from remarkably poor productivity relative to other sectors. Global labor-productivity growth in construction has averaged only 1% per year over the past two decades (and was flat in most advanced economies). Contrasted with growth of 2.8% in the world economy and 3.6% in manufacturing, this clearly indicates that the construction sector is underperforming. While many sectors including agriculture and manufacturing have increased productivity 10–15 times since the 1950s, the productivity of construction remains stuck at the same level as 80 years ago. Current measurements find that there has been a consistent decline in the industry's productivity since the late 1960s (McKinsey Global Institute, 2017). Construction remains largely manual, while manufacturing and other industries have made significant progress in the use of digital, sensing, and automation technologies.

In the past few years, many of us have witnessed the wide availability of 3D printers in consumer markets using various types of plastic filaments based on fused deposition modeling technology. 3D printers using metals are becoming commonplace in advanced industries. 3D printing was made famous by the former US President Barak Obama during his 2013 State of the Union Address as a technology that has "the potential to revolutionize the way we make almost everything." These events have generated significant interest among construction researchers and construction industry innovators in attempting 3D printing using concrete. They envision the 3D printing technology as a way of introducing much-needed automation in construction.

Unlike the conventional approach of casting concrete into a mold (formwork), 3D printing will combine digital technology and new insights from materials technology to allow free-form construction without the use of formwork. Recent advances in numerical control technology, sensing technology, and automatic driving systems with improved functions and accuracy are new enablers for 3D concrete printing. While the technology enthusiasts interested in 3D printing are optimistic, the technology is still in its infancy. The early researchers have identified the key challenges faced in making this technology commercially viable and attempted to solve by innovative techniques. This book presents these key techniques in its 19 chapters detailing the advances made in this field during the past few years. We hope the book will be valuable to those who want to quickly grasp the viability of this technology, the challenges faced, the future research and development needs, and the future directions where this technology can be commercially deployed for maximum benefits.

The main economic driver is the cost of formwork. Formwork contributes approximately 60% of the total cost of concrete construction and represents a significant source of waste, given that all of it is discarded sooner or later, contributing to a generally increasing amount of waste worldwide. Chapter 17, Construction 3D Printing, illustrates this point by comparing the cost of building a wall using 3D concrete printing technology which is estimated at only 22% of a wall constructed using conventional concrete using formwork systems. If these cost reductions become a commercial reality, 3D concrete printing will be a major disruptive technology for the construction sector. However, the road to success for this technology is riddled with many challenges which are discussed in this book.

Conventional concrete has been progressively optimized over more than a century where concrete is cast in formwork. The concrete for 3D printing has different requirements and a new form of concrete needs to be developed for this application. Chapter 1, 3D Concrete Printing for Construction Applications, summaries the current 3D printing technologies applicable in the construction industry. In Chapter 2, Performance-Based Testing of Portland Cement Concrete for Construction-Scale 3D Printing, performance-based test methods to develop Portland cement systems are described, including various new requirements such as shape stability, print quality, and printability window. Another challenge, most obvious immediately, is the reinforcements. While there are some technologies developed to place reinforcements in-line with the layers, there are not satisfactory techniques available for reinforcements perpendicular to the layers. As such, many researchers have focused on fiber reinforcements in concrete as an alternative to circumvent this problem. Chapters 3–5 focus on the fiber reinforcement technologies for 3D concrete printing. The machine design, particularly, the nozzle design, to suit the concrete placement is also an important factor where the research is still in progress (Chapter 6: 3D Concrete Printing: Machine Design, Mix Proportioning, and Mix Comparison Between Different Machine Setups).

The use of admixtures to satisfy the demands placed on the concrete are new frontiers where significant research needs to be done (Chapter 7: Investigation of Concrete Mixtures for Additive Construction). While extrusions-based 3D printing is the most popular so far, powder-based systems have also been developed. Chapter 8 gives methods for the enhancement of buildability and bending resistance of 3D printable tailing mortar. The powder-based systems can achieve significantly higher resolution than the extrusion-based method. These and other issues are described in Chapters 9–11 and 13. Chapter 19, Industrial Adoption of 3D Concrete Printing in the Australian Market: Potentials and Challenges, gives the potential and challenges that 3D construction printing might face in the market based on the strengths, weaknesses, opportunities, and threats available. The 3D printed concrete will have different structural properties to conventional concrete. Understanding these differences are important so that the structural engineers can design the 3D printing structures with confidence. These issues are dealt with in many of the chapters focusing on both the powder-based and extrusion-based methods.

This book can provide the reader with a quick overview of the state-of-the-art of the 3D concrete printing technology. This technology is moving fast and many new innovations are revealed on a regular basis. We hope the book will create an interest among researchers and industry innovators. We believe this is a "green field" opportunity for investors who want to participate in a technology which, we believe, will transform the construction industry during the coming decades.

Jay Sanjayan, Ali Nazari and Behzad Nematollahi

ACKNOWLEDGMENTS

We would like to express our gratitude and appreciation to Elsevier for their ongoing support in publishing this book. Working with Elsevier was a great journey from the beginning to the end. We received endless support from the team who managed us effortlessly in every publishing milestone. We also appreciate all the book's contributors who shared their experience and knowledge through writing invaluable book chapters. All of the contributors significantly invested their time to make the dream of publishing the first book on 3D concrete printing a reality.

CHAPTER 1

3D Concrete Printing for Construction Applications

Jay G. Sanjayan and Behzad Nematollahi
Centre for Sustainable Infrastructure, Faculty of Science, Engineering and Technology, Swinburne University of Technology, Hawthorn, VIC, Australia

1.1 INTRODUCTION

Construction is one of the largest sectors of the global economy with construction-related spending at $10 trillion globally, equivalent to 13% of GDP. However, construction has shown remarkably poor productivity gains relative to other sectors. Under these conditions, the global infrastructure and housing construction industry will lag behind and not meet the global demand [1]. This situation is further exacerbated in Australia, for example, by inefficiencies and difficulties of delivering appropriate infrastructure in remote areas. The cost of building an average house in remote communities in Australia is, on average, USD 600k [2], which is much more expensive than in urban regions.

Since the discovery of modern concrete in the 19th century, many researchers have sought to automate concrete construction without much success. Thomas Edison's attempt to create a machine to build concrete houses in a single pour, which he patented in 1917 (Fig. 1.1), was a well-documented failure due to technological challenges in concrete. The great inventor is said to have spent as much time with his concrete house project as with his other inventions, but the complexity of concrete eluded him. Concrete as a construction material appears deceptively simple, but has many hidden challenges. Many advancements in concrete construction technologies have been made since then, including innovative development in concrete pumping technologies and admixture technologies. However, it is also commonplace to find construction sites transporting, placing, compacting, and curing concrete using technologies that are more than 100 years old. Concrete construction remains labor intensive, costly, and highly accident prone.

3D Concrete Printing Technology
DOI: https://doi.org/10.1016/B978-0-12-815481-6.00001-4

Figure 1.1 Thomas Edison with his single pour concrete house.

Annual production of concrete is reaching nearly 30 billion tons worldwide [3], making it the most widely used construction material. However, the concrete itself plays only a partial role since the formwork represents 35%–60% of the overall cost of concrete construction [4]. Formwork is the temporary structure and mold for pouring wet concrete into and is typically built with timber. Formwork represents a significant source of waste, given that all formwork is discarded sooner or later, contributing to the generally increasing amount of waste worldwide. According to a study from 2011 [5], 80% of the total worldwide waste is generated in the construction industry, with significant contributions from formwork timber which has limited reuse value. Further, pouring concrete into formworks limits the creativity of architects to build in various geometries unless very high costs are paid for bespoke formworks. Unlike the conventional approach of casting into a mold, 3D concrete printing (3DCP) is an emerging technology that combines digital technologies and new insights from materials technologies to allow free-form construction without the use of formwork. 3DCP is a type of additive manufacturing technique where the construction is through layer-by-layer addition of material.

Two state-of-the-art processes currently leading the 3DCP field are [6]: (1) The Single Deposition Nozzle Concrete Printer which is similar to fused deposition modeling. Contour Crafting is another technology where concrete is extruded against trowel; and (2) powder deposition process where the "ink" is deposited on a powder bed. There is no clear winner at this stage. Further research in materials and structural forms will eventually decide the direction of the technology for the future. It is, therefore, important for researchers to experiment with both leading 3DCP technologies.

The lack in underpinning concrete materials technology research has been identified as hampering progress in 3DCP [7]. The four key characteristics of the fresh concrete relevant to 3D printing identified by Le et al. [8] are: (1) Pumpability: the ease and reliability with which material is moved through the delivery system; (2) Printability: the ease and reliability of depositing material through a deposition device; (3) Buildability: the resistance of deposited wet material to deformation under load; and (4) Open time: the period where these properties are consistent within acceptable tolerances.

1.2 3D PRINTING IN CONSTRUCTION

Pegna [9] was the first successful researcher who tried to adopt additive manufacturing in construction applications. The process was in the form of making structures with sand and then using cement as an adhesive. When compared with conventional construction processes, the application of 3D printing techniques in concrete construction may offer excellent advantages including:

1. Reduction of construction costs by eliminating formwork.
2. Reduction of injury rates by eliminating dangerous jobs (e.g., working at heights), which would result in an increased level of safety in construction.
3. Creation of high-end, technology-based jobs.
4. Reduction of onsite construction time by operating at a constant rate.
5. Minimizing the chance of errors by precise material deposition.
6. Increasing sustainability in construction by reducing wastages of formwork.
7. Increasing architectural freedom, which would enable more sophisticated designs for structural and esthetic purposes.

8. Enabling the potential of multifunctionality for structural/architectural elements by taking advantage of the complex geometry [10,11].

1.3 EXTRUSION-BASED 3D CONCRETE PRINTING

The extrusion-based 3DCP is similar to the fused deposition modeling used in polymer and metal technologies. Contour crafting is one of the proprietary terminologies used for the layered fabrication technology that has been under development for almost 15 years [12]. It is based on extruding a cement-based concrete against a trowel that allows a smooth surface finish created through the build-up of subsequent layers. The current deposition head is capable of laying down material to create a full-width structural wall (Fig. 1.2). Contour Crafting technology allows a variety of materials (such as mortar, concrete, cement pastes and fiber-reinforced concrete) to be investigated (Fig. 1.2). Contour Crafting was selected by NASA for its Innovative Advanced Concepts (NIAC) to explore the use of building a Lunar Settlement infrastructure [13].

Another fused deposition modeling type technology that was introduced during the early stages is led by the 3DCP group at Loughborough University which conducted a series of early trials on this technology [6].

1.3.1 Current Examples of Extrusion-Based 3D Concrete Printing Elements/Structures

In 2014, the Chinese Winsun company claimed to have built 10 basic houses in less than a day, with the area and cost of each being about

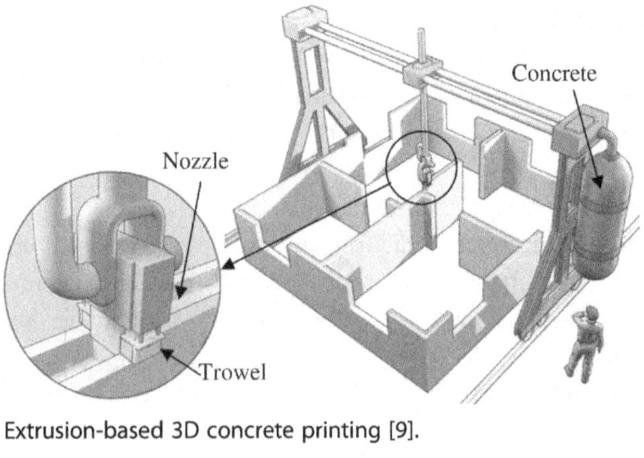

Figure 1.2 Extrusion-based 3D concrete printing [9].

195 m^2 and US$4800, respectively. The company used a large extrusion-based 3D printer to manufacture the basic house components separately offsite before they were transported and assembled on site [14]. In 2015, the company built a five-story apartment building with an area of about 1100 m^2, currently the tallest 3D printed structure. The company also claimed to have built a stand-alone concrete villa with interior fittings for a cost of about US$160,000. The company claimed to 3D print the walls and other components of the structure offsite and then assembled them together onsite [15].

The Chinese Huashang Tengda company in Beijing has recently claimed to 3D print an entire 400 m^2 two-story villa onsite within 45 days (see Fig. 1.3A). Unlike the Winsun company, the Huashang Tengda company used a unique process allowing to print an "entire house," "onsite" in "one go." The frame of the house, including conventioinal steel reinforcements and plumbing pipes, were first erected. Then, ordinary Class C30 concrete-containing coarse aggregates was extruded into the framework and around the rebars through the use of a novel nozzle design and a gigantic 3D printer [16]. The Huashang Tengda project seemingly eliminated one of the major challenges of 3DCP which is the incorporation of conventional steel reinforcements when structural concrete is to be 3D printed. The company claimed that the two-story villa is durable enough to withstand an earthquake measuring 8.0 on the Richter scale. Their giant 3D printer has a sort of forked nozzle (see Fig. 1.3B) that simultaneously lays concrete on both sides of the rebars, swallowing it up, and encasing it securely within the walls [16].

The researchers at the University Federico II of Naples, Italy used a 4 m high BIGDELTA WASP (World's Advanced Saving Project) printer to build the first modular, reinforced-concrete beam of about 3 m long

Figure 1.3 (A) The two-story villa 3D printed by Huashang Tengda company; and (B) the novel nozzle of the giant 3D printer [16].

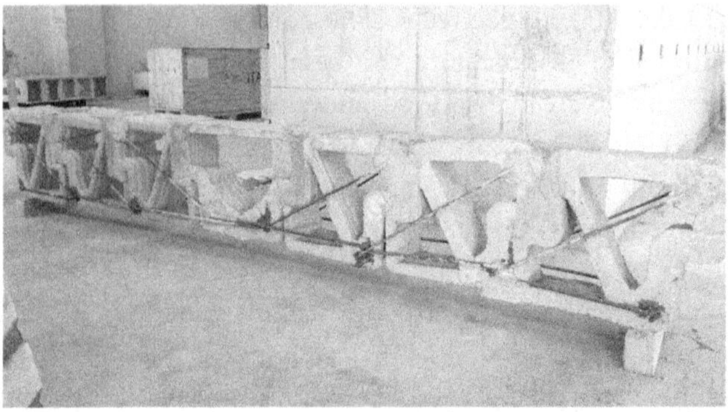

Figure 1.4 The first 3D printed modular reinforced concrete beam of about 3 m [17].

Figure 1.5 The Y-Box Pavilion, 21st-century Cave 3 m tall structure [18].

(see Fig. 1.4). With this WASP printer, the researchers have developed a system to produce concrete elements that can be assembled with steel bars and beams or can compose pillars in reinforced concrete [17].

As a result of collaboration between Supermachine Studio and the Siam Cement Group (SCG), a 3 m tall cave structure called the "Y-Box Pavilion, 21st-century Cave" was built in Thailand using the 4 m high BIGDELTA WASP printer (see Fig. 1.5). The components of the pavilion were 3D printed offsite at the SCG factory and then all the components were assembled together. The cost of manufacture of the pavilion was reported to be about US$28,000 [18].

In December 2016, the Apis Core company announced to have built the first onsite house in Russia using a mobile 3D concrete printer in just

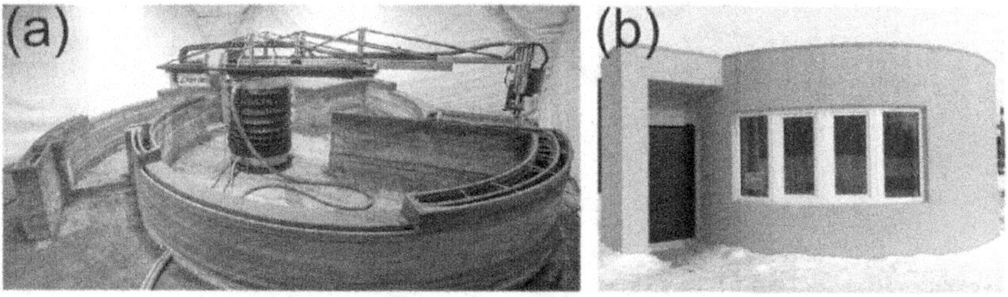

Figure 1.6 Onsite 3D printed house by Apis Core. (A) Construction using a mobile 3D concrete printer; (B) house exterior [19].

24 hours (see Fig. 1.6). The entire 38 m^2 house was 3D printed onsite. The total construction cost was claimed to be US$10,134 [19].

1.4 POWDER-BED-BASED 3D CONCRETE PRINTING

In the powder-bed process, a thin layer of powder is spread over the powder bed surface first. Once a layer is completed, binder droplets are selectively applied on the powder layer by a print-head causing powder particles to bind each other. By repeating these steps, the built part is completed and removed after a certain setting time. The unbound powder is then removed by using an air blower.

The D-shape 3D printing construction technology was introduced by Enrico Dini in Italy [20]. This process uses a powder deposition process by which the powder is selectively hardened using a binder in much the same way as the Z-Corp 3D printing process [21]. Each layer of building material is laid to the desired thickness, compacted and then the nozzles mounted on a gantry frame deposit the binder. The binders are deposited only in places where the building material should become solid and the rest is kept loose and removed at a later stage. Once a part is complete it is then dug out of the loose powder bed. The process has been used to create 1.6 m high architectural pieces called "Radiolaria" (Fig. 1.7).

1.5 EMERGING OBJECTS

The Emerging Objects technology developed in the United States uses the powder-based technique to selectively harden a proprietary cement composite formulation by deposition of a binding agent [22]. The technology

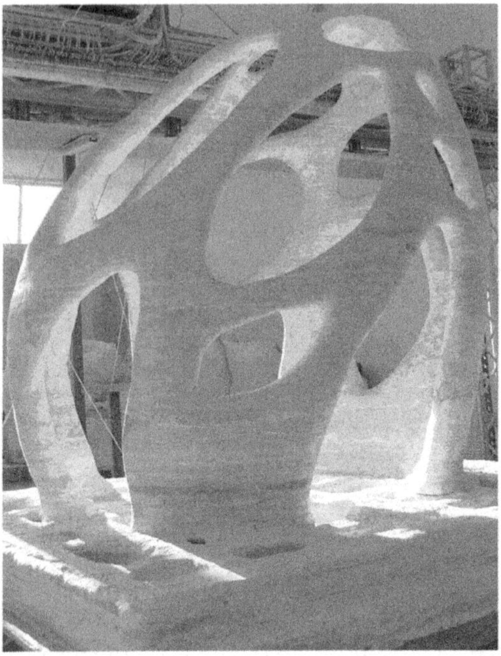

Figure 1.7 Sculpture by D-shape process [20].

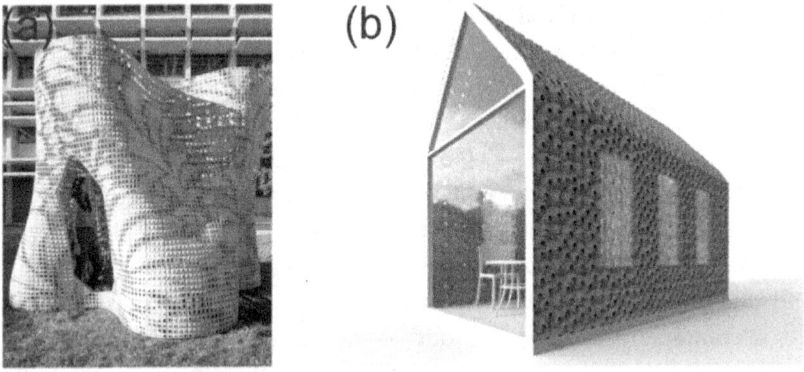

Figure 1.8 (A) Bloom [22] and (B) Shed [23] printed by Emerging Objects.

was used to manufacture Bloom (see Fig. 1.8A). Bloom is a 2.74 m tall, freestanding tempietto with a footprint that measures approximately 3.66 m by 3.66 m and is composed of 840 customized 3D printed blocks. Each block is printed using a farm of 11 powder 3D printers with a proprietary cement composite formulation comprised chiefly of iron oxide-free

OPC. The blocks are held in place using stainless steel hardware and assembled into 16 large, lightweight, prefabricated panels that can be assembled in just a few hours. The technology was also used to manufacture Shed (see Fig. 1.8B). Shed is a small 3D printed prototype building constructed with Picoroco Blocks, modular 3D printed building blocks for wall fabrication printed from sand measuring 0.3 × 0.3 × 0.3 m [23].

1.6 FUTURE DIRECTIONS AND CHALLENGES

Researchers and industry professionals currently working on 3D printing using concrete envision the technology to be a major disruptor to the construction industry. The vision for futuristic house construction (Fig. 1.9) and multistory buildings using 3DCP technology (Fig. 1.10) are some examples of the aspirations of those who are currently working in this field.

Western countries, in general, have higher levels of safety regulations in construction work than developing countries, with a significant cost to the construction industry. Despite this, the injury rate in the construction sector is still one of the highest when compared to other sectors. According to the 2014 data from the Australian Bureau of Statistics, 52 workers in 1000 were injured in construction sector. Similarly, in the United States this number was 40 in 1000. One of the main benefits of automated construction is its potential to meaningfully decrease the number of injuries and deaths in the construction sector by avoiding many of the dangerous and laborious tasks.

Figure 1.9 A vision for house construction.

Figure 1.10 A vision of multistory construction.

REFERENCES

[1] McKinsey Global Institute, Reinventing construction: a route to higher productivity, <http://www.mckinsey.com/>, 2017.
[2] The Auditor-General, Audit Report No. 12 2011-12 Performance Audit, Australian National Audit Office.
[3] CEMBUREAU Activity Report 2013, The European Cement Association, 44 p.
[4] D.W. Johnston, Design and construction of concrete formwork, in: G. Nawy Edward (Ed.), Concrete Construction Engineering Handbook, CRC Press, New Jersey, Boca Raton, 2008, pp. 7.1–7.49.
[5] C. Llatas, A model for quantifying construction waste in projects according to the European waste list, Waste Manage. 31 (6) (2011) 1083–1426.
[6] S. Lim, R.A. Buswell, T.T. Le, S.A. Austin, A.G.F. Gibb, T. Thorpe, Developments in construction-scale additive manufacturing processes, Autom. Constr. 21 (1) (2012) 262–268.
[7] L. Ding, R. Wei, H. Che, Development of a BIM-based automated construction system, Procedia Eng. 85 (C) (2014) 123–131.
[8] T.T. Le, S.A. Austin, S. Lim, R.A. Buswell, A.G.F. Gibb, T. Thorpe, Mix design and fresh properties for high-performance printing concrete, Mater. Struct. 45 (8) (2012) 1221–1232.
[9] J. Pegna, Exploratory investigation of solid freeform construction, Autom. Constr. 5 (5) (1997) 427–437.
[10] R.A. Buswell, R. Soar, A.G. Gibb, A. Thorpe, Freeform construction: mega-scale rapid manufacturing for construction, Autom. Constr. 16 (2007) 224–231.
[11] C. Gosselin, R. Duballet, Ph Roux, N. Gaudillière, J. Dirrenberger, Ph Morel, Large-scale 3D printing of ultra-high performance concrete – a new processing route for architects and builders, Mater. Des. 100 (2016) 102–109.
[12] J. Zhang, B. Khoshnevis, Optimal machine operation planning for construction by contour crafting, Autom. Constr. 29 (2013) 50–67.

[13] B. Khoshnevis, A. Carlson, N. Leach, M. Thanavelu, Contour crafting simulation plan for lunar settlement infrastructure build-up, Earth and Space 2012@ Engineering, Science, Construction, and Operations in Challenging Environments, ASCE, Pasadena, CA, 2012, pp. 1458–1467.
[14] L. Wang, Chinese company assembles 10 3D-printed concrete houses in a day for less than $5,000 each. On-line: <http://inhabitat.com/chinese-company-assembles-ten-3d-printed-concrete-houses-in-one-day-for-less-than-5000-each/>, (accessed 20.03.17).
[15] B. Sevenson, Shanghai-based Winsun 3D prints 6-story apartment building and an incredible home. On-line: <https://3dprint.com/38144/3d-printed-apartment-building/>, (accessed 20.03.17).
[16] C. Scott, Chinese construction company 3D prints an entire two-story house on-site in 45 days. On-line: <https://3dprint.com/138664/huashang-tengda-3d-print-house/>, (accessed 20.03.17).
[17] WASP, Concrete beam created with 3D printing. On-line: <http://www.waspro-ject.it/w/en/concrete-beam-created-with-3d-printing/>, (accessed 20.03.17).
[18] Alec, Thai cement maker SCG develops an elegant 3m-tall 3D printed 'pavilion' home, 21st C. Cave. On-line: <http://www.3ders.org/articles/20160427-thai-cement-maker-scg-develops-an-elegant-3m-tall-3d-printed-pavilion-home-21st-c-cave.html>, (accessed 20.03.17).
[19] Apis Core, The first on-site house has been printed in Russia. On-line: <http://apis-cor.com/en/about/news/first-house>, (accessed 20.03.17).
[20] E. Dini, D_Shape, vol. 2014. <http://www.d-shape.com>, 2014.
[21] Z-Corp, Sls-based 3D printers. <http://www.zcorp.com/en/Products/3DPrinters/spage.aspx>.
[22] R. Ronald Rael, V. San Fratello, Bloom. On-line: <http://www.emergingobjects.com/project/bloom-2/>, (accessed 20.03.17).
[23] R. Ronald Rael, V. San Fratello, Shed. On-line: <http://www.emergingobjects.com/project/shed/>, (accessed 20.03.17).

CHAPTER 2

Performance-Based Testing of Portland Cement Concrete for Construction-Scale 3D Printing

Ali Kazemian[1,2,3], Xiao Yuan[3], Ryan Meier[1] and Behrokh Khoshnevis[1,3,4]
[1]The Sonny Astani Department of Civil and Environmental Engineering, University of Southern California, Los Angeles, CA, United States
[2]Department of Computer Science, University of Southern California, Los Angeles, CA, United States
[3]Contour Crafting Corporation, El Segundo, CA, United States
[4]Department of Industrial and Systems Engineering, University of Southern California, Los Angeles, CA, United States

2.1 INTRODUCTION

Almost every major industry, except construction, has already adopted automation. Aerospace, automobile, retail, and manufacturing are examples of these industries which have reaped the benefits of automation over the past few decades. The construction industry could be described as the last frontier for automation. It should be noted, however, that there have been earlier attempts to use robotic systems for in situ construction. For instance, the Big Canopy system was developed in 1980s by Japanese engineers in an attempt to develop an in situ robotic construction system. The system consisted of a 13-ton rack and pinion gondola-type construction lift for vertical material delivery and automated overhead cranes for horizontal delivery and structural element orientation and positioning [1]. Even though this system resulted in some improvement in productivity for a few structures, it was unable to defeat traditional construction methods. Based on reviews from the Japan Construction Mechanization Association, the failure of Big Canopy and similar projects in 1980s and 1990s could be attributed to the inability to recover the research, development, and manufacturing costs, as well as the inability to significantly reduce onsite labor requirements [1].

A more recent major attempt toward automated construction is based on the novel idea of scaling up additive manufacturing techniques. Additive manufacturing is defined as "the process of joining materials to

3D Concrete Printing Technology
DOI: https://doi.org/10.1016/B978-0-12-815481-6.00002-6

make objects from 3D model data, usually layer upon layer, as opposed to subtractive manufacturing methodologies." [2] It should be noted that additive manufacturing technologies have previously been used for concept modeling in architecture [3]. However, use of this technique for full-scale, in situ, automated building construction is in the process of emerging in the construction industry. A well-developed automated layer-by-layer construction process would present numerous advantages, including design freedom, superior construction speed, minimal waste of construction materials, and a higher degree of customization.

Contour crafting (CC), invented in 2004 by Dr. Behrokh Khoshnevis of University of Southern California, is a pioneering additive fabrication technology that uses computer control to exploit the surface-forming capability of troweling to create smooth and accurate planar and free-form surfaces out of extruded materials [4]. CC is commonly recognized as the first viable construction-scale additive manufacturing process for construction [5–7]. Some important advantages of CC include unprecedented surface quality of printed elements, increased fabrication rate, and a vast choice of materials [8]. The prototype CC machine (Fig. 2.1) has work envelope dimensions of $5\,m \times 8\,m \times 3\,m$, corresponding to a $120\,m^2$ printing zone. CC technology has been considered to be a viable method for building immediate infrastructures on the surface of the moon and Mars for potential colonization purposes [9,10].

Figure 2.1 Contour crafting machine.

With respect to the construction material, Portland cement concrete has been found to be the most viable option for widespread use in automated construction processes in the near future [6,11]. Concrete is well-understood and has unique fresh and hardened properties as well as an extensive variety of readily available admixtures to customize its performance.

Limited research has been carried out on properties of printing concrete. In 2016, Perrot et al. [11] studied the time-dependent, structural build-up of cementitious materials in layerwise construction. The time required to harden is important because during the layer-by-layer construction process, the previously deposited layers need to be able to withstand the load caused by following layers. Based on a comparison of the vertical stress acting on the first printed layer with the critical stress related to the plastic deformation, a theoretical framework was proposed. Assuming linear evolution of yield stress over time, these researchers defined a critical failure time (t_f) as a function of concrete-specific weight, concrete yield stress with no time at rest, structuration rate, construction rate, and a geometric factor (α_{geom}). Finally, layerwise construction of a 70 mm diameter column with building rates of 1.1–6.2 m/h was used to validate the findings. Except for the smallest building rate, 1.13 m/h, the experimentally measured t_f values were highly correlated with values calculated based on proposed expressions [11].

While limited past studies have provided an initial understanding of some of the desirable properties of printing concrete, extensive research and experimental data is still needed. Specifically, characterization of the fresh-state behavior of a printing mixture requires deeper investigation. The traditional definition for workability of fresh concrete as "a measure of ease by which fresh concrete can be placed and compacted" [12] does not seem to be applicable to construction-scale 3D printing and, thus, new measures should be developed for describing the fresh-state behavior of a printing mixture.

2.2 RESEARCH OBJECTIVES

Printing concrete, as the latest special concrete, has no relevant guidelines or proposed procedures for mixture evaluation, or any set of well-defined acceptance criteria. Few prior studies have focused on specific properties of printing mixtures such as shape stability (also called shape retention and green strength) [11,13–16]. However, a comprehensive list of performance requirements and test methods for a printing mixture has not yet been developed.

The goal of this chapter is to present and examine a framework for performance-based laboratory testing and evaluation of printing mixtures. It should be noted that only the fresh properties of a printing mixture are considered herein; further research is needed to investigate the structural requirements for hardened printing mixtures. Development of a comprehensive framework for laboratory testing of printing concrete would be a starting point for systematic investigation on this special concrete by researchers and would provide a basis for future specifications and guidelines. Establishing universal acceptance criteria for printing concrete would be possible only after many relevant studies have been carried out and a reasonable amount of data is available on the performance of different printing mixtures used in actual construction projects.

Next, a proposed framework for laboratory testing of fresh printing concrete is introduced and relevant details are discussed. Then, an experimental program that was carried out to test the proposed framework is discussed. The proportions of four printing mixtures designed to study effects of nanoclay, silica fume, and fiber are also presented. Finally, the results of some conventional tests as well as several proposed new tests are presented and used to briefly discuss the performance of different mixtures.

2.3 EXPERIMENTAL PROGRAM

2.3.1 Proposed Framework

The proposed framework for laboratory testing of printing concrete in fresh state is presented in Fig. 2.2. The testing procedure is designed based on properties of printed layers (rather than parameters directly related to the employed pumping or extrusion mechanism). As such, the proposed framework is applicable to different 3D concrete printing systems.

Based on experience and other research [13,17–19], printing concrete can be characterized by high powder content, no coarse aggregate, increased paste fraction, and use of viscosity modifying admixture (VMA). In fact, the reported mixture proportions of successful printing mixtures could be used as a starting point. After designing an initial mixture, fulfilling print quality requirements is the first step in the proposed iterative mixture assessment and modification process. With respect to acceptable print quality, three requirements must be satisfied: (1) surface quality; (2) squared edges; and (3) dimension conformity and consistency.

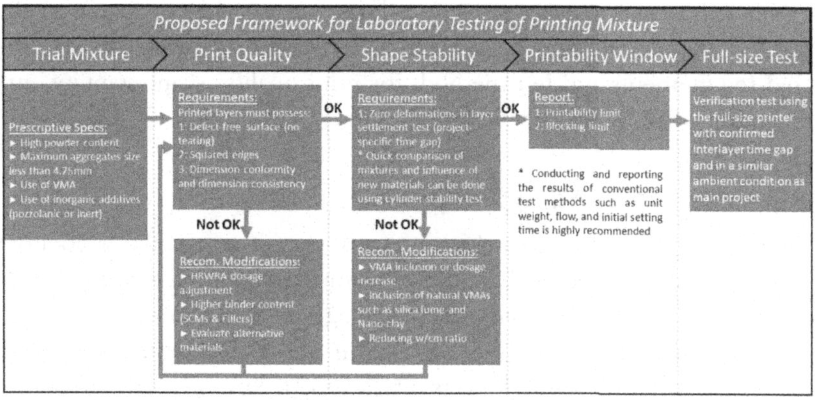

Figure 2.2 Proposed framework for laboratory testing of printing mixture in fresh state.

Next, the shape stability of a mixture must be examined and necessary adjustments made. In this regard, the cylinder stability test is proposed for rapid evaluation and comparison of the influence of different materials (additives or admixtures) on the shape stability of a mixture. However, the decision to accept or reject a mixture must be based on the layer settlement experiment where concrete layers are printed on top of each other with the same extrusion mechanism as the full-scale concrete printer. Further, no visible deformations should occur when the target interlayer time gap is used.

The third step in laboratory testing of a printing mixture refers to the printability window of a mixture. Two important parameters related to the printability window include the printability limit and blockage limit. Printability limit is the longest period during which a mixture can be printed with acceptable print quality. Blockage limit refers to the longest period of time when a mixture can remain in the nozzle before the concrete hardens and blocks the extrusion. Both limits should be measured and reported for each specific mixture. Finally, when the laboratory testing is aimed at developing a mixture for a specific construction project, the proposed framework suggests a verification test. This test should be carried out using the full-size printer in a similar ambient temperature and humidity as the intended project. Also, the use of the same concrete batching, mixing, and transporting equipment as the actual project is highly recommended. The main function of the proposed verification test is to subject the designed mixture to actual jobsite-based assessment.

Next, the experimental program which was carried out based on the proposed framework will be presented. In addition, the details of proposed requirements and test methods for print quality, shape stability, and printability window will be discussed.

2.3.2 Concrete Printing Setup

A linear concrete printing machine was constructed to simplify laboratory testing of various mixtures (Fig. 2.3A). Compared to implementing a full-size printer (e.g., CC machine), the advantages of using such a laboratory-scale printer include saving time, cost, and material. The developed system is capable of printing up to 10 concrete layers, 1.2 m in length. The nozzle uses an extrusion mechanism like the CC machine to print 38.1 mm × 25.4 mm concrete layers. The control system was developed using a combination of Arduino Mega (based on ATmega1280) and Arduino Uno (based on ATmega328) microcontrollers. The machine is able to print concrete at different linear speeds and deposition rates. The implemented feedback control system (closed-loop) for the extruder ensures the consistent material deposition rate.

In addition, Bluetooth communication with the printing setup enables the user to conveniently control the machine using an Android application (Fig. 2.3B) on a smartphone or tablet. This application enables the user to select the parameters (such as linear speed and deposition rate) for concrete printing process and provides several predefined settings for convenient operation. For the purposes of this study, a linear printing speed of 60 mm/s was used in all experiments.

2.3.3 Materials and Mixture Proportions

In this study, ASTM C150 Type II Portland cement was used to produce printing mixtures. A commercially available manufactured sand with nominal maximum aggregate size of 2.36 mm was used as fine aggregate. The sieve analysis results and other physical properties of used sand are presented in Table 2.1.

A polycarboxylate-based, high-range water reducing admixture (HRWRA) was used to achieve the required flowability for the mixtures. In addition, to increase the plastic viscosity and cohesion of printing mixtures, a commercially available VMA for antiwashout concrete was used. Polypropylene fiber was also used as a shrinkage reinforcement for a printing mixture as it inhibits and controls the formation of plastic and drying

(A)

(B)

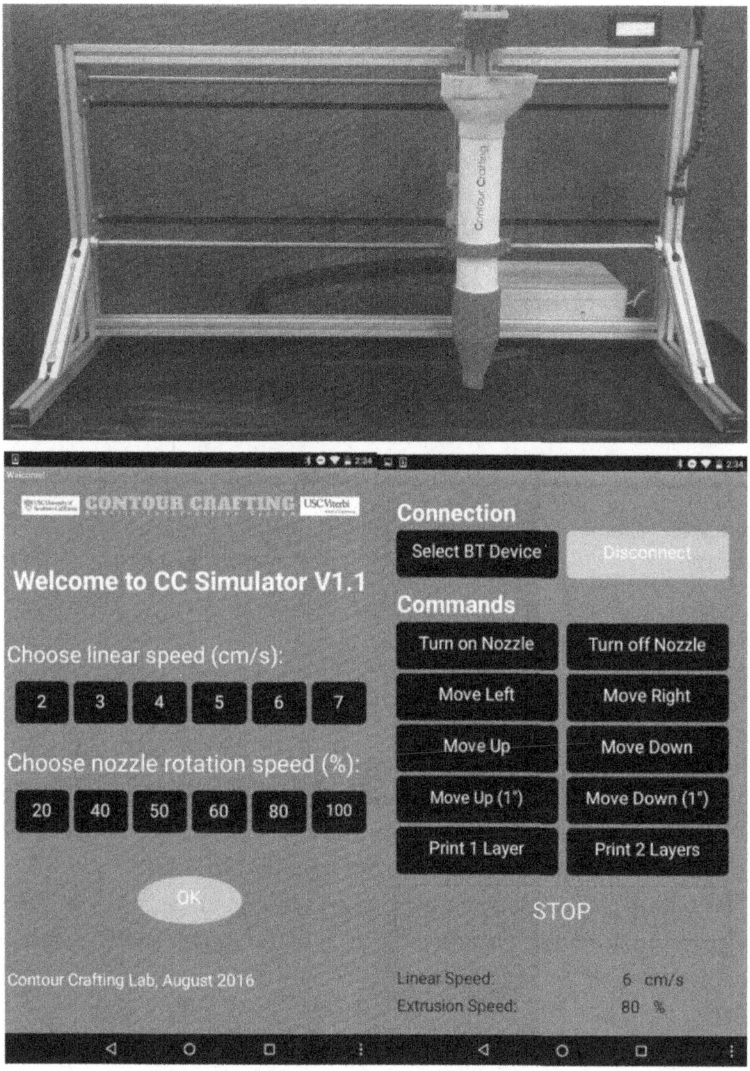

Figure 2.3 (A) Linear concrete printing machine for laboratory testing of printing mixtures. (B) Android application developed for remote control of concrete printing setup.

shrinkage cracking in concrete [20]. This is particularly important for printing concrete since a higher rate of water evaporation is anticipated for a printed structure given that there is no formwork covering the surface of freshly printed elements. The length of fiber used in this study is 6 mm and tensile strength is 415 MPa. Furthermore, densified silica fume

Table 2.1 Sieve analysis and other properties of fine aggregate

Sieve		Percent passing
Number	**Size (mm)**	
4	4.75	100
8	2.36	83
16	1.18	62
30	0.6	38
50	0.3	19
100	0.15	8
200	0.075	3.5
Specific gravity		2.6
Absorption capacity (%)		1.3
Fineness modulus		2.9

Table 2.2 Properties of used chemical admixtures, fiber, and silica fume

Property	HRWRA	VMA	Polypropylene fiber	Silica fume
Specific gravity	1.1	1.02	0.91	2.2
pH	5.6	6	—	—
Aspect ratio	—	—	29	—

was used as a supplementary cementitious material. It is well-established that the addition of silica fume improves cohesion of fresh concrete and the mechanical strength and impermeability of hardened concrete [21,22]. Properties of mentioned materials are presented in Table 2.2.

A highly purified attapulgite clay with average particle length of 1.75 μm and average particle diameter of 3 nm was used in this study. The specific gravity of this clay is 2.29 and the average length divided by average diameter of clay particles is 583, indicating a high aspect ratio [23]. Therefore, it may form a highly entangled gel even at a small addition rate, provided proper dispersion is achieved. This nanoclay is commercially available and several studies [23–25] have investigated the influence of its addition on different properties of other special concretes, such as the formwork pressure of self-consolidating concrete.

Proportions within different printing mixtures are presented in Table 2.3. These mixtures were designed to demonstrate possibilities with respect to an array of materials available in 3D concrete printing. For comparison purposes, the total cementitious materials content and water/cementitious materials ratio were kept constant in all mixtures at

Table 2.3 Mixture proportions of printing mixtures (20 L of entrapped air per cubic meter concrete is assumed)

Mixture ID	Fine aggregate (SSD) (kg/m^3)	Portland cement (kg/m^3)	Free water (kg/m^3)	Silica fume (kg/m^3)	Fiber (kg/m^3)	Nanoclay (%)[a]	HRWRA (%)[a]	VMA (%)[a]
PPM	1379	600	259	0	0	0	0.05	0.11
SFPM	1357	540	259	60	0	0	0.16	0
FRPM	1379	600	259	0	1.18	0	0.06	0.10
NCPM	1379	600	259	0	0	0.30	0.15	0

[a]The percentages are reported by cementitious materials mass.

600 kg/m^3 and 0.43, respectively. With respect to mixing procedure, a drum mixer was used to blend the ingredients for eight minutes. For NCPM mixture preparation, nanoclay was initially mixed with water and introduced as a suspension to the mixer.

2.4 TESTING AND RESULTS

In this section, different test methods used to characterize the printing mixtures and their results are presented. Initially, some conventional test methods and corresponding results are discussed. These conventional tests serve as standard characterization of mixtures and provide the opportunity for other researchers to compare these results with other studies. Then, the workability of printing mixtures is described and evaluated in terms of print quality, shape stability, and printability window.

2.4.1 Conventional Mixture Characterization

To characterize the printing mixtures per conventional concrete technology, unit weight, flow, and compressive strength at 7- and 28-day ages were measured. Flow of each mixture was determined using a flow table (ASTM C1437-15 [26]), which involves a mold being filled with mortar and then compacted. The mold is subsequently lifted away from mortar and the table is immediately dropped 25 times in 15 seconds. The flow is the resulting increase in average base diameter of the mortar mass expressed as a percentage of the original base diameter. Compressive strength was measured according to ASTM C109 [27], using 50 mm cubes (cured in a water tank) at a loading rate of 1200 N/s. The results are presented in Table 2.4.

As observed in Table 2.4, flow of the developed mixtures is in the range of 113%−119%, and the minimum flow was measured for mixture

Table 2.4 Unit weight, flow, and strength results for printing mixtures

Mixture ID	Unit weight (kg/m^3)	Flow (%)	7-Day compressive strength (MPa)	28-Day compressive strength (MPa)
PPM	2250	119	32.9 [0.7][a]	44.7 [1.3]
SFPM	2210	116	35.2 [1.6]	49.9 [1.3]
FRPM	2265	118	31.0 [1.9]	45.1 [1.1]
NCPM	2250	113	31.8 [1.2]	45.9 [1.5]

[a]Values in brackets are standard deviations of the strength measurements for each set of three cube specimens (in MPa).

with nanoclay (NCPM). In fact, a considerable increase in viscosity and cohesion of mixture was observed because of a small addition of nanoclay (1.8 kg/m^3). With respect to compressive strength, all mixtures except silica fume addition (SFPM) demonstrated almost similar strength values after 7 days and 28 days of water curing. Compared to PPM, a 7% increase in 7-day strength and a 12% increase in 28-day strength was measured because of SFPM. This could be attributed to the pozzolanic behavior of silica fume, where formation of secondary hydration products (mainly C-S-H) around the silica fume particles fills the large capillary pores with a microporous, low-density material [21,28].

2.4.2 Print Quality

In this section, "print quality" refers to the properties of a printed layer, such as surface quality and dimensional conformity/consistency, when using a specific printing mixture. A printing mixture could be considered acceptable when three requirements are satisfied: (1) The printed layer must be free of surface defects, including any discontinuity due to excessive stiffness and inadequate cohesion; (2) the layer edges must be visible and squared (vs round edges); and (3) dimension conformity and dimension consistency must be satisfied by the printed layer. Based on these three proposed criteria, the print quality of a mixture can be evaluated and an acceptance decision can be made.

Fig. 2.4 presents a case where a mixture with poor workability is rejected by the first print quality requirement due to observation of discontinuity. With respect to the third print quality requirement, dimension conformity guarantees that the dimensions of the printed layers are within an acceptable range of the target dimensions (Fig. 2.5A), while dimension consistency refers to changes in width of a printed layer and

Figure 2.4 Discontinuity (tearing) in the printed layer due to excessive stiffness of the mixture.

(A)
(B)

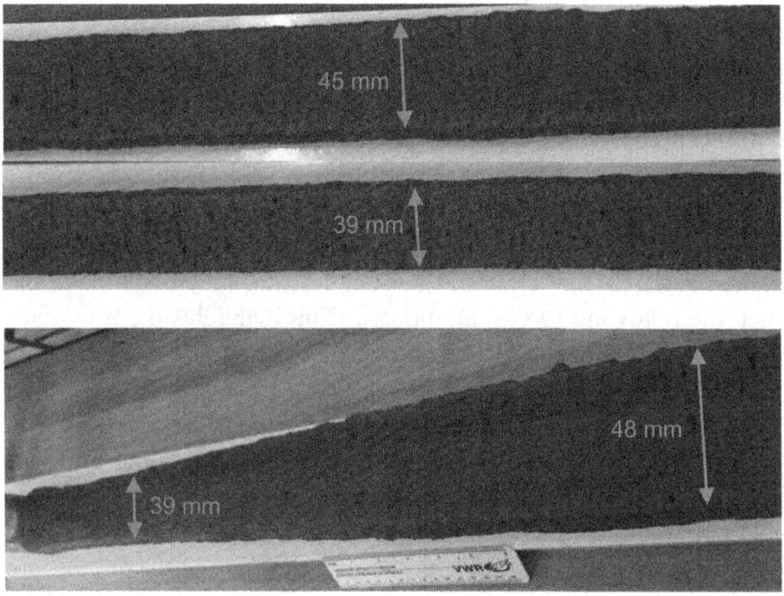

Figure 2.5 (A) Variations in width of printed layer using different mixtures at the same printing speed (dimension conformity). (B) Variations in width of a single layer (dimension consistency).

acceptable variations (Fig. 2.5B). It should be noted that these dimensional limitations are set for a fresh concrete layer and do not consider variations caused by shrinkage over time. In this study, a width of 38.1 mm was designed for each layer. After running a large number of experiments, it was concluded that a 10% error in the target width was a reasonable range for accepting or rejecting printed layers. In other words, the width of printed layers using all the accepted printing mixtures was in the range of

38.1–42 mm. It should be noted that five measurements were done along each printed layer to assess the dimension conformity and, for each mixture, the experiment was carried out four times (four replicates of each mixture). The print quality of a mixture was considered "acceptable" only if the three requirements were satisfied by all four mixture replicates. The printing mixtures presented in Table 2.3 were selected based on the three print quality requirements. A trial-and-error approach was adopted for this purpose, as currently there is no guideline or suggested procedure for designing and testing printing mixtures.

2.4.3 Shape Stability

Shape stability is a critical property of fresh concrete printing, which refers to the concrete's ability to resist deformations during layerwise concrete construction. More specifically, there are three main sources of deformation which apply to a deposited layer: (1) self-weight; (2) weight of the following layer(s) to be printed on top of it; and (3) the extrusion pressure. Based on observations, a mixture with acceptable print quality produces a layer without visible deformations due to self-weight. However, the two latter parameters could possibly lead to undesirable deformations when the following layer(s) are printed. This highlights the importance of laboratory testing of the shape stability of the printing concrete during the mixture design stage.

A 2012 study on the hardened properties of high-performance printing concrete [29] investigated the influence of the printing time gap between consecutive layers on the mechanical bond strength of layers. The results indicated that a longer interlayer time gap leads to lower bond strength, which is undesirable in terms of the structural properties of the printed structure or element. For instance, between a 30-minute and 7-day time gap the average bond strength of printed specimens was 53% and 77% lower than the conventionally cast specimens, respectively [29]. This signifies the importance of reducing the interlayer time gap in order to minimize bond strength loss. However, this is practical only if previously deposited layers possess high shape stability. Ideally, a printed layer should withstand the printing of following layers right after it is deposited, which means no deformations should occur when the time gap is zero.

In this study, in order to obtain a realistic notion of the time gap required between deposition of consecutive layers, the layer-by-layer construction of a one-story, 108 m^2 house with two bedrooms and one bathroom designed by an architectural and design company [30] is considered. Based on the structure's plan, the nozzle traveling distance for each layer was measured as 67 m. Considering the linear printing speed of 60 mm/s, an interlayer time gap of 19 minutes was obtained. The calculated 19-minute time gap as well as the worst-case scenario of a zero time gap were used for shape-stability assessment of the four developed mixtures. Two different test methods, namely the layer settlement and cylinder stability tests, were developed and carried out to study the shape stability of printing concrete. The test procedures and obtained results are provided next.

The layer settlement test was developed to enable quantitative evaluation of shape stability. In this test, two concrete layers were printed on top of each other with a specific time gap. A camera was placed in front of the printed layers, a ruler was placed next to the layers as a scale, and photos were taken before and after the second layer was printed. Then, ImageJ software, a public domain Java-based image processing program [31], was used to analyze the photos and to measure layer settlement. The average of five readings for a printed layer was reported as a test result, while the average of three tests (three printed layers) was used as the final result for a printing mixture. For each experiment, the bottom layer was printed as soon as the 8-minute mixing procedure was complete.

The layer settlement test results for both scenarios, namely, zero and 19-minute time gaps, are presented in Table 2.5. As anticipated, shape stability of a printed layer improves over time and smaller deformations were measured for all mixtures when the time gap was 19 minutes. The results indicate that all mixtures except PPM possess high shape stability when there is a 19-minute time gap between layers, denoted by no visible deformations of the bottom layer. An average 1.5 mm deformation,

Table 2.5 Layer settlement test results (mm)

Mixture ID	Time gap: 0				Time gap: 19 min			
	Test 1	Test 2	Test 3	Average reading	Test 1	Test 2	Test 3	Average reading
PPM	Collapse	Collapse	Collapse	–	1.9	1.1	1.6	1.5 [0.3]
SFPM	2.2	1.8	1.5	1.8 [0.3][a]	0	0	0	0
FRPM	2.8	3.3	2.5	2.9 [0.3]	0	0	0	0
NCPM	2.0	1.1	1.6	1.6 [0.4]	0	0	0	0

[a]Values in brackets are standard deviations of the deformation measurements for each set of three layers (in mm).

equivalent to 5.9% of the layer height, was measured for the layers printed with PPM.

For the PPM mixture with a zero time gap (see Fig. 2.6), considerable deformations occurred after the second layer was deposited. Considering the significant changes in both width and height of the layer in this case, the result of testing was reported as "collapse." For the zero time gap, the lowest deformations were measured for SFPM and NCPM mixtures, where the average layer height reduced by approximately 6.7%. However, considering the standard deviation of the obtained results, there is no statistically significant difference between shape stability of these two mixtures. Acceptable print quality, as defined in this chapter, was not found to guarantee high shape stability as the four printing mixtures with acceptable print quality showed different levels of shape stability. As such, this property must be separately evaluated during mixture design.

It should be noted that yield stress of fresh concrete is the main parameter determining the shape stability before setting. Yield stress increases over time in the absence of agitation and shear stress [11,32]. This is due to the nucleation of cement grains at their contact point by C-S-H formation during the dormant period before the setting time [33]. Several researchers have reported a linear increase in yield stress during the dormant period [11,32,34], which suggests a corresponding linear increase of shape stability with time. Another important consideration for a fundamental study of shape stability is thixotropy, which is defined as the

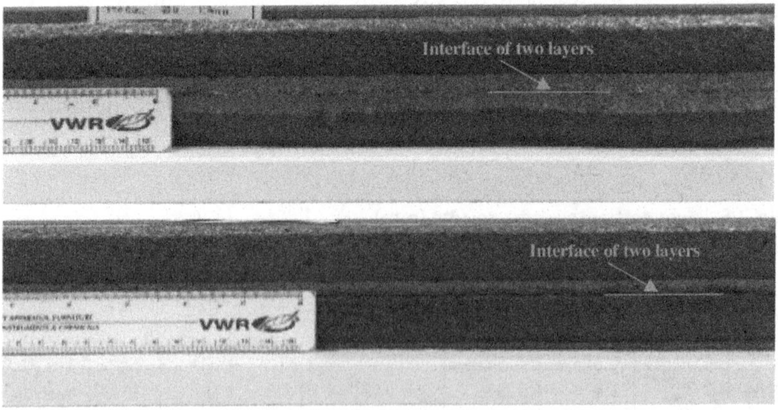

Figure 2.6 A double-layer specimen printed with zero time gap, reported as "collapse" (top) and another specimen with a 19-min time gap and no visible deformations (bottom).

build-up and breakdown of internal the 3D structure within the cementitious paste. This phenomenon happens due to flocculation or coagulation and dispersion of cement particles that, in turn, result from interparticle forces and chemical connections [35,36]. Build-up and breakdown of the internal structure causes an increase and reduction in viscosity of fresh paste, respectively [37]. Considering the process of 3D concrete printing, where the cementitious mixture undergoes considerable shear stress and agitation before being deposited as a layer, these changes in the internal structure and the consequent influence on shape stability need to be considered. For example, Kawashima et al. [24] reported a higher build-up rate of the internal structure after shear-induced breakdown for nanoclay incorporated cementitious mixtures, especially within the first few minutes. This faster "structuration at rest" of nanoclay included mixtures could explain the enhanced shape stability of the NCPM mixture in this study. Further fundamental research on rheology of printing concrete and more experimental data are needed for a deeper understanding of the shape stability of a printing mixture.

Additionally, in order to examine the scalability of obtained results, five-layer specimens were printed using NCPM and SFPM mixtures and an interlayer time gap of 19 minutes (see Fig. 2.7). For both mixtures, no deformations were detected (either by visual inspection or using ImageJ software) during, and after, the printing process.

The cylinder stability test was developed as another test method to evaluate shape stability of different mixtures. The test apparatus includes a frame, two semicylinder shells (with sealed vertical joints), a tamping rod,

Figure 2.7 A five-layer printed specimen with interlayer time gap of 19 min.

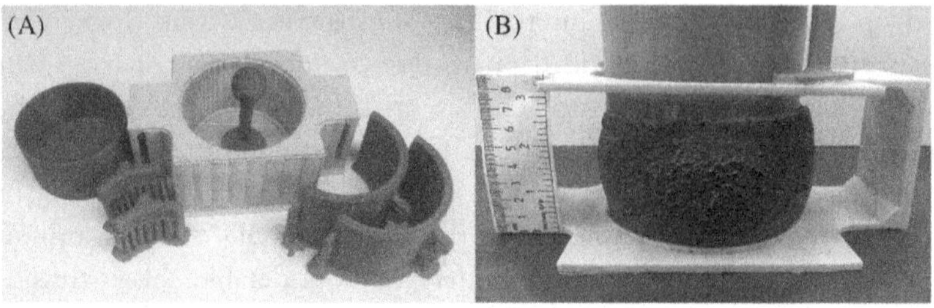

Figure 2.8 (A) 3D printed parts for cylinder stability test. (B) Cylinder stability test.

two loading guides, and a container for uniform application of load to the fresh concrete specimen. These parts were designed and 3D printed using ABS plastic (Fig. 2.8). There are five steps in the cylinder stability test procedure: (1) the semicylinders are fixed in place and locked, and a concrete layer of 40 mm is placed; (2) using the tamping rod, the layer is consolidated by rodding 15 times and evenly distributed; (3) the same procedure is repeated for the second layer and excessive concrete is removed from the top (a well-compacted fresh concrete cylinder with a total height of 80 mm is achieved); (4) the two semicylinders are unlocked and gently removed and any possible change in height as a result of self-weight is measured and recorded; and, finally, (5) a load of 5.5 kg (equivalent to a 4.77 kPa stress) is applied and the resulting deformation in the fresh concrete cylinder is measured in terms of change in height. The main advantage of this test (compared to the layer settlement test) is eliminating the need to print concrete layers, leading to saving time during the mixture design phase. The measured deformations in this test method would be primarily reliable for comparison purposes, while the layer settlement test result is a realistic indicator of mixture performance in layer-based construction. It should be noted Perrot et al. [11] used a similar idea to simulate the load acting on the first deposited layer in layerwise construction. In their experiment, a 35 mm $\times$ 60 mm fresh concrete cylinder was placed between two metal plates and the upper plate was then loaded in 1.5 N increments. The combination of surface fracture occurrence and upper plate displacement was used as indications of plastic deformations of the concrete cylinder over time.

In this study, the cylinder stability test was carried out after the mixing procedure was finished. The obtained results are presented in Table 2.6.

The largest deformation in fresh concrete cylinder was observed for the PPM mixture, followed by FRPM. The NCPM mixture exhibited the highest shape stability, while the SFPM mixture resulted in slightly

Table 2.6 Cylinder stability test results (mm)

Mixture ID	Test 1	Test 2	Test 3	Average reading
PPM	41	37	38	38.7 [1.7][a]
SFPM	15	15	14	14.7 [0.5]
FRPM	34	29	31	31.3 [2.1]
NCPM	12	15	11	12.7 [1.7]

[a]Values in brackets are standard deviations of measurements from three replicate tests (in mm).

larger deformation. Considering the similar performance of SFPM and NCPM, the experimental results indicate the cylinder stability test ranks the shape stability of mixtures in a similar order as the layer settlement test. This implies that this simple test can be used for comparing the shape stability of different printing mixtures as well as studying the influence of new materials added and chemical admixtures. Further research and experimental data are required to investigate any correlation between the values given by the two test methods.

2.4.4 Printability Window

The printability window is the third property of a concrete mixture considered in this chapter. This window is the period of time during which the printing mixture could be extruded by the nozzle with acceptable quality, considering the workability loss that occurs over time. The timing of material delivery to the nozzle is essential to the operation of a full-size building printer, such as the CC machine. In this study, two time limits are introduced to define the printability window of a mixture and the printability and blockage time limits. The printability limit refers to the time when the quality of printed layer is affected as a result of workability loss, recognized by the triple print quality requirements, whereas the blockage limit is the time when the concrete cannot be guided out of the printing nozzle at all, so further delay would result in mixture solidification and damage to the nozzle.

An important related concept is the initial setting time of concrete. Per ASTM C125 [38], the initial setting time is defined as the elapsed time after cement–water contact required for the mortar sieved from the concrete to reach a penetration resistance of 3.5 MPa. In this study, a concrete penetrometer (based on ASTM C403 [39]) was used to measure the initial setting time of developed mixtures. Since the setting times of the four mixtures were similar (325–350 minutes), three other mixtures

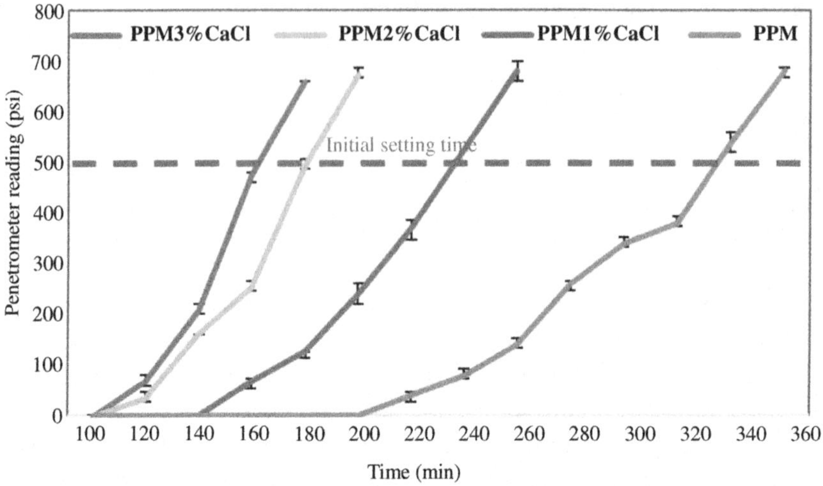

Figure 2.9 Concrete penetrometer readings for the three PPM $CaCl_2$ mixtures.

(based on PPM) were developed and studied. Calcium chloride ($CaCl_2$) is commonly known as an effective accelerator for cementitious mixtures as it acts like a catalyst for C_3S reactions, which results in larger hydration heat liberation [21]. Thus, three different dosages of analytical grade $CaCl_2$ (1%, 2%, and 3% of Portland cement mass) were added to the PPM mixture and the resulting mixtures were labeled PPM1%CaCl, PPM2%CaCl, and PPM3%CaCl, respectively. For each mixture, the measurements using concrete penetrometer were carried out three times. The variations in readings are depicted with error bars in Fig. 2.9.

Based on penetrometer readings, the initial setting time of PPM, PPM1%CaCl, PPM2%CaCl, and PPM3%CaCl was measured as 335, 237, 181, and 163 minutes, respectively. As anticipated, adding $CaCl_2$ resulted in accelerated hydration reaction and shorter setting times. The results for printability window parameters, as well as initial setting time are presented in Table 2.7. To determine the printability limit, a single layer was printed every 5 minutes, beginning 20 minutes after the initial water–cement contact. The earliest time when print quality criteria were not satisfied was recorded as the printability limit. Similarly, the earliest time when concrete could not be extruded out of the nozzle was recorded as the blockage limit.

The results indicate nozzle blockage could happen long before the initial setting time of each mixture, while setting time could not be used as an alternative indicator. Nozzle blockage can result in significant time loss,

Table 2.7 Initial setting time, printability limit and blockage limit of mixtures (min)

Mixture ID	Initial setting time	Printability limit	Blockage limit
PPM	335	55	85
PPM1%CaCl	237	40	75
PPM2%CaCl	181	40	60
PPM3%CaCl	163	45	55

Table 2.8 Flow of printing mixtures at different times (%)

Mixture ID	Time (min)						
	8	20	40	60	70	80	90
PPM	119	111	100	79	69	61	55
PPM1%CaCl	112	110	102	63	52	48	42
PPM2%CaCl	114	113	97	54	46	37	31
PPM3%CaCl	107	115	95	47	45	34	28

nozzle damage, and extra cost during construction. As such, measuring the blockage limit for each mixture is recommended during mixture design and laboratory testing.

Finally, to investigate the relationship between changes in workability over time and proposed time limits, workability loss of the four mixtures was studied. Flow at different times, from 8 to 90 minutes after initial water–cement contact, were measured for different mixtures (see Table 2.8). The results indicate the influence of the $CaCl_2$ addition on workability loss appears only after 60 minutes and becomes distinct at longer times. This finding justifies the similar printability limits which were measured for the three $CaCl_2$-incorporated mixtures (Table 2.7). Also, increasing the accelerator dosage from 2% to 3% did not have a significant influence on the workability loss during first 90 minutes.

As for conventional concrete, setting time and workability loss measurements could be used to study the effects of different chemical admixtures on fresh-state behavior of printing concrete. However, it is apparent that neither setting time measurements nor workability loss measurements could replace the direct measurement of printability and blockage limits. Also, both proposed limits are directly dependent on the specific extrusion mechanism (i.e., auger, piston, etc.) used by the 3D concrete printer. Therefore, finding a printer-independent relationship between conventional parameters, such as setting time and printability window limits, seems unlikely.

2.5 CONCLUSIONS AND FUTURE WORK

In this chapter, a framework is proposed for performance-based laboratory testing of fresh printing mixtures. An iterative laboratory testing procedure is described to assess and modify the print quality, shape stability, and printability window of a printing mixture. The proposed testing procedure is based only on different properties of already extruded layers; therefore, the procedure is not dependent on an extrusion mechanism and could be applied to different 3D concrete printing systems. To realize the proposed framework, a laboratory-scale linear concrete printer machine capable of printing 38.1 mm × 25.4 mm layers was constructed. Four different printing mixtures were developed and the results for a combination of conventional and new test methods were reported. Acceptable print quality, as defined in this study, was not found to guarantee high shape stability, because the four printing mixtures with acceptable print quality showed different levels of shape stability. Experimental data revealed that inclusion of silica fume and nanoclay (a highly purified attapulgite clay) enhanced shape stability of fresh printing mixture, while minor improvement was observed from polypropylene fiber addition. The experimental results also suggest that the cylinder stability test could be considered for rapid shape stability assessment, while the layer settlement test result was a more realistic indicator of shape stability. Based on the printability window, adding $CaCl_2$ resulted in accelerated hydration reaction and shorter setting times. Further, nozzle blockage was found to occur long before the initial setting time of each mixture, hence it was concluded that direct measurement of blocking limit is necessary.

There are numerous unexplored areas in the emerging field of construction-scale 3D concrete printing. Some major topics and unresolved challenges that need to be further investigated include shrinkage of 3D printed concrete, structural performance and durability of 3D printed structures, robustness of printing concrete as well as the printing process, and quality monitoring of the construction process.

REFERENCES

[1] M. Taylor, S. Wamuziri, I. Smith, Automated construction in Japan, Proc. Inst. Civil Eng. (2003).
[2] ASTM F2792-12a (withdrawn), American Society of Testing and Materials, 2012.
[3] R.A. Buswell, R.C. Soar, A.G. Gibb, A. Thorpe, Freeform construction application research, Adv. Eng. Struct. Mech. Constr. (2006) 773–780.

[4] B. Khoshnevis, Automated construction by contour crafting-related robotics and information technologies, Autom. Constr. 13 (1) (2004) 5–19.

[5] R. Naboni, I. Paoletti, Advanced Customization in Architectural Design and Construction, Springer, 2015.

[6] T. Wangler, E. Lloret, L. Reiter, N. Hack, F. Gramazio, M. Kohler, et al., Digital concrete: opportunities and challenges, RILEM Tech. Lett. 1 (2016) 67–75.

[7] R. Wolfs, 3D Printing of Concrete Structures, Eindhoven University of Technology, 2015.

[8] B. Khoshnevis, M.P. Bodiford, K.H. Burks, E. Ethridge, D. Tucker, W. Kim, et al., Lunar contour crafting—a novel technique for ISRU-based habitat development, in: 43rd AIAA Aerospace Sciences Meeting and Exhibit, Reno, Nevada, 2005.

[9] B. Khoshnevis, X. Yuan, B. Zahiri, Z. Jing, B. Xia, Construction by contour crafting using sulfur concrete with planetary applications, Rapid Prototyping J. 22 (5) (2016) 848–856.

[10] B. Khoshnevis, M. Thangavelu, X. Yuan, J. Zhang, Advances in contour crafting technology for extraterrestrial settlement infrastructure buildup, in: AIAA SPACE 2013 Conference and Exposition, San Diego, CA, 2013.

[11] A. Perrot, D. Rangeard, A. Pierre, Structural built-up of cement-based materials used for 3D-printing extrusion techniques, Mater. Struct. 49 (4) (2016) 1213–1220.

[12] Specification and guidelines for self-compacting concrete, European Federation of National Associations Representing Producers and Applicators of Specialist Building Products for Concrete (EFNARC), 2002.

[13] T.T. Le, S.A. Austin, S. Lim, R.A. Buswell, A.G.F. Gibb, T. Thorpe, Mix design and fresh properties for high-performance printing concrete, Mater. Struct. 45 (2012) 1221–1232.

[14] S. Shah, Design and Application of Low Compaction Energy Concrete for Use in Slip-form Concrete Paving, Infrastructure Technology Institute, Northwestern University, 2008.

[15] T. Voigt, J. Mbele, K. Wang, S. Shah, Using fly ash, clay, and fibers for simultaneous improvement of concrete green strength and consolidatability for slip-form pavement, J. Mater. Civil Eng. 22 (2) (2010) 196–206.

[16] A. Kazemian, X. Yuan, R. Meier, E. Cochran, B. Khoshnevis, Construction-scale 3D printing: shape stability of fresh printing concrete, in: 12th International Manufacturing Science and Engineering Conference (MSEC 2017), Los Angeles, CA, 2017.

[17] V. Nerella, M. Krause, M. Nather, V. Mechtcherine, Studying printability of fresh concrete for formwork free concrete on-site 3D printing technology (CONPrint3D), in: 25th Conference on Rheology of Building Materials, Regensburg, 2016.

[18] L. Anell, Concrete 3D Printer, Lund University, Sweden, 2015.

[19] G. Ma, L. Wang, A critical review of preparation design and workability measurement of concrete material for largescale 3D printing, Front. Struct. Civil Eng. (2017) 1–19.

[20] BASF, [Online]. Available from: <https://www.master-builders-solutions.basf.us/en-us/products/masterfiber/1649> (accessed 01.11.16).

[21] P.K. Mehta, P.J.M. Monteiro, Concrete: Microstructure, Properties, and Materials, fourth ed., McGraw-Hill Education, 2013.

[22] A.A. Ramezanianpour, A. Kazemian, M. Nikravan, M.A. Moghadam, Influence of a low-activity slag and silica fume on the fresh properties and durability of high performance self-consolidating concrete, in: International Conference on Sustainable Construction Materials & Technologies (SCMT3), Kyoto, Japan, 2013.

[23] S. Kawashima, M. Chaouche, D.J. Corr, S.P. Shah, Influence of purified attapulgite clays on the adhesive properties of cement pastes as measured by the tack test, Cem. Concr. Compos. 48 (2014) 35–41.
[24] S. Kawashima, M. Chaouche, D.J. Corr, S.P. Shah, Rate of thixotropic rebuilding of cement pastes modified with highly purified attapulgite clays, Cem. Concr. Res. 53 (2013) 112–118.
[25] J. Kim, M. Beacraft, S.P. Shah, Effect of mineral admixtures on formwork pressure of self-consolidating concrete, Cem. Concr. Compos. 32 (9) (2010) 665–671.
[26] ASTM C1437-15: Standard Test Method for Flow of Hydraulic Cement Mortar, American Society of Testing and Materials (ASTM), 2015.
[27] American Society of Testing and Materials (ASTM), [Online]. Available from: <https://www.astm.org/Standards/C109.htm> (accessed 08.16).
[28] A.A. Ramezanianpour, A. Kazemian, M. Sarvari, B. Ahmadi, Use of natural zeolite to produce self-consolidating concrete with low portland cement content and high durability, J. Mater. Civil Eng. 25 (5) (2013).
[29] T.T. Le, S.A. Austin, S. Lim, R.A. Buswell, R. Law, A.G.F. Gibb, et al., Hardened properties of high-performance printing concrete, Cem. Concr. Res. 42 (3) (2012) 558–566.
[30] FreeGreen, [Online]. Available from: <https://www.houseplans.com/plan/1160-square-feet-2-bedroom-1-bathroom-0-garage-modern-39050> (accessed 15.06.16).
[31] ImageJ, [Online]. Available from: <https://imagej.nih.gov/ij/>.
[32] L. Josserand, O. Coussy, F. de Larrard, Bleeding of concrete as an ageing consolidation process, Cem. Concr. Res. 36 (9) (2006) 1603–1608.
[33] N. Roussel, G. Ovarlez, S. Garrault, C. Brumaud, The origins of thixotropy of fresh cement pastes, Cem. Concr. Res. 42 (1) (2012) 148–157.
[34] N. Roussel, A thixotropy model for fresh fluid concretes: theory, validation and applications, Cem. Concr. Res. 36 (2006) 1797–1806.
[35] J. Wallevik, Thixotropic investigation on cement paste: experimental and numerical approach, J. Non-Newtonian Fluid Mech. 132 (2005) 86–99.
[36] J. Wallevik, Rheological properties of cement paste: thixotropic behavior and structural breakdown, Cem. Concr. Res. 39 (1) (2009) 14–29.
[37] G. Heirman, Modelling and Quantification of the Effect of Mineral Additions on the Rheology of Fresh Powder Tupe Self-Compacting Concrete, Arenberg Doctoral School of Science, Engineering and Technology, 2011.
[38] ASTM C125-15b: Standard Terminology Relating to Concrete and Concrete Aggregates, American Society of Testing and Materials (ASTM), 2015.
[39] ASTM C403/C403M-08: Standard Test Method for Time of Setting of Concrete Mixtures by Penetration Resistance, American Society of Testing and Materials (ASTM), 2008.

FURTHER READING

A.A. Ramezanianpour, A. Kazemian, M.A. Moghaddam, F. Moodi, A.M. Ramezanianpour, Studying effects of low-reactivity GGBFS on chloride resistance of conventional and high strength concretes, Mater. Struct. 49 (7) (2016) 2597–2609.
E.P. Koehler, D.W. Fowler, ICAR 108-2F: Aggregates in Self-Consolidating Concrete, International Center for Aggregates Research (ICAR), The University of Texas at Austin, 2007.
S. Lim, R.A. Buswell, T.T. Le, R. Wackrow, S.A. Austin, A.G. Gibb, et al., Development of a viable concrete printing process, in: International Association for Automation and Robotics in Construction (IAARC), 2011.

R.A. Buswell, A.G.F. Gibb, R.C. Soar, S.A. Austin, A. Thorpe, Applying future industrialised processes to construction, in: CIB World Building Congress 'Construction for Development', Cape Town, 2007.

J. Irizarry, M. Gheisari, B.N. Walker, Usability assessment of drone technology as safety inspection tools, J. Inf. Technol. Constr. 17 (2012) 194–212.

J. Irizarry, D.B. Costa, Exploratory study of potential applications of unmanned aerial systems for construction management tasks, J. Manage. Eng. 32 (3) (2016).

S. Lim, T. Le, J. Webster, R. Buswell, A. Austin, A. Gibb, Fabricating construction components using layered manufacturing technology, in: Global Innovation in Construction, Loughborough University, Leicestershire, UK, 2009.

P. Feng, X. Meng, J. Chen, L. Ye, Mechanical properties of structures 3D printed with cementitious powders, Constr. Build. Mater. 93 (2015) 486–497.

S. Kubba, Handbook of Green Building Design and Construction, second ed., Butterworth-Heinemann, 2017.

H. Penttilä, Describing the changes in architectural information technology to understand design complexity and free-form architectural expression, Electron. J. Inf. Technol. Constr. 11 (2006) 395–408.

O. Davtalab, Benefits of real-time data driven BIM for FM departments in operations control and maintenance, in: ASCE International Workshop on Computing in Civil Engineering (IWCCE 2017), Seatle, 2017.

H. Wei, S. Zheng, L. Zhao, R. Huang, BIM-based method calculation of auxiliary materials required in housing construction, Autom. Constr. 78 (2017) 62–82.

CHAPTER 3

Building Applications Using Lost Formworks Obtained Through Large-Scale Additive Manufacturing of Ultra-High-Performance Concrete

N. Gaudillière[1,2], R. Duballet[1,3], C. Bouyssou[1], A. Mallet[1], Ph. Roux[1], M. Zakeri[1] and J. Dirrenberger[1,4]
[1]XtreeE, Rungis, France
[2]Laboratoire GSA, Ecole Nationale Supérieure d'Architecture Paris-Malaquais, Paris, France
[3]Laboratoire Navier, UMR 8205, Ecole des Ponts, IFSTTAR, CNRS, UPE, Paris, France
[4]Laboratoire PIMM, Arts et Métiers-ParisTech, Cnam, CNRS UMR 8006, Paris, France

3.1 INTRODUCTION

Until recently, additive manufacturing (AM) techniques were confined to high value adding sectors such as the aeronautical and biomedical industries, mainly due to the steep cost of primary materials used for such processes. In the past decade, there has been development of large-scale AM in such domains as design, construction, and architecture by using various materials such as polymers, metals, and cement [1].

Historically, the first attempt at cement-based AM was made in 1997 [2] using an intermediate process between the classical powder bed and inkjet head 3D printing (3DP) [3] and fused deposition modeling (FDM) [4] in order to glue sand layers together with a Portland cement paste. Many groups have been involved with the development of large-scale AM for construction applications, all of which have been using processing routes derived from FDM or 3DP, although these vary depending on the chosen material and targeted application.

The aim of this chapter is to shed a new light on the perspective of 3D concrete printing in the construction industry, specifically using the lost formwork technique, by describing three case studies based on actual

3D Concrete Printing Technology
DOI: https://doi.org/10.1016/B978-0-12-815481-6.00003-8

building projects performed using the XtreeE[1] 3D concrete printing system. In Section 3.2, an introduction on designing structures for large-scale AM is given. In Section 3.3, the concrete formwork 3D printing technique is presented, followed by three case studies demonstrating the potential advantages of this technique for the construction industry. In Section 3.4, conclusions are drawn regarding the results obtained.

3.2 DESIGN FOR ADDITIVE MANUFACTURING

Generating and modeling shapes for AM follows specific rules from both processing constraints and functional objectives. According to Ref. [1], the concept of freeform previously used in the literature is not adequate or sufficient for describing 3D concrete printing. For a given printing process and automation complexity, one can attain specific types of topologies within a given time-frame and performance criterion for the structure. Although out of the scope of this chapter, design conditions for large-scale AM depend on many other parameters than merely the properties of extruded cementitious materials, such as the printing spatial resolution, overall size of parts to be printed, the environment, and the presence of assembling steps, etc. A tentative classification of such relationships between geometrical complexity, processing, and design is proposed in Ref. [5].

Processing constraints depend mostly on the fresh material properties in their viscous state as well as early-age behavior in the interaction with the building strategy and the stiffness of the structure being built, as reviewed in Ref. [11]. On the other hand, functional requirements will depend on the properties of the hardened material, the structural geometry for effective stiffness, and other functional properties such as thermal and sound insulation. See Ref. [1] for a geometrically induced thermal insulation case study. Both types of constraints have to be considered at the design stage.

Printing path generation is a critical step during the design phase. There are two main approaches to tool-path generation in the context of 3D printing: (1) 3D-to-2D slicing, by far the most common method adopted, yields planar layers of equal thickness built on top of each other. This approach is not optimal from the design and structural viewpoints as it will induce cantilevers when two consecutive layers have different sizes

[1] http://www.xtreee.com/.

and limit the attainable geometries; and (2) the tangential continuity method introduced in Ref. [1] in order to optimize the structure being built by creating layers of varying thickness. These layers exhibit a maximized surface area of contact between each other, hence stabilizing the overall structure. Moreover, this method actually exploits the possibilities of the process in terms of printing speed and flow for generating variations in the layer thickness. As of 2017, the tangential continuity method is no longer available in commercial software packages.

Along with the paradigm shift of AM comes the possibilities enabled by topology optimization, which aims at attaining the most efficient structure geometrically possible for a given set of requirements. Optimality in terms of industrial design has become more and more critical due to the scarcity of material resources and the need for lightweight structures. This technique has become well-established in the field of structural mechanics, especially when associated with finite element simulation. Classical methods, such as Solid Isotropic Material with Penalization (SIMP [6]) rely on node-based values to evaluate and optimize the geometry, that is, the number of design variables is equal to the number of elements available in the model at initialization. Then, the optimization procedure consists of determining at each element whether it should either remain as a material element or become a void element, that is, be removed. This technique has been applied to different scales; for instance with regards to the design of efficient building structures [7], or as a tool for designing micro- and nanoarchitectured materials [8].

A driving force for AM is its ability to produce more complex 3D shapes in comparison with casting or subtractive processes. This complexity allows us to design optimal structures based on topology optimization techniques. One of the main challenges currently is to modify optimization algorithms in order to account for the AM constraints, especially with regards to the processing parameters and structural stability while printing. A possible solution to these challenges would be to consider the multiphysics phenomenon aspect of 3D printing which involves the elastic stability of the overall structure being built, the kinetics of hydration, the evolving viscoplasticity of fresh cement, and the evolution of temperature within the printing environment, etc. All these physical problems, at multiple time and space scales, can be modeled on their own, but coupling them generates complexity and uncertainty regarding the process of 3D printing. Therefore, efforts should be concentrated on understanding and modeling the printing process for its multiple physical aspects, only then can optimization will be fully integrated with the processing, which would change the way 3D-printed structures are conceived today.

In order to fully exploit the potential of AM technologies in construction, various alternative ways have been explored regarding the building process—as done, for instance, in Refs. [1,9] for varying layer thickness and multifunctionality, and in Ref. [10] for the case of 3D-printing-aided robotic assembly for spatial structures. In this chapter, the path of concrete formwork 3D printing is explored in the light of several applications in architecture and construction.

3.3 CONCRETE FORMWORK 3D PRINTING

The process of large-scale 3D concrete printing developed by XtreeE has been presented in Ref. [1]. The overall process is summarized in Fig. 3.1.

Based on this process, a construction strategy can be derived for concrete formwork 3D printing. The principle consists in only 3D printing the formwork that is necessary for casting another structural material, such as ultra-high-performance concrete (UHPC) for fiber-reinforced concrete, or insulation material such as foamed concrete, in order to build multifunctional components, as shown on Fig. 3.2. The printed formwork is left inplace and becomes a *lost formwork*. An optimal tradeoff has to be taken into account from the early design steps concerning the ratio of printed material within the built part, which can be critical for reaching economic viability. Depending on the application considered, concrete formwork 3D printing can be more efficient than either all-3D concrete printing, or traditional building techniques, from both an economic and/or building strategy viewpoint. This assertion is demonstrated in the next three sections using different case studies.

3.4 CASE STUDY 1: POST IN AIX-EN-PROVENCE, FRANCE

3.4.1 Context

This 4 m-high freeform pillar is placed in the sports facilities of a school in Aix-en-Provence, France. It supports a concrete awning covering part of the playground, as shown in Fig. 3.3. The sports facilities project was mandated by the Aix-Marseille Metropolis. The pillar, part of this larger project, was handled by: Marc Dalibard as the architect (also the architect for the whole sports facilities building), Artelia as the structural engineering office, AD Concept as the construction company, LafargeHolcim as

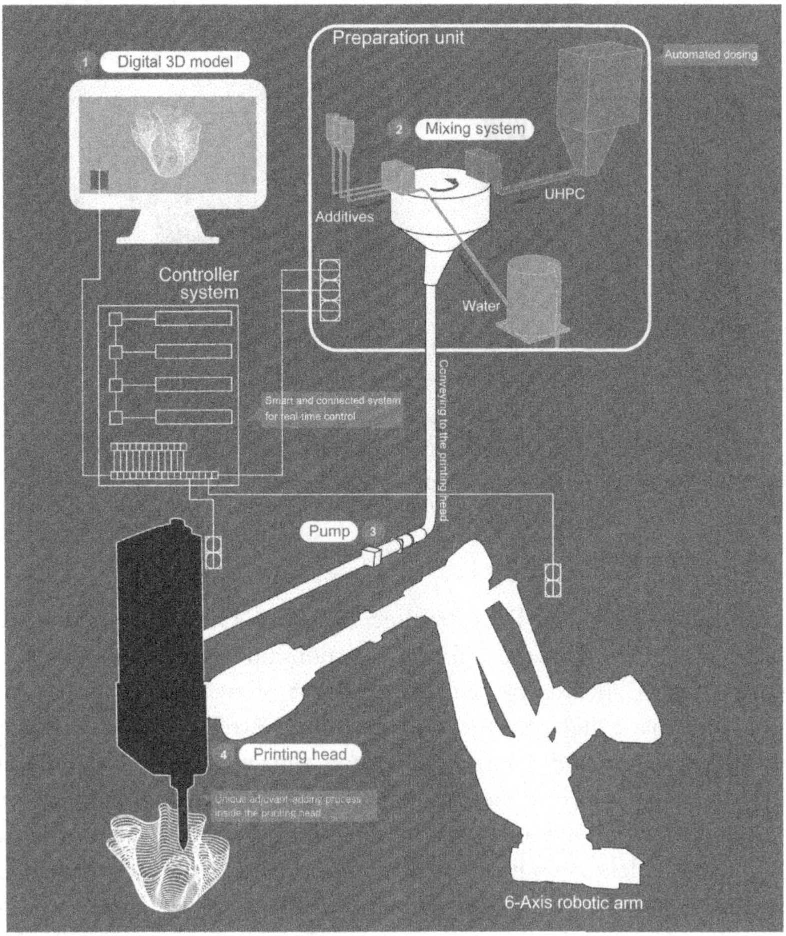

Figure 3.1 Workflow of the large-scale concrete 3D printing process.

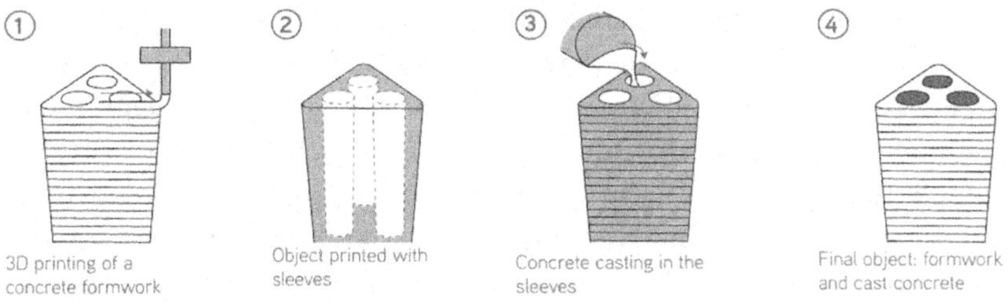

Figure 3.2 Schematic view of the concrete formwork 3D printing.

Figure 3.3 Post in Aix-en-Provence. *Photo by Lisa Ricciotti.*

the material supplier, and Fehr Architectural as the UHPC concrete caster. For the construction of the pillar, the responsibilities of the actors were divided: Marc Dalibard was also the manager of the overall project and defined the shape and placement of the pillar. Artelia supported XtreeE both during design and construction phases and was tasked with defining the load case on which to base the topological optimization and verifying the strength and stability of the printed pillar in accordance with the load case. LafargeHolcim supplied a specific 3D printing concrete, developed with XtreeE in an earlier collaboration. Fehr Architectural casted UHPC inside the pillar, a task requiring a specific license. Each one of the key players supported XtreeE in the definition of the fabrication strategy adopted for the pillar by providing inputs regarding their field of expertise.

XtreeE identified a fabrication strategy for the post and adapted the printing system developed earlier (presented in Section 3.3) according to the fabrication strategy and its requirements. During the design stages, XtreeE codefined the load case for the topological optimization with structural engineer Artelia and designed an exact shape for the pillar through the optimization. In the fabrication stages, XtreeE coded the manufacturing files for the printing system and performed the manufacturing before cosupervising the placing of the pillar on site with Dalibard.

3.4.2 Design

In the initial project designed by Dalibard, a complex truss-shaped pillar supporting the roof was already planned as shown on Fig. 3.4. But, although the idea of a complex truss-shaped pillar was featured, no viable design for the pillar existed at this stage of the project. XtreeE came in at this point and took over the design of the pillar, based on the sketches provided by Dalibard.

The design of the pillar is based as much on the formal intention highlighted in the sketches as on the constraints fixed by the building regulations in effect at the time and by the 3D printing manufacturing method.

As no building regulation existed regarding 3D-printed items integrated into buildings at the time of construction and in order to stick to the projected schedule, the choice was made to use the lost formwork manufacturing method, as introduced in Section 3.3. Instead of having to validate the pillar and its manufacturing method by using an Experimental technical appreciation (ATEx), a long and expensive procedure for experimental construction in France, the lost formwork made it possible to rely on existing building regulations for UHPC concrete.

For the 3D printing system developed by XtreeE, the main limitation encountered for complex truss shapes, such as the pillar, is the maximal inclination attainable for a given geometry. In the case of the Aix-en-Provence pillar, this issue was avoided by printing supports at the same time as the sought-after geometry, in order to enable any angle to be printed.

To define the precise shape of the pillar, we relied on a topological optimization method to ensure optimal use of matter in the truss. The entire circular volume containing the pillar is used as research space for

Figure 3.4 Initial design of the overall structure.

the truss to develop, and the applied load case included the weight of the concrete awning supported by the pillar as well as site-specific constraints (wind, etc.). The resulting final shape is shown in Fig. 3.5. A more thorough examination of the possibilities offered by topology optimization in the context of 3D concrete printing is available in Ref. [10].

Given the selected approach of lost formwork printing, the pillar was made of an outer shell that was 3D-printed and later filled with UHPC. The 3D-printing system in place at the time at XtreeE did not allow for printing the outer shell in one piece. Therefore, the pillar was divided in three smaller parts (cf. Fig. 3.6), each one to be filled with concrete and then assembled together to form the whole element. During casting, metallic female connectors were inserted in the concrete at each end of the parts. Male plugs were then used to connect the different parts of the pillar and assemble them.

3.4.3 Manufacturing

The manufacturing of the pillar included three stages: 3D-printing of the outer shell at XtreeE's headquarters in the south of Paris, France; casting the UHPC and integrating the connectors at the Fehr Architectural production facility in the north of France; and final assembly onsite in Aix-en-Provence. 3D-printing the outer shell inside the facility allows, like for UHPC casting, a precise control and monitoring of the environment, to ensure ideal temperature and humidity conditions for the concrete to behave as expected.

As a precaution, after running trials on smaller geometries similar to the pillar, it was decided to 3D-print the formwork in four parts rather than three. The concrete formwork took 15 hours to print, that is, approximately 3 hours and 45 minutes for each part of the post. Setting time was typically comparable to the setting time of C60 concrete, that is, about 2 hours. Once the formwork was successfully 3D-printed, an assembly trial was conducted at our facility to ensure the results were as precise as expected before shipping the parts to the site.

The casting of UHPC in each part of the pillar was then operated by the team from Fehr Architectural. To resist the hydrostatic pressure resulting from the casting, supports printed with the pillar were left in place until the UHPC set. The supports were then removed, as shown in Fig. 3.7, and the parts were shipped to the site in Aix-en-Provence. The definitive assembling of the parts was performed there, before installing

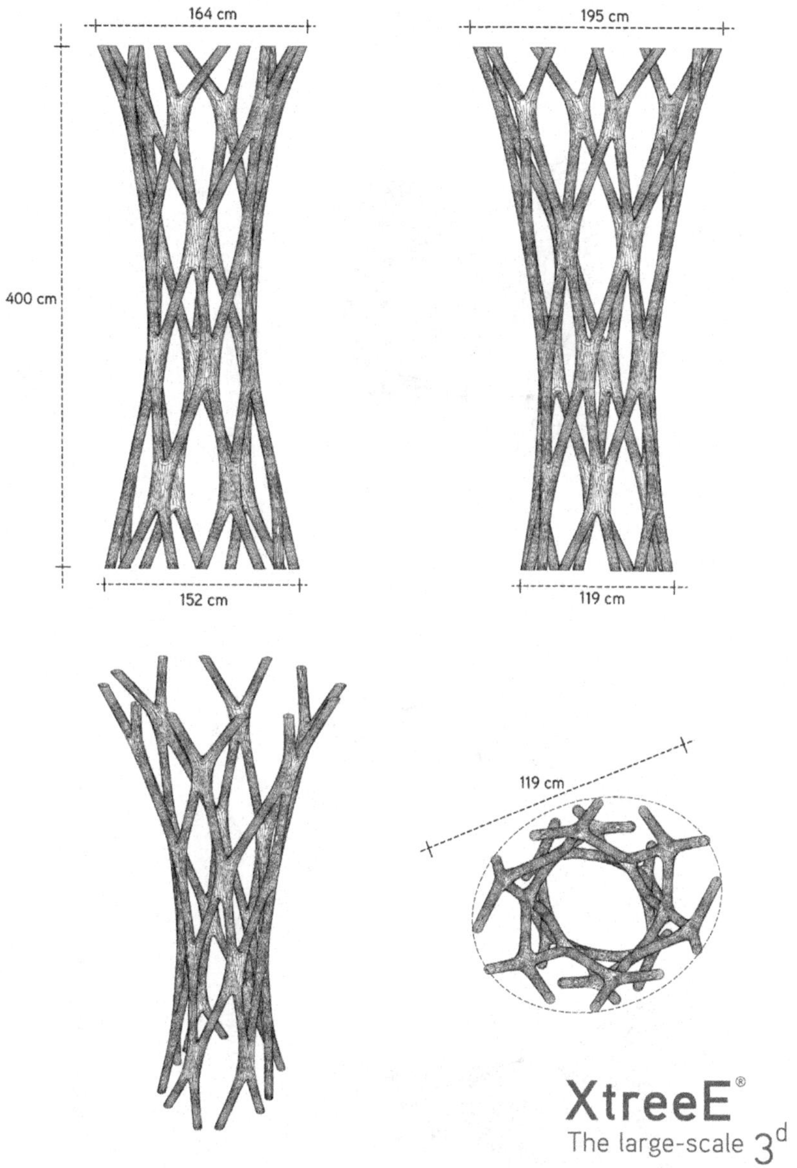

Figure 3.5 Final design of the pillar after topology optimization.

the pillar in place with sliding supports on top and at its feet. Finally, on request of the architect, the pillar was given a smooth finish by coating it to cover the line pattern specific to 3D-printed objects. The difference of surface roughness is shown in Fig. 3.8.

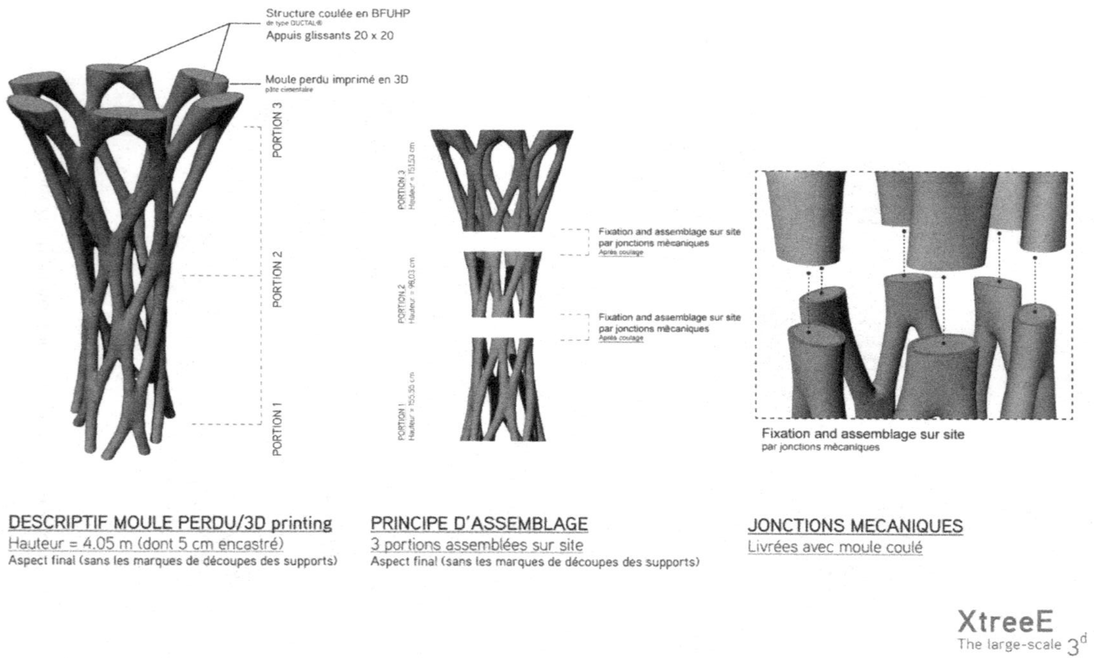

Figure 3.6 Splitting and assembling principles for the 3D-printed pillar system.

Figure 3.7 Cast and lost formwork assembly from which the printing supports were cut.

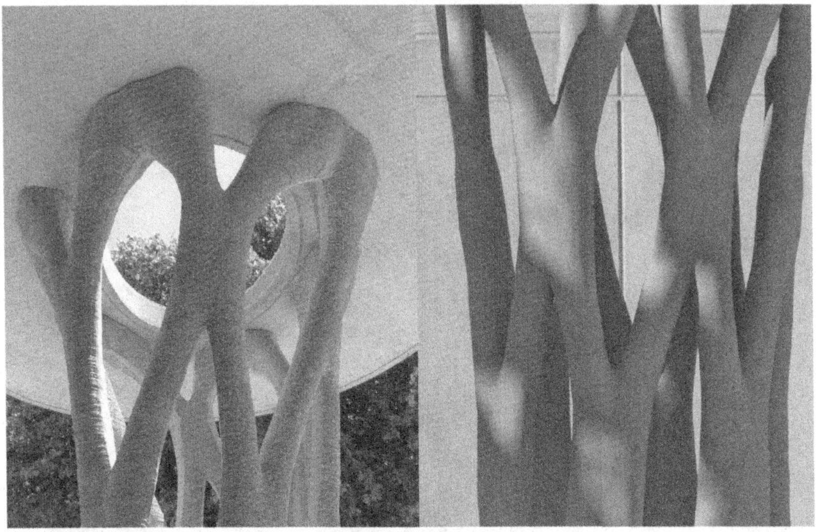

Figure 3.8 Before (left) and after (right, photo by Lisa Ricciotti) surface smoothing obtained through manual coating.

3.4.4 Comparison to Standard Building Methods

After completing the Aix-en-Provence pillar project, we conducted a study to compare our process for complex truss-shaped pillars fabrication to standard building methods. Two types of pillars were compared: a traditionally built pillar with a complex shape and geometry comparable to the Doha Convention Center pillars as shown in Fig. 3.9, but in a smaller scale; and a more complex pillar, in terms of shape, built with the AM lost formwork technique, such as the Aix-en-Provence pillar. The building process and gain in terms of material, build time, and workforce for both types of pillar is illustrated in Fig. 3.10. The data for 3D printing is based on the experience gained from the Aix-en-Provence pillar project, while the data for the traditional casting technique using steel molds is based on realistic values for implementation in the same socioeconomic environment, that is, Western Europe. It is noteworthy that the actual cost of steel molds has not been considered although it is likely to be the most costly aspect of a traditional casting technique. Even without considering the molds, concrete formwork 3D printing emerges as a cost-effective alternative for pillar construction.

Figure 3.9 Doha Convention Center, with truss-shaped pillars.

XtreeE (3D-printing)		*Traditional method* (*Steel mold*)	
Material consumption			
3D-printable concrete	650 kg	*UHPC concrete (casting)*	*2250 kg*
UHPC concrete (casting)	1600 kg		
Steel reinforcement (fibers)	50 kg	*Steel reinforcement (fibers)*	*50 kg*
(+ 3D-printable concrete for supports)		*+ Steel for the mold*	
Concrete consumption: −0 kg/−0.0%			
Steel consumption (mold): −100.0%			
Manufacturing process			
1. Digital design optimization and process design	3.0 days	*1. Digital design optimization and process design*	*3.0 days*
2. 3D-printing of the outer shell in concrete	2.0 days	*2. Creation of a mannequin*	*3.0 days*
3. Setting time	1.0 days	*3. Creation of a steel mold*	*3.0 days*
4. Casting of concrete inside the shell	1.0 days	*4. Casting of concrete inside the mold*	*1.0 days*
5. Setting time	5.0 days	*5. Setting time*	*5.0 days*
6. Transportation to site	0.5 days	*6. Transportation to site*	*0.5 days*
7. Installation and assembly	1.0 days	*7. Installation*	*1.0 days*
8. Coating	2.0 days	*8. Coating*	*2.0 days*
		(+ discard and recycle the mannequin and mold)	
Total	**15.5 days**	***Total***	***18.5 days***
Manufacturing time: −3.0 days/−16.2 %			
Workforce			
Designer/engineer for toolpath design	1 pers. - 3.0 days	*Designer/engineer for mold design*	*1 pers. - 3.0 days*
3D printing supervisor	1 pers. - 2.0 days	*Steel mold fabrication operator*	*2 pers. - 5.0 days*
3D printing operator	2 pers. - 2.0 days	*Concrete casting operator*	*1 pers. - 1.0 days*
Concrete casting operator	1 pers. - 1.0 days	*Transportation*	--- external ---
Transportation	--- external ---	*Assembly operator*	*2 pers. - 1.0 days*
Assembly operator	2 pers. - 1.0 days	*Coating operator*	*1 pers. - 2.0 days*
Coating operator	1 pers. - 2.0 days	*Mold and mannequin recycling*	--- external ---
Total (FTE)	**14.0 man.days**	***Total (FTE)***	***18.0 man.days***
Workforce use: −4.0 man.days/−22.2%			

Figure 3.10 Table of comparison between concrete formwork 3D printing versus traditional concrete casting in steel molds.

It could be interesting, as well, to further compare 3D-printed pillars with lost formwork to traditionally built pillars of semistandard shapes, such as the Nantes train station pillars depicted in Fig. 3.11.

The table presented in Fig. 3.10 highlights the potential gain from several viewpoints, some being more significant than others. Another element of comparison can be given by considering the total production price of the Aix-en-Provence pillar in relation to the total

Figure 3.11 Example of semistandard shaped pillar on the construction site of the new train station of Nantes, France.

production price of a traditionally built complex pillar: a 62.5% total price gain is obtained, based on our information for the price of the Aix-en-Provence pillar and quotes obtained for a traditional manufacturing. One of the main reasons for this price difference (on top of the gain identified on time, materials, and workforce) is the absence of a specific material and shaping for the mold, hence eliminated by using the lost formwork method. Furthermore, comparison with the Nantes train station pillars, cost brings to light the fact that prices become comparable when building at least 18 identical pillars with a traditional-mold manufacturing method.

The Aix-en-Provence pillar has also provided input on possible improvements for the lost formwork method, including getting rid of the supports by advancing the development of a 3D-printing system, gathering each of the manufacturing stages (i.e., 3D-printing, casting, assembling) at the same place to reduce transportations, as well as setting up construction regulations for 3D-printed concrete structural parts.

3.5 CASE STUDY 2: YRYS CONCEPT HOUSE IN ALENÇON, FRANCE

The YRYS Concept House was developed by Maisons France Confort (one of the largest individual house constructors in France) to support and implement various innovations in building methods. The YRYS Concept House has been designed to experiment new building solutions for a more environmentally friendly, evolutive, and adaptive housing. Among the 18 partners taking part in the construction of the project, XtreeE was tasked with designing and 3D-printing four pillars and an interior separation wall for the house.

The four pillars were placed in the exterior and support part of the upper floor, as shown in Fig. 3.12. The design and manufacturing method is similar to the Aix-en-Provence pillar, with adjustments made in accordance to the size of the pillars, which are smaller than the Aix-en-Provence pillar, to the load case, which is heavier than the Aix-en-Provence pillar, and in accordance with the inputs provided by the Aix-en-Provence project for the improvement of the lost formwork method.

The four pillars were each divided into two parts for easier transportation and for more rapid printing. The eight formwork parts were printed in a row, next to each other, as shown in Fig. 3.13. As the pillars were only divided into two parts and are significantly smaller in diameter than the Aix-en-Provence post, with less assembly points, they were not assembled with metallic connectors and/or plugs. The two 3D-printed shells of each pillar were stacked before UHPC casting, which yielded structural continuity for the pillar (cf. Fig. 3.14). The assembled pillars were finally brought onsite and placed.

The wall is an interior separation partition between the stair's circulation area and the main room. The use of 3D-printed lost formwork allows for a perforated design with a nonrepetitive pattern. This

Figure 3.12 Exterior pillars supporting part of the upper floor on the YRYS Concept house in Alençon, France.

Figure 3.13 Concrete formwork 3D printing for the 4 pillars (2 parts per pillar).

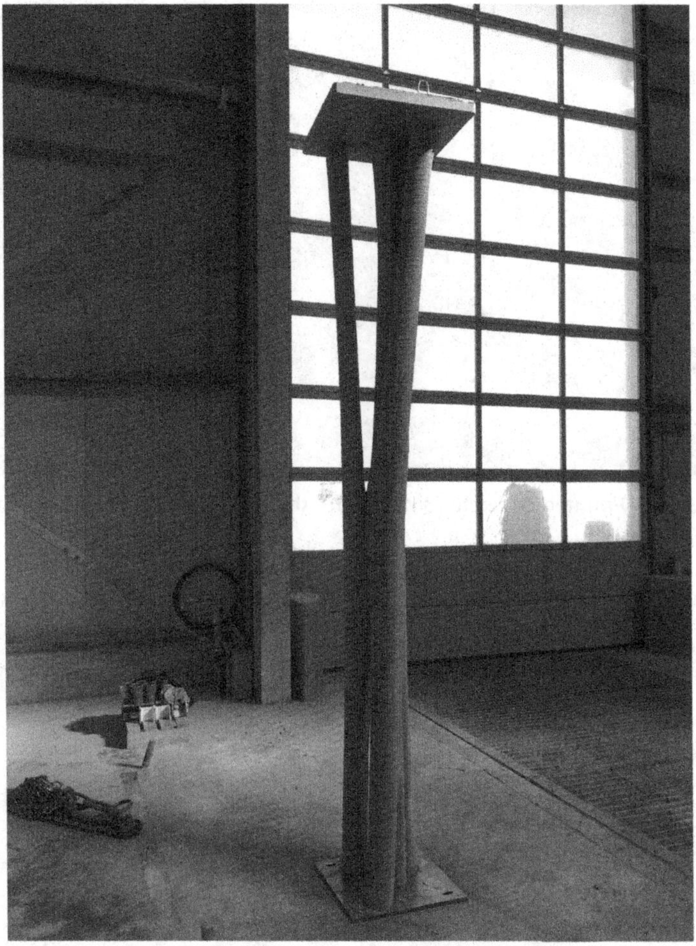

Figure 3.14 Assembled 3D-printed concrete formwork filled with UHPC casting.

manufacturing method also allowed for the efficient production of such perforated concrete partitions as only the contour was printed (cf. Fig. 3.15), and the body itself was then cast inside the contour, used as lost formwork. The same manufacturing stages that were implemented for the pillars were used to produce the wall: 3D-printing the formwork in our facility, shipping the parts on site, casting concrete, finishing, and finally placing the elements on the construction site, as shown in Fig. 3.16.

The YRYS Concept house pillars and interior wall produced by the concrete formwork 3D printing technique present comparable time,

Figure 3.15 3D-printed concrete formwork for the YRYS Concept house interior wall.

Figure 3.16 Interior wall made of the 3D-printed concrete formwork filled with cast concrete after finishing.

material, and workforce efficiency potential as the Aix-en-Provence pillar. Moreover, it has also been an opportunity for implementing some of the improvements considered following the previous project. In particular, casting onsite has allowed a significant reduction in transportation costs and the geometry of the pillars allowed for their printing without supports to be removed afterwards.

3.6 CASE STUDY 3: RAIN COLLECTOR IN LILLE

As part of a long-term collaboration with Point P TP, the French leader in construction materials distribution, we developed a process for the design and manufacturing of rain collectors. As rain collectors should be implemented underground on the city's drainage network, rapidity of execution is a priority for cities in order to avoid blocking the streets for a long period of time due to construction. Nevertheless, manufacturing rain collectors in the traditional way requires blocking the circulation, digging into the ground to take measurements, building adequate casing, casting, curing, setting the concrete, and, finally, rehabilitating the road. This lengthy process can be drastically improved by using the 3D-printing, lost formwork technique.

This first rain collector, measuring $2.15 \times 2.2 \times 2.6$ m, was manufactured in partnership with Point P TP and Sade, a utility network-specialized company located in Lille, France. XtreeE developed the design with Point P TP in order to fit the site constraints, such as the placing of pipes connected to the collector, as well the 3D-printing constraints. The rain collector features a 3D-printed concrete shell with a sinusoidal element inside. The sine part, developed in former projects as a way to

Figure 3.17 3D printing of the rain collector structure.

reduce thermal bridges (see Ref. [1]), also works as a stiffener element for the two external parts of the shell. The whole structure was printed within 9 hours, as shown in Fig. 3.17, directly on a reinforced concrete slab designed to support the lifting and placing operations of the collector onsite (cf. Fig. 3.18).

A comparison between the rain collector produced by XtreeE and the traditionally produced rain collectors was performed on three aspects based on the data supplied by Point P TP: material, time, and workforce-gain potential. The details and results are presented in Fig. 3.19.

Figure 3.18 On-site lifting and placing operations for the rain collector.

XtreeE (3D-printing)		Traditional method (On-site casting)	
Material consumption			
3D-printable concrete	2200 kg	*Standard concrete (casting)*	*3500 kg*
		Steel reinforcement (bars)	*200 kg*
(+ standard concrete for slabs and reinforcement)		*(+ wood for casing)*	
Concrete use: −1300 kg/−40.5%			
Production process			
1. Road excavation *	1.0 days	*1. Road excavation **	*1.0 days*
2. 3D scanning of drainage network intersection *	0.5 days	*2. Construction of the casing **	*3.0 days*
3. Digital design of the collector (including optimisation and toolpath)	2.0 days	*3. Concrete casting **	*1.0 days*
4. 3D printing of the outer shell and slabs	2.0 days	*4. Setting time **	*5.0 days*
5. Setting time	1.0 days	*5. Casing removal **	*1.0 days*
6. Transportation and crane	0.5 days	*6. Road rehabilitation **	*2.0 days*
7. Implementation (including connection to drainage network) *	1.0 days	*(+ discard and recycle the casing)*	
8. Road rehabilitation *	2.0 days		
Total	**10.0 days**	***Total***	***13.0 days***
Manufacturing time: −3.0 days/−23.0%			
* On-site production time: −8.5 days/−65.0%			
Workforce			
Road excavation and rehabilitation operators	3 pers. – 3.0 days	*Road excavation and rehabilitation operators*	*3 pers. – 3.0 days*
Operator for 3D scanning	1 pers. – 0.5 days	*Specialized worker for casing construction*	*2 pers. – 3.0 days*
Designer/engineer for design and toolpath	1 pers. – 2.0 days	*Concrete casting operators*	*2 pers. – 1.0 days*
3D printing supervisor	1 pers. – 2.0 days	*Material transportation*	--- external ---
3D printing operators	2 pers. – 2.0 days	*Casing discarding/recycling*	--- external ---
Transportation	--- external ---		
Concrete casting operator	1 pers. – 1.0 days		
Total (FTE)	**16.5 man.days**	***Total (FTE)***	***17.0 man.days***
Workforce use: −0.5 man.days/−76.0%			

Figure 3.19 Table of comparison between concrete 3D printing versu. traditional on-site casting for rain collectors.

3.7 CONCLUSIONS AND OUTLOOK

The various advantages of large-scale AM of ultra-high-performance concrete, as well as the concrete formwork 3D printing technique, are reviewed in this chapter based on the analysis of three case studies taken from industrial construction projects in France which were performed using the XtreeE 3D-printing technology. The multiple aspects of the potential socioeconomic gain for relying on AM are threefold: (1) materials saving by using the right amount of matter where needed, given that a multiobjective topology optimization computational framework is available; (2) time-saving by reducing the number of steps in the construction process, as well as being building information model-compatible for construction-planning strategies; and (3) workforce saving by limiting onsite manual building steps, therefore, enhancing safety on the construction site.

Although the lost formwork strategy allowed to get around experimental technical certification, further work should have to be conducted with certification authorities for the construction industry in order to define a legal and regulatory framework for 3D-printed structures. The technological feats presented in this work are commercially available, but a legal framework and an anaylsis of the economic market need to be developed in order for the 3D-printing technology to be transferred into the mainstream construction industry.

REFERENCES

[1] C. Gosselin, R. Duballet, P. Roux, N. Gaudillière, J. Dirrenberger, P. Morel, Large-scale 3d printing of ultra-high performance concrete — a new processing route for architects and builders, Mater. Des. 100 (2016) 102–109.

[2] J. Pegna, Exploratory investigation of solid freeform construction, Autom. Constr. 5 (1997) 427–437.

[3] E. Sachs, J. Haggerty, M. Cima, P. Williams, Three-Dimensional Printing Techniques, 1993.

[4] S. Crump, Apparatus and Method for Creating Three-Dimensional Objects, 1992.

[5] R. Duballet, O. Baverel, J. Dirrenberger, Classification of building systems for concrete 3d printing, Autom. Constr. 83 (2017) 247–258.

[6] M. Bendsøe, O. Sigmund, Topology Optimization, Springer, 2004.

[7] C. Cui, H. Ohmori, M. Sasaki, Computational morphogenesis of 3d structures by extended eso method, J. Int. Assoc. Shell Spatial Struct. 44 (1) (2003) 51–61.

[8] S. Zhou, Q. Li, Design of graded two-phase microstructures for tailored elasticity gradients, J. Mater. Sci. 43 (2008) 5157–5167.

[9] R. Duballet, C. Gosselin, P. Roux, Additive manufacturing and multi-objective optimization of graded polystyrene aggregate concrete structures, in: M. Thomsen, M. Tamke, C. Gengnagel, B. Faircloth, F. Scheurer (Eds.), Modelling Behaviour-Design Modelling Symposium, 2015, Springer, 2016.

[10] R. Duballet, O. Baverel, J. Dirrenberger, Humanizing Digital Reality, Chapter Design of Space Truss Based Insulating Walls for Robotic Fabrication in Concrete, Springer, 2018, pp. 453–461.

[11] R.A. Buswell, W.R. Leal de Silva, S.Z. Jones, J. Dirrenberger, 3D printing using concrete extrusion: a roadmap for research, Cement and Concrete Research 112 (2018) 37–49.

CHAPTER 4

Fiber-Reinforced Cementitious Composite Design with Controlled Distribution and Orientation of Fibers Using Three-Dimensional Printing Technology

Young Jun Nam, Young Kwang Hwang, Ji Woon Park and Yun Mook Lim
School of Civil and Environmental Engineering, College of Engineering, Yonsei University, Seoul, Republic of Korea

4.1 INTRODUCTION

In order to understand the behavior of composite materials, such as fiber-reinforced cementitious composites (FRCCs), it is important to understand the behavior of the matrix, fiber, and fiber–matrix interface. Interface behaviors have a great effect on the transformation of FRCCs with multiple cracks from quasi-brittle to ductile behavior (or strain hardening) [1]. Fig. 4.1 illustrates how enhanced interfacial properties between fibers and matrix can lead to advanced properties of FRCCs. To understand the interface behavior, we have to know the interfacial properties derived from the interfacial property test, the so-called fiber pull-out test. Conventional research usually provides the bond stress-slip curve resulting from the fiber pull-out test of a single fiber. However, in the case of FRCCs generally, the interfacial behavior of each fiber can interactively affect numerous fibers in the vicinity because a large number of fibers are mixed with the matrix. The random characteristics of the fiber distribution—which is affected by the casting process, material rheology, and configuration, among others—multiplies the gap between the pullout response of fiber and the real behavior of the composite [2]. Thus, if we know exactly how the fibers are positioned and oriented prior to the

3D Concrete Printing Technology
DOI: https://doi.org/10.1016/B978-0-12-815481-6.00004-X

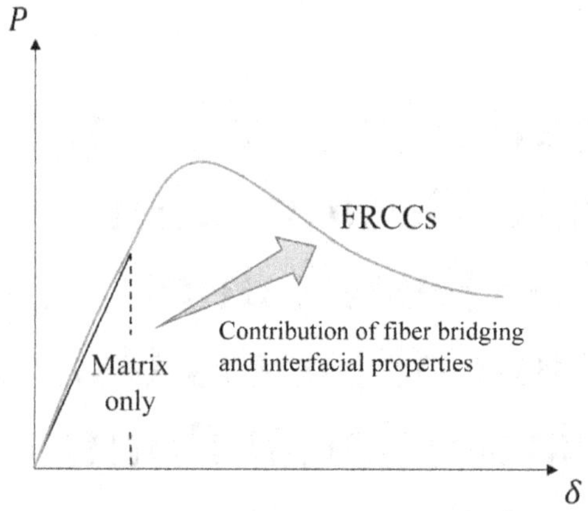

Figure 4.1 Advanced load-carrying capacity of FRCC due to the contribution of interfacial properties.

experiment, the interfacial data can be achieved through inverse analysis. Therefore, beyond the conventional method (single or aligned fiber pullout test), numerical and experimental research on the fiber–matrix interface behavior are crucial.

To figure out the interfacial behavior between the matrix and fibers, the following research can be considered. Firstly, material properties of an individual matrix and fiber, P-δ relations are plotted by using the direct tensile test or flexural test. Next, through a numerical analysis model with high accuracy, a virtual experiment is performed with the interfacial characteristics. In this kind of simulation, it is necessary to implement exactly the same conditions as the real experiment. In order to do this, geometric information (both ends of each fiber) from the distribution and orientation state of the fibers is required. Through 3D image scan technology, such as computer tomography (CT), the exact distribution of fibers inside the FRCC can be extracted. Information on how the fibers are distributed can be represented in the form of a point cloud in 3D space, as shown in Fig. 4.2A. By selecting a constraint condition (the volume or length of the element) on the individual content, the initial information including voids and other noises can be filtered out, as shown in Fig. 4.2B. For this kind of filtered extraction, it is necessary to develop an adequate segmentation algorithm which automatically extracts single fibers from tomographic 3D data [3–5]. However, depending on the characteristics of the CT equipment and material, the resolution may not be sufficient to gain information on a section of the

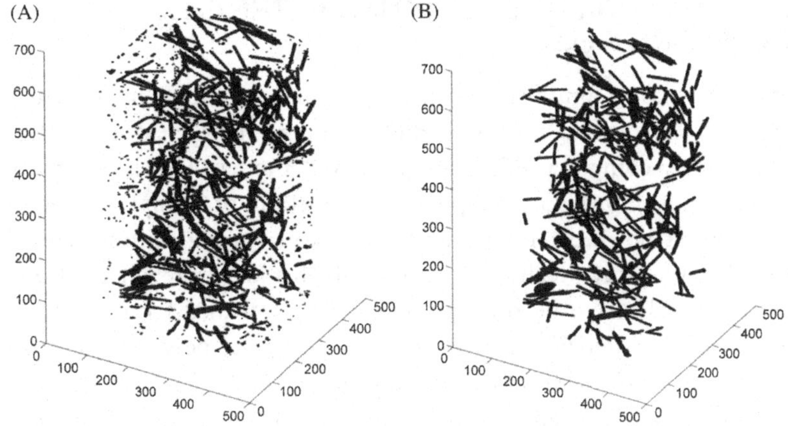

Figure 4.2 3D information for fiber contents from CT images. (A) Extracted image; and (B) Filtered image.

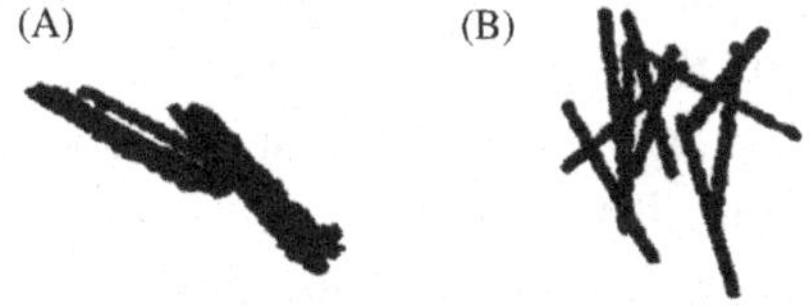

Figure 4.3 Geometrically intricate form of fibers observed from CT images. (A) Fibers concentrated in a certain region; and (B) Intertwined fibers.

fiber. Additionally, as illustrated in Fig. 4.3, in case of fibers clustered and intertwined with each other, it is very difficult to implement an algorithm that extracts the geometric information of individual fibers. Thus, because of the inaccuracies in the spatial information of the fibers when using this methodology, numerically implementing the experimental behavior of the fibers is quite difficult.

In order to overcome this problem and implement identical experimental conditions for both laboratory and virtual (numerical) experiments, this study proposes utilizing 3D printing technology to study the interface and whole behavior of the FRCC. With this methodology researchers can proactively determine the volume fraction, distribution, and orientation of fibers and their interfacial properties. The geometrical data can then be imported to the numerical analysis model to accurately predict the behavior represented in the experiment.

4.2 APPLICATION OF 3D PRINTING IN THE CIVIL ENGINEERING FIELD

Three-dimensional printing, formally known as additive manufacturing (AM), is a layer-based manufacturing technique that create a 3D solid object from a computer-aided design (CAD) model. This layer-by-layer manufacturing allows for the precise production of complex structures which cannot be made using traditional manufacturing routes. With the use of this innovative technology, materials that were previously deemed as non-printable have become printable. Currently, AM can produce a wide range of materials including metallic, ceramic, polymers, and their combinations in the form of composites [6].

Whether in the manufacturing process, engineering test, or design process, numerous industries utilize AM and benefit from its advantages [7]. In the field of civil engineering, however, the utilization of AM is less effective compared to other engineering fields because of the large-scale of infrastructures being built—such as buildings, bridges, tunnels, harbors, and dams—along with their unique characteristics and the condition of the construction sites. Consequently, the application of AM in this field has been rather slow and challenging [8].

The need to automate the construction process arose in the 1990s to effectively manage the use of raw materials and solve problems such as low labor efficiency; more than 40% of raw materials are consumed globally in the civil engineering field [9,10]. As a result, a cementitious material AM process called contour crafting (CC) was introduced ([11]; Fig. 4.4A). CC is a free-form construction method that constructs concrete buildings through extrusion of cementitious material. Lim et al. [12] pointed out some limitations of CC regarding its excessive steps, including molding and installing reinforcement. Thereafter, another free-form construction method called concrete printing was developed ([13]; Fig. 4.4C). Dini [14], on the other hand, invented a largescale powder-based 3D printer named D-shape (Fig. 4.4B), which is similar to stereolithography except that Dini took a different approach by using other materials. The number of communities researching AM for construction showed an enormous increase from 2012 [15–17].

In short, largescale AM has been developed as a need for such the method, especially with concrete material, in construction increased [18]. However, it should be noted that the application of AM to the various reinforcements inside of the matrix is still at the beginning stages.

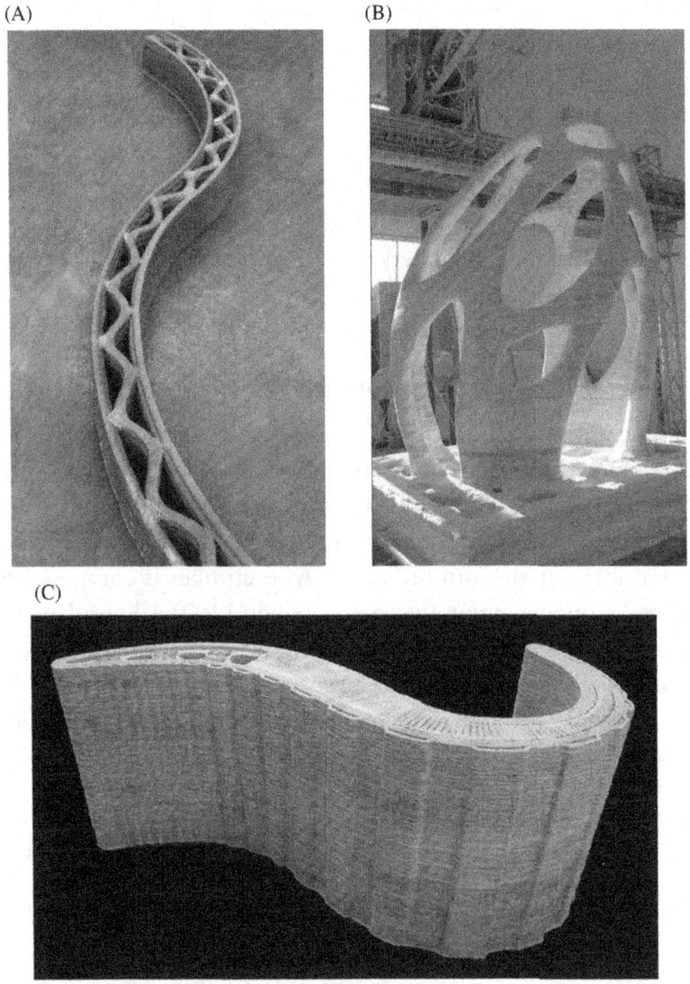

Figure 4.4 Examples of largescale additive manufacturing processes. (A) Contour crafting [11]; (B) D-shape [14]; and (C) Concrete printing [13].

4.3 3D MODELING OF REINFORCEMENTS

Concrete is a quasi-brittle material that can exhibit a failure without yielding. While strong in compression, concrete is weak in tension. To overcome its structural imbalances and to improve its strength and overall ductility, concrete has many types of reinforcement added to the crack-initiating section (lower part of the concrete specimen), which is why four different types of fibers were fabricated for 3D printing and tested in this study. The four different fiber structures are:

1. Regular mesh (RM)
2. Functional mesh (FM)
3. Randomly distributed fibers (RFs)
4. Functionally distributed fibers (FFs)

The obtained structures can mainly be classified as either (1) mesh or (2) fiber. Each category exhibited two different orientations between the specimen's bottom and top.

4.3.1 Generating Fiber Coordinates

Two considerations were made when generating points. Firstly, a boundary condition-setting process was conducted to prevent the possibility of some fibers protruding from the surface of the specimen. As shown in Fig. 4.5A, the boundary was created inside the concrete mold at a distance of half of a fiber length from the mold. As each fiber is composed of two coordinates, a middle point was generated inside the boundary using a random function. Two end points of fibers were generated in accordance to specified variables of length and number of fibers, and random variables of azimuthal angle (θ) and polar angle (φ). By providing constraints on directional angles (θ, φ), closely orienting the fibers in direction of the

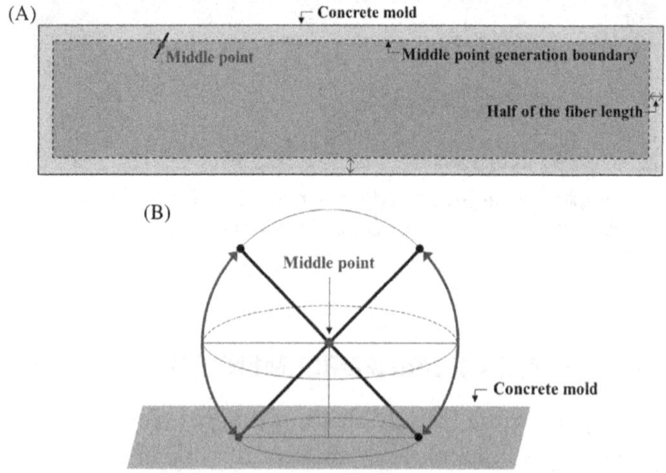

Figure 4.5 Two different fiber distributing methodology. (A) Random distribution of the middle points with random orientation; and (B) Fibers generation considering angular boundary.

bottom surface was possible. However, because the middle point was placed inside the boundary, placing the fibers closer to the center of the bottom surface of the beam where tensile strength is at its weakest was difficult.

To supplement the first method, a second method attempted to distribute the fibers without using a physical boundary condition. As shown in Fig. 4.5B, after generating the middle point, if a fiber had a possibility of protruding the surface, the end points were generated within the angular boundary. The fibers created from the second method, compared to the first method, were closely located to each other and to the surface of the concrete mold. Therefore, the 3D modeling of fibers used the second method.

4.3.2 Adjustment for Self-Sustaining Support

Through the fiber coordinates generating process, fibers are distributed throughout the entire domain. However, because the fibers cannot sustain their formation and structure through distribution alone, some kind of support was necessary to hold the fibers in desired locations. Thus, connections between fibers were employed for its self-sustainability. Concerning the different behavior when two fibers are connected by a support, for this study connection size was determined to be half the size of the fibers with consideration of printer resolution.

As a methodology to connect all the fibers, distances between the fibers were measured and two fibers with the shortest distance between them were joined. Accordingly, the fiber structure was built using a 3D CAD program (Rhinoceros 5.0). Fig. 4.6 shows some parts of the fibers that were extracted for better visualization of how the fibers are interconnected. Fig. 4.7 shows printed results of the mesh structures.

(A) (B)

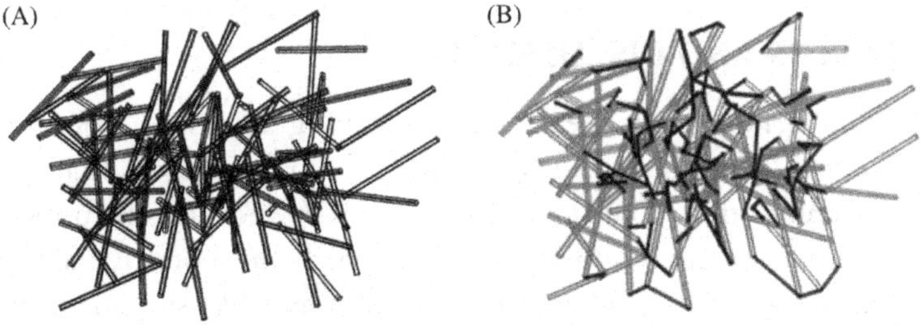

Figure 4.6 Example parts of fibers showing: (A) Original fibers distribution; and (B) Completed fiber structure by connecting fibers.

Figure 4.7 3D-printed fiber reinforcements: (A) RFs; and (B) FFs.

4.3.3 Mesh

Two different types of mesh structures with different distances between the nodes were generated. Before producing the meshes, their layers were divided into three sections. The first mesh (shown in Fig. 4.8A) was composed of the elements with even sizes and identical distances between the layers. In the case of the second mesh (shown in Fig. 4.8B), the layers were skewed downward in order to locate more fibers on the tension part of the specimen. Fig. 4.9 shows printed results of the two mesh structures.

4.3.4 Summary of the Whole Procedure

The process starts with a model designed from the 3D CAD program. The model is translated into a standard tessellation language (STL) file containing information on each layer. With the file, 3D printer manufactures the layers from bottom to top. As both fiber and mesh models have their own structure, ordinary mixing procedures are unnecessary. Therefore, the printed fiber reinforcements are preplaced in the molds, thereafter cement-based slurry is infiltrated through the structures much alike the process taken when making slurry-infiltrated fibrous concrete (SIFCON), as shown in Fig. 4.10 [19]. The whole process of making the fiber structure is shown in the Fig. 4.11.

4.4 RESULTS AND DISCUSSION

Three-point bending tests were performed using an Instron 3369 instrument (Fig. 4.12). In Fig. 4.13, the experimental responses of each type of specimens are plotted. The strengths of the RFs and FFs structures were lower than that of the control specimen which did not have any reinforcement inside; on the other hand, the RM and FM specimens exhibited higher strengths compared to the control one. Moreover, the specimens (FFs and FM) with functional reinforcements showed lower

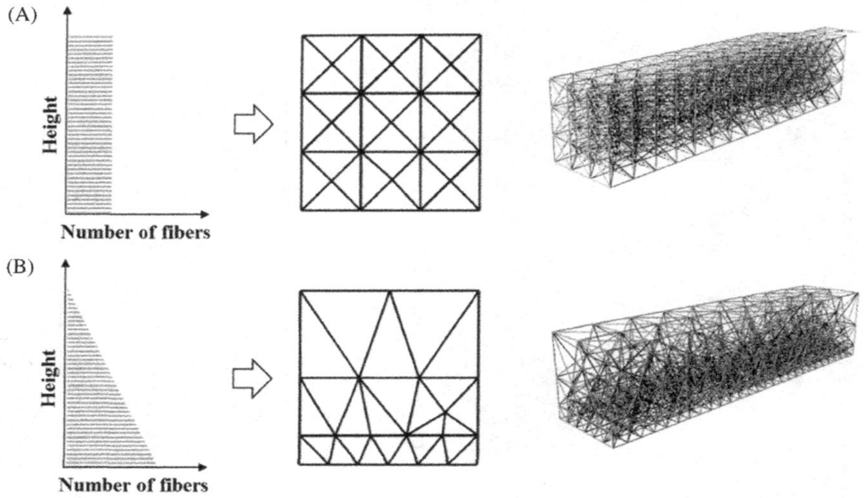

Figure 4.8 Schematic illustration of modeling mesh structures. Side and overall view of: (A) RM; and (B) FM.

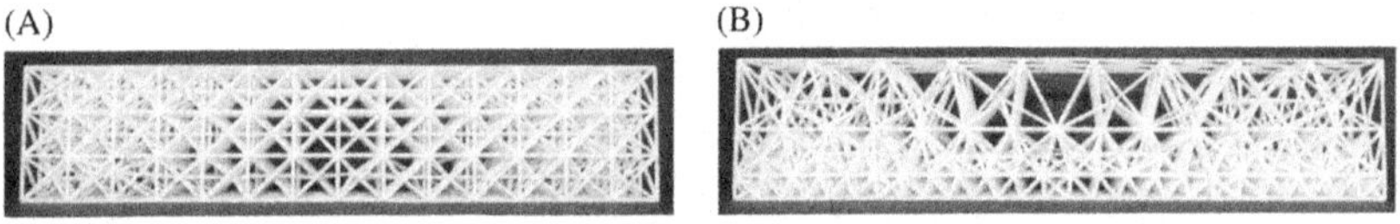

Figure 4.9 3D-printed mesh reinforcements: (A) RM; and (B) FM.

Figure 4.10 Fabrication of the specimens following slurry-infiltrated fibrous concrete process.

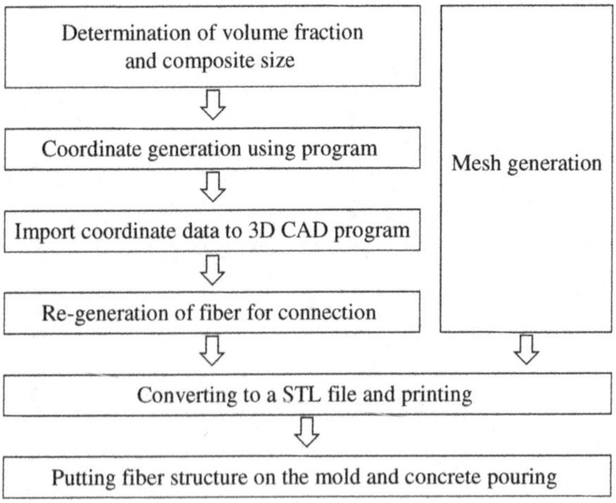

Figure 4.11 Flow chart of creating fiber-reinforced cementitious composites.

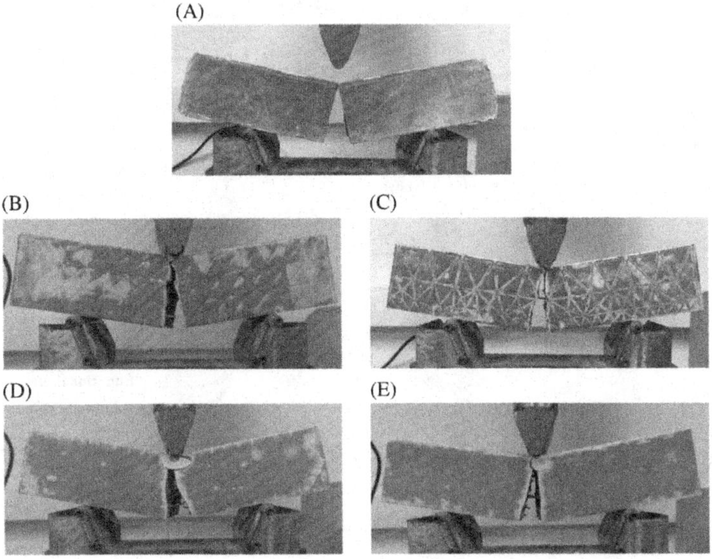

Figure 4.12 Results of the three-point bending tests performed: (A) Control specimen, (B) RM, (C) FM, (D) RFs, and (E) FFs.

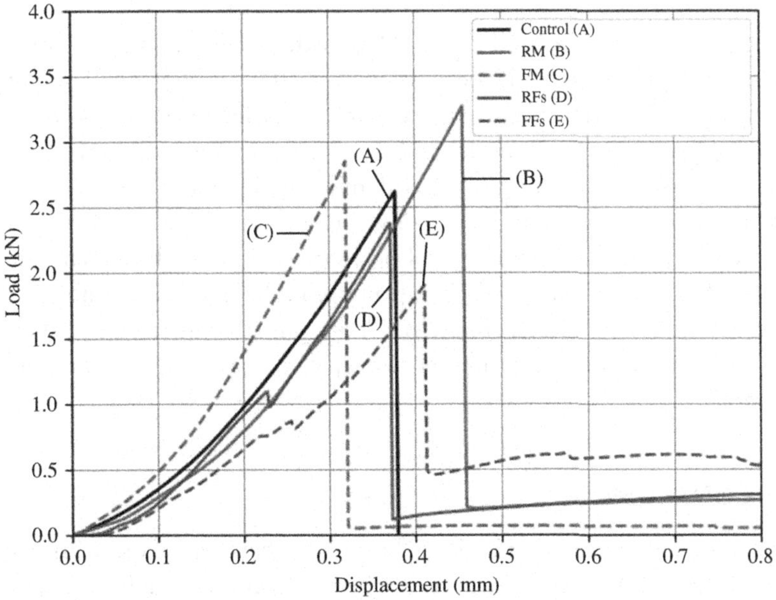

Figure 4.13 The corresponding load-displacement curves [20].

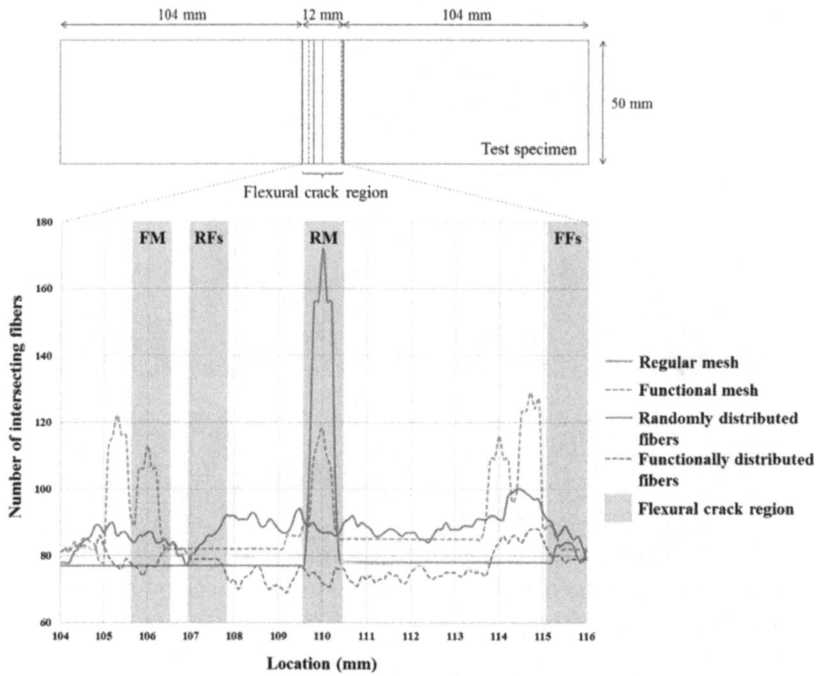

Figure 4.14 Locations of the crack planes from each specimen and the corresponding numbers of intersecting fibers [20].

strength than the ones (RFs and RM) with no functional distribution. These strength order could be explained by the number of fibers intersecting near the crack plane. Fig. 4.14 roughly describes the locations of cracks obtained for each model and the corresponding numbers of intersecting fibers. Unlike the fiber-based models, the mesh models contain several points where many fibers converge, and therefore the corresponding distribution plots exhibit several peaks (they illustrate how many fibers cross each particular plane which explains why different values of strengths were obtained for different models). The number of intersecting fibers at the crack plane of each model was proportional to the composite strength.

4.5 CONCLUSION

Generally, during the construction of FRCCs, fibers are mixed with matrices at a variety of volume fractions of fibers, which causes difficulty in controlling the fiber distribution and orientation. The proposed methodology on FRCCs fabrication exhibits a feasibility to control the location and orientation of the fibers. The material used in this study was to examine whether AM can produce the complex structure required. According to the results, due to the low tensile strength of the 3D printed reinforcement and its brittle characteristics, neither strain-softening nor strain-hardening were exhibited. However, along with the adoption of higher strength materials, such as metal and polyvinyl alcohol, and with the development of related technologies in AM, the performance of FRCCs is expected to be maximized with controlling fibers' distribution and orientation. In addition, it would be possible to modify the surface of the printed fiber to control the interfacial properties in near future.

ACKNOWLEDGMENTS

This work represents an ongoing research study supported by the Education-research Integration through Simulation on the Net (EDISON) Program through the National Research Foundation of Korea (NRF) funded by the Ministry of Science, ICT and Future Planning (grant No. NRF-2014M3C1A6038855).

This work was supported by the Korea Institute of Energy Technology Evaluation and Planning (KETEP) and the Ministry of Trade, Industry and Energy (MOTIE) of the Republic of Korea (No. 20171510101910)

REFERENCES

[1] T. Kanda, V.C. Li, Effect of fiber strength and fiber-matrix interface on crack bridging in cement composites, J. Eng. Mech. 125 (3) (1999) 290–299.

[2] J. Kang, K. Kim, Y.M. Lim, J.E. Bolander, Modeling of fiber-reinforced cement composites: discrete representation of fiber pullout, Int. J. Solids Struct. 51 (10) (2014) 1970–1979.

[3] C. Redon, L. Chermant, J.L. Chermant, M. Coster, Automatic image analysis and morphology of fibre reinforced concrete, Cem. Concr. Compos. 21 (5) (1999) 403–412.

[4] M. Teßmann, S. Mohr, S. Gayetskyy, U. Haßler, R. Hanke, G. Greiner, Automatic determination of fiber-length distribution in composite material using 3D CT data, EURASIP J. Adv. Signal Process. 1 (2010) 545030.

[5] G. Zak, C.B. Park, B. Benhabib, Estimation of three-dimensional fibre-orientation distribution in short-fibre composites by a two-section method, J. Compos. Mater. 35 (4) (2001) 316–339.

[6] S.A. Tofail, E.P. Koumoulos, A. Bandyopadhyay, S. Bose, L. O'Donoghue, C. Charitidis, Additive manufacturing: scientific and technological challenges, market uptake and opportunities, Mater. Today (2017).

[7] M.W. Hyer, Stress Analysis of Fiber-reinforced Composite Materials, updated ed., DEStech Publications, Lancaster, Pennsylvania, 2009.

[8] Y.J. Nam, J.W. Park, S.H. Yoon, Y.M. Lim, 3D printing of fibers and reinforcements for cementitious composites to maximize the fracture-resisting performance of FRCCs, in: Proceedings of the 9th International Conference on Fracture Mechanics of Concrete Structures, 2016, pp. 1–7.

[9] D.M. Roodman, N.K. Lenssen, J.A. Peterson, A Building Revolution: How Ecology and Health Concerns are Transforming Construction, Worldwatch Institute, Washington, DC, 1995 (pp. 11-11).
[10] A. Warszawski, R. Navon, Implementation of robotics in building: current status and future prospects, J. Constr. Eng. Manage. 124 (1) (1998) 31−41.
[11] B. Khoshnevis, R. Dutton, Innovative rapid prototyping process makes large sized, smooth surfaced complex shapes in a wide variety of materials, Mater. Technol. 13 (2) (1998) 53−56.
[12] S. Lim, T. Le, J. Webster, R. Buswell, S. Austin, A. Gibb, et al., Fabricating construction components using layer manufacturing technology, Global Innovation in Construction Conference, Loughborough University, Loughborough, UK, 2009.
[13] S. Lim, R.A. Buswell, T.T. Le, S.A. Austin, A.G. Gibb, T. Thorpe, Developments in construction-scale additive manufacturing processes, Autom. Constr. 21 (2012) 262−268.
[14] Dini, E. D-shape printers. 2007, <http://d-shape.com/d-shape-printers>.
[15] G. Cesaretti, E. Dini, X. De Kestelier, V. Colla, L. Pambaguian, Building components for an outpost on the Lunar soil by means of a novel 3D printing technology, Acta Astronaut. 93 (2014) 430−450.
[16] R.J.M. Wolfs, T.A.M. Salet, L.N. Hendriks, 3D printing of sustainable concrete structures, in: Proceedings of the International Association for Shell and Spatial Structures (IASS) Symposium 2015, Amsterdam, 2015.
[17] L. Feng, L. Yuhong, Study on the status quo and problems of 3D printed buildings in China, Glob. J. Hum. −Soc. Sci. Res. 14 (5) (2014).
[18] P. Wu, J. Wang, X. Wang, A critical review of the use of 3-D printing in the construction industry, Autom. Constr. 68 (2016) 21−31.
[19] Y. Farnam, M. Moosavi, M. Shekarchi, S.K. Babanajad, A. Bagherzadeh, Behavior of slurry infiltrated fibre concrete (SIFCON) under triaxial compression, Cem. Concr. Res. 40 (11) (2010) 1571−1581.
[20] Y.J. Nam, Y.K. Hwang, J.W. Park, Y.M. Lim, Feasibility study to control fiber distribution for enhancement of composite properties via three-dimensional printing, Mech. Adv. Mater. Struc. (2018).

FURTHER READING

M. GuoWei, W. Li, J. Yang, State-of-the-art of 3D printing technology of cementitious material-An emerging technique for construction, Sci. China Technol. Sci. (2017) 1−21.
M. Yossef, A. Chen, Applicability and limitations of 3D printing for civil structures, in: Civil, Construction and Environmental Engineering Conference Presentations and Proceedings, vol. 35, 2015, pp. 237−246.

CHAPTER 5

Properties of 3D-Printed Fiber-Reinforced Portland Cement Paste

Manuel Hambach, Matthias Rutzen and Dirk Volkmer
Chair of Solid State and Materials Chemistry, University of Augsburg, Augsburg, Germany

5.1 INTRODUCTION

Regular Portland cement-based construction materials exhibit a high compressive strength (around 20–60 MPa for general usage) [1], but they fall short in terms of tensile and flexural strength values (3–10 MPa) for plain cement pastes [2,3]. As a common solution, steel-reinforcement is placed in the formworks in order to improve flexural strength of the cementitious composite. However, steel-reinforcement results in time- and material-consuming labor costs during the construction process since the steel has to be placed and fixed by hand in the construction molds. To avoid the disadvantages of steel-reinforcement, mortars and concretes containing high-performance synthetic fibers (e.g., glass or carbon fibers) were introduced in the 1960s in the scientific literature [4]. The resulting composite materials show a remarkable increase in flexural and tensile properties leading to an ultimate flexural strength of up to 50 MPa, one order of magnitude higher than the corresponding value for plain concrete (without further reinforcement) [5–7]. Since the late 1990s, efforts have been made on extruding fiber-reinforced cement pastes in a simple extrusion process in order to increase the density of the cement paste and to influence, to a certain extent, the orientation of fibers in the cementitious matrix [8–11]. A manual nozzle-injection process realizes an extraordinarily high degree of fiber orientation in reinforced cement pastes. If the mean length of the reinforcing fibers is greater than the nozzle diameter, orientation of the fibers parallel to the path of movement through the nozzle takes place. With three vol.% fiber content, a flexural strength of 120 MPa can be realized as reported recently [12]. Since extrusion techniques forcing fiber alignment are capable of producing

3D Concrete Printing Technology
DOI: https://doi.org/10.1016/B978-0-12-815481-6.00005-1

high-performance fiber-reinforced cementitious composites, the next technological step should head toward a fully automated layer-based fabrication, known as additive manufacturing (AM) or 3D printing, and has already been introduced into the manufacturing of cement-based materials [13–15]. Similar developments towards 3D-printing can also be observed in related material classes, for example, biomedical materials [16,17], polymer composites [18,19], or bone replacement materials [20–24].

3D printing is a technique first introduced in the late 1980s and is gaining more and more importance in production processes during recent years [25] since it enables the fabrication of complex, multiscale structures through computer-aided design (CAD) [26,27]. Continuous extrusion or fused deposition modeling (FDM) is based on extruding a (semi-)liquid paste, typically a molten thermoplastic polymer, and depositing the material by a computer-controlled extrusion system [28]. In this chapter we apply FDM 3D-printing on cementitious materials by storing the ready-mixed paste in a reservoir and extruding the material by a movable piston system of a dispenser in order to form layered structures. However, the need for steel reinforcement hampers the fully automated processing of free-formed 3D construction structures made from Portland cement-based materials. As a possible solution, common steel-reinforced concrete might be substituted by a 3D-printable short carbon fiber-reinforced cement exhibiting high flexural strength, thus minimizing the content of reinforcement steel in load-bearing structures. In the next section we describe the required materials development in conjunction with a suitable 3D printing approach, which point in this direction.

5.2 EXPERIMENTAL PROCEDURE

Fiber treatment: To provide good fiber dispersion [29] and fiber bonding to the cementitious matrix [30,31], all fibers were heat-treated to remove the fibers' sizing and to ensure a hydrophilic fiber surface. For carbon fibers a 400°C and for glass and basalt fibers a 500°C thermal treatment ensures the removal of the polymer sizing, as illustrated in supplementary Fig. A5.1.

Cement paste preparation: For all specimens 61.5% by weight of type I 52.5 R Portland cement (Schwenk Zement KG, Ulm, Germany) was used along with 21% by weight of silica fume (Microsil, Elkem, Oslo, Norway), 15% by weight of water and 2.5% by weight of a water-reducing agent (Glenium ACE 430, BASF SE, Ludwigshafen). The water cement ratio was 0.3 (including water-reducing agent). To avoid thickening of the cement paste during 3D printing 0.3% by weight of a hydration inhibitor (PANTARHOL 85, Ha-Be, Hameln, Germany) was added. The infill mortar for hierarchical specimens was prepared from 40.0% by weight of type I 52.5 R Portland cement and 60.0% by weight of sand (Gebenbacher Sand) along with a water cement ratio of 0.4. All solid components, apart from reinforcement fibers, were mixed in dry state. Water, water-reducing agent, and hydration inhibitors were added and mixed with a rotary mixer at 600 RPM until a homogeneous mixture was obtained. Finally, the reinforcement fibers were added and the mixture was stirred again at 50 RPM until the fibers became uniformly dispersed. Characteristic properties of the reinforcement fibers employed in our studies are presented in Table 5.1.

Nozzle-injection technique: The reinforced cement paste was extruded through a disposable syringe (B. Braun Melsungen AG, Melsungen, Germany, volume: 20 mL, nozzle diameter: 2 mm) by hand. After injecting the mixture into PTFE molds of 60 × 13 × 3 mm (specimens for flexural strength) or 60 × 50 × 15 mm (specimens for compressive strength) it was densified by putting it on a plate vibrator for 60 seconds. Specimens intended for compressive strength tests were cut into cubes of 15 × 15 × 15 mm with a low-speed saw after 7 days of storage [12].

3D-printing system and specimen preparation: 3D objects fitting the desired specimen dimensions were sketched in FreeCAD (build 0.15) CAD-software and a 3D-print preparation software (Cura, build 15.14.02) [32] was used to create g-code print paths. For 3D printing a DeltaWASP 20 × 40 3D-printer (WASProject, Ravenna, Italy) was used along with the WASP Clay Extruder Kit 2.0 having a nozzle 2 mm in diameter

Table 5.1 Properties of reinforcement fibers used for the present study

Fiber	Diameter (μm)	Tensile strength (MPa)	Young's modulus (GPa)	Length (mm)	Supplier
Carbon fiber (HT C261)	7	3950	230	3	Toho Tenax
Glass fiber (AR Force D-6)	20	3500	72	6	DuraPact
Basalt fiber (BS 13 0064 12)	13	4200	93	6	Incotelogy

attached to the extruder. After the fiber-reinforced cement paste was mixed, it was immediately filled into the storage container of the 3D-printing system, a pressure of 3 bar was applied on the piston of the storage container and kept constant at this value during specimen printing, which was typically accomplished within 3 hours. Best results were obtained by adjusting a layer height of 1.5 mm and a print speed of 30 mm/s.

Specimen storage: After preparation the samples were stored for 24 hours in a desiccator providing 100% relative humidity followed by placing the samples for another 6 days in a water bath (about 0.5 L for five specimens). The samples were stored for another 21 days in a desiccator over-saturated sodium bromide solution (at 58% relative air humidity). All mechanical tests were immediately performed after a storage period of 28 days. Prior to testing, all specimens were polished in order to obtain clean and parallel object surfaces required for mechanical tests.

Flexural strength measurements: Flexural strength measurements were carried out using a 3-point bending test setup. The testing machine was a Zwick/Roell BT1-FR05TN.D14 with a 500 N or 5 kN load cell attached (depending on specimen dimensions and expected maximum load). By measuring the maximum force, F, the flexural strength, fs, can be calculated by Eq. (5.1):

$$fs = \frac{3}{2} \times \frac{F \times l}{w \times h^2} \tag{5.1}$$

where l represents the distance between the supports (50 ± 0.1 mm for specimens 60 mm in length and 100 ± 0.1 mm for specimens 120 mm in length), w is the specimen's width, and h is the specimen's height. The specimen's dimensions were determined prior to the measurement within an accuracy of about 0.1 mm.

Compressive strength measurements: For compressive strength measurements, at least five specimens for each testing series were tested. A Zwick/Roell BZ1-MM14640.ZW03 testing machine with a 50 kN load cell attached was used. By measuring the maximum force, F, the compressive strength, cs, can be calculated by the following Eq. (5.2):

$$cs = \frac{F}{l \times w} \tag{5.2}$$

l represents the length and w the width of the sample. Again, the specimen's dimensions were determined prior to the measurement within an accuracy of 0.1 mm.

Preparation of thin sections: To quantify fiber alignment in each sample, thin sections were prepared and characterized by optical microscopy. Specimens of 0.5, 1.5, and 2.5 mm thickness were produced, glued to a microscopy slide by means of an epoxy resin (Akepox 1005, AKEMI GmbH, Germany) and wet polished down to a thickness of 80−100 μm. The microscope used was an Olympus BX 51 model in transmission mode with a QImaging MicroPublisher 5.0 RTV imaging system.

Specimen density and helium porosity measurements: The surfaces of each specimen were polished in order to get even and parallel borders required for accurate volume measurement. After storing the specimens for 12 hours in an oven at a constant temperature of 130°C, the specimen density was calculated by determining the specimen's volume (vol_{spec} = length × width × height) within an accuracy of 0.01 mm as well as the specimen's weight within an accuracy of 0.1 mg. The specimens were then crushed with a Kinematica Polymix PX-MFC 90 D mill in order to measure the specimen's bulk volume (vol_{bulk}) by helium pycnometry with a Micromeritics AccuPyc II 1340. Helium porosity p can be calculated by Eq. (5.3):

$$p = \left(1 - \frac{vol_{bulk}}{vol_{spec}}\right) \times 100\ (vol.\%) \tag{5.3}$$

5.3 RESULTS AND DISCUSSION

5.3.1 Analysis of Fiber Orientation in Samples Prepared by Nozzle-Injection

The specific alignment of the carbon fibers is investigated by preparing thin sections of the samples as described in Chapter 2, Performance-Based Testing of Portland Cement Concrete for Construction-Scale 3D Printing, and analyzing them using transmission light microscopy. Fig. 5.1 demonstrates the basic principle behind the nozzle-injection method and shows the degree of fiber orientation that can be obtained. Using the open-source image analysis software *imageJ/Fiji* with the plugin *directionality*, a semiquantitative breakdown of fiber orientation in the micrographs can be given [12].

The principle on how to obtain an orientation histogram is shown in Fig. 5.2: micrographs of thin-sections with a thickness of about 80 μm are recorded and intentionally overexposed. This leaves the carbon fibers as easily recognizable dark lines, contrasting with the white, translucent background. As a result, automatic orientation mapping can be carried

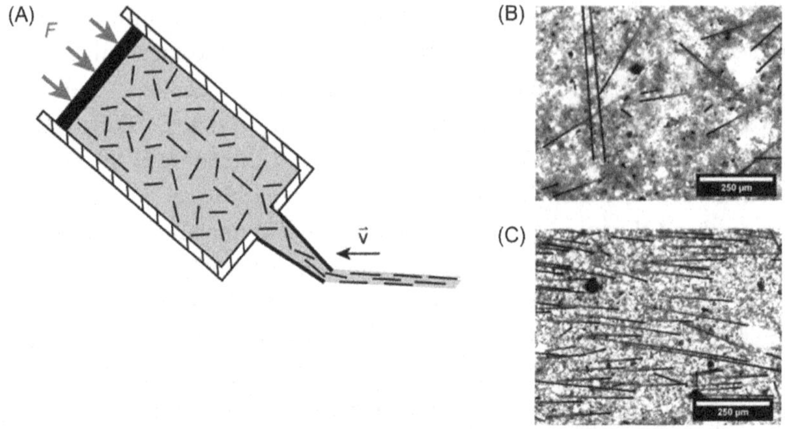

Figure 5.1 (A) Schematic diagram of the nozzle-injection technique: by pressing the fiber-reinforced paste through a syringe, fibers are aligned parallel to the movement of the nozzle. Thin sections of specimens with (B) randomly distributed, and (C) aligned carbon fibers. *Reproduced from M. Hambach, H. Möller, T. Neumann, D. Volkmer, Portland cement paste with aligned carbon fibers exhibiting exceptionally high flexural strength (>100 MPa), Cem. Concr. Res. 89 (2016) 80–86, slightly modified.*

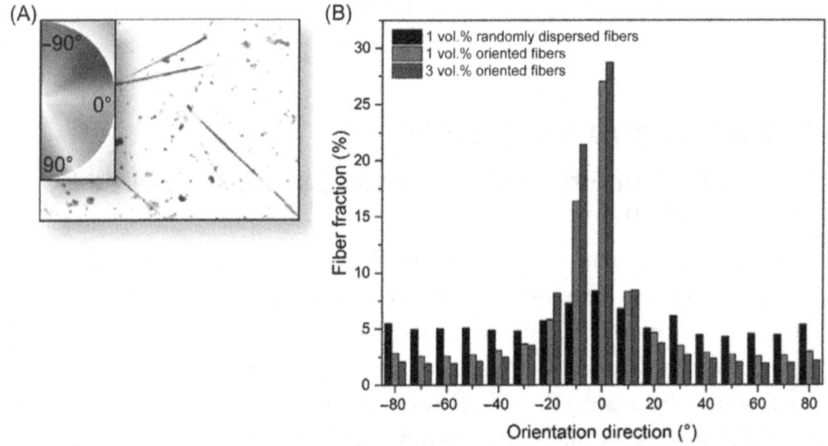

Figure 5.2 The process of thin section analysis: (A) Using the image-analysis software Fiji, fiber orientation is mapped using an overexposed micrograph; and (B) from the raw data, a histogram is generated. *Reproduced from M. Hambach, H. Möller, T. Neumann, D. Volkmer, Portland cement paste with aligned carbon fibers exhibiting exceptionally high flexural strength (>100 MPa), Cem. Concr. Res. 89 (2016) 80–86, slightly modified.*

out. The orientation distribution in Fig. 5.2 shows the average from 18 distribution histograms per sample tested. The original distribution histograms can be found in the Appendix, Figs. A5.2–A5.10.

The mold-cast sample with randomly distributed fibers (*black bars*) shows a relatively even distribution for all angles. A small peak at 0 degree is seen and is probably due to border effects when pouring the cement paste into the PTFE mold. For the nozzle-injection method at a fiber content of 1 vol.% (*red bars* (light gray in print version)) and 3 vol.% (*blue bars* (dark gray in print version)), a strong peak around 0 degree is observed. If the fiber-volume-fractions oriented at 0 ± 20 degrees are added up, 62% of fibers of the sample containing 1 vol.% and 71% of fibers of the sample containing 3 vol.% can be described as unidirectionally oriented. Higher fiber content seems to induce a slightly higher degree of orientation.

5.3.1.1 Flexural and Compressive Strength of Cement Paste With Aligned Carbon Fibers

Due to the nearly unidirectional fiber alignment, samples with fiber-orientation can be tailored to withstand a specific kind of load. Fig. 5.3 shows the typical setups for strength testing with the gray lines indicating fiber alignment. The samples were characterized by their flexural and compressive strength. To illustrate the influence of the aligned fiber, plain cement paste and a mold-cast sample with random fiber alignment are included in the tests. Fig. 5.3 shows the stress-deflection curves for the 3-point bending test; Fig. A5.11 in the supplementary material shows the

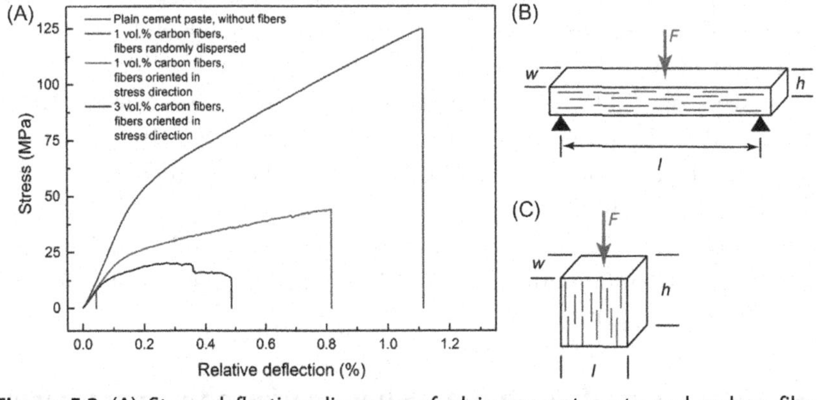

Figure 5.3 (A) Stress-deflection diagrams of plain cement paste and carbon fiber-reinforced cement composites in a 3-point-bending setup. Fiber orientation in specimens for (B) flexural strength and (C) compressive strength is indicated by gray dashed lines. *Reproduced from M. Hambach, H. Möller, T. Neumann, D. Volkmer, Portland cement paste with aligned carbon fibers exhibiting exceptionally high flexural strength (>100 MPa), Cem. Concr. Res. 89 (2016) 80–86 slightly modified.*

Table 5.2 Flexural and compressive strength of cement paste and carbon fiber cement composites with varying fiber content [12]

Sample	Flexural strength (MPa)	Compressive strength (MPa)
Plain cement paste, without fibers	8.3 ± 1.1	101.8 ± 19.1
1 vol.% carbon fibers, fibers randomly dispersed	20.3 ± 2.5	87.4 ± 21.5
1 vol.% carbon fibers, fibers oriented in stress direction	46.5 ± 4.3	87.3 ± 11.6
3 vol.% carbon fibers, fibers oriented in stress direction	119.6 ± 7.6	83.8 ± 9.0

stress–strain curves for uniaxial compression. Table 5.2 gives a comprehensive overview of all the results concerning the strength testing.

Upon admixing 1 vol.% of carbon fibers with random alignment into the cement paste, the flexural strength rises from 8.3 (±1.1) MPa to 20.3 (±2.5) MPa. As can be seen in Fig. 5.3, the toughness of the material also improves by a huge margin. This behavior is bolstered when the same amount of fibers is aligned; flexural strength rises to 46.5 (±4.3) MPa. Samples containing 3 vol.% aligned fibers reach a staggering 119.6 (±7.6) MPa, which corresponds to a 1340% increase in flexural strength in comparison to plain cement paste.

All samples with aligned fibers show deflection-hardening-behavior, which suggests the formation of multiple microcracks with limited crack size, instead of the development of a single large crack which leads to sudden material failure.

As seen in Table 5.2, compressive strength decreases from 101.8 (±19.1) MPa for plain cement paste to 87.3 (±11.6) MPa for a fiber content of 1 vol.% or 83.8 (±9.0) MPa for a fiber content of 3 vol.%, respectively. A likely cause is introduction of macroporosity with increasing fiber content. However, the compressive strength still complies with the standardized minimum of 52.5 MPa for the highest classification of Portland cement [12].

5.3.2 Properties of 3D-Printed Fiber-Reinforced Cement Paste

Using a DeltaWasp 20 × 40 printer, a small-scale setup (specimen dimensions up to several centimeters, Fig. 5.4A) is used to investigate the influence of layer orientation, density, and porosity on the strength of 3D-printed Portland cement paste. In addition to carbon fibers, glass and

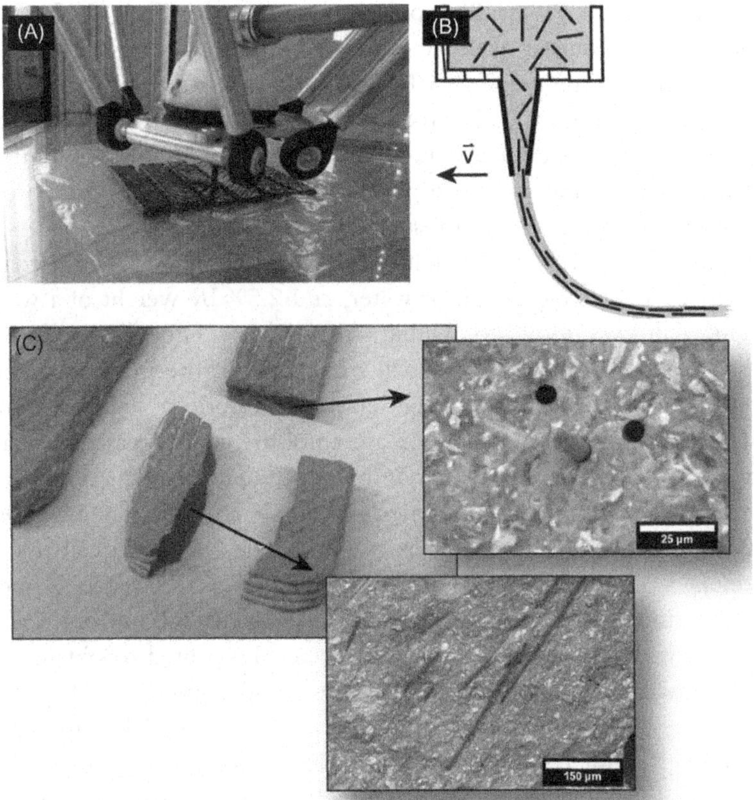

Figure 5.4 (A) Photograph of 3D printing of beam test specimens used in 3-point bending tests (dimensions: 6 × 12 × 60 mm). (B) Schematic illustration of fiber alignment during 3D printing. (C) Photograph of a sample containing aligned reinforcement fibers and corresponding ESEM micrographs which show the fiber orientation (perpendicular and parallel to the fracture surface of the specimen).

basalt fibers were used to determine the influence of fiber reinforcement on distinct mechanical properties, namely flexural and compressive strength. Since filler alignment in 3D-printed structures has previously been employed successfully for polymer composites [18], fiber alignment is here investigated as an efficient means for fiber reinforcement of Portland cement paste. As described for the nozzle-injection method, a nozzle diameter of 2 mm was used to align fibers with a mean length of 3 or 6 mm. ESEM micrographs of 3D-printed cement pastes containing reinforcement fibers prove that our prototypic 3D-printing setup leads to strongly aligned carbon, glass, or basalt fibers in the printed specimens, as shown in Fig. 5.4C and in the supplementary material (Fig. A5.12). Thus, the process of alignment of the fibers is used in this study to spatially control mechanical properties, especially flexural/tensile strength, by automatically adjusting the print path directions using the GNU-licensed printing software, Cura. Besides mechanical properties, 3D printing of Portland cement pastes can also influence porosity [13], consequently the porosity of 3D-printed samples is also investigated in this report.

A mixture of 61.5% by weight of Portland cement, 21% by weight of Microsilica, 15% by weight of water, and 2.5% by weight of a water-reducing agent was found to exhibit a viscosity low enough to ensure no blockage of the printing nozzle will occur during the printing process. Nevertheless, the viscosity was high enough to fabricate stable structures by 3D printing that retained their shape until the cement paste was hardened for both plain and fiber-reinforced cement pastes. Attempts to increase the fiber volume content (exceeding about 1.5 vol.% fibers) caused frequent blocking of the extrusion nozzle, 1 vol.% of fiber reinforcement was, thus, used for all specimens' preparations. Cubic specimens (18 mm edge length) and beam specimens (60 mm in length, 12 mm in width, and 6 mm in height) were 3D-printed to obtain samples for uniaxial compressive strength tests and 3-point bending tests, respectively. Two different print patterns, namely a parallel shaped (print path A for 3-point bending tests, print path C for uniaxial compressive strength tests) and a crosshatch-shaped pattern (print path B for 3-point bending tests, print path D for uniaxial compressive strength tests) were employed (Fig. 5.5A and B) to investigate the influence of different fiber orientations along predetermined print paths. For print paths A and C (parallel shape) each layer was printed identically to the layer below and above, for print paths B and D (crosshatch shape) each layer was twisted by 90 degrees with respect to the layers beneath and above, resulting in a sandwich-like structure, as illustrated in Fig. 5.5B and D. Besides fiber-reinforced samples, a fiber-free reference was also produced for all print paths reported in this study. Additionally, 3-point bending tests were performed perpendicular to layer orientation, simulating the most-common load case for 3D-printed constructions, as illustrated in Fig. 5.5E. Compressive strength tests were performed in two directions, perpendicular (test direction I) and longitudinal (test direction II) to the layer orientation, in order to simulate two different load cases for the anisotropic specimens which are obtained in a 3D-printing fabrication process (Fig. 5.5F and G).

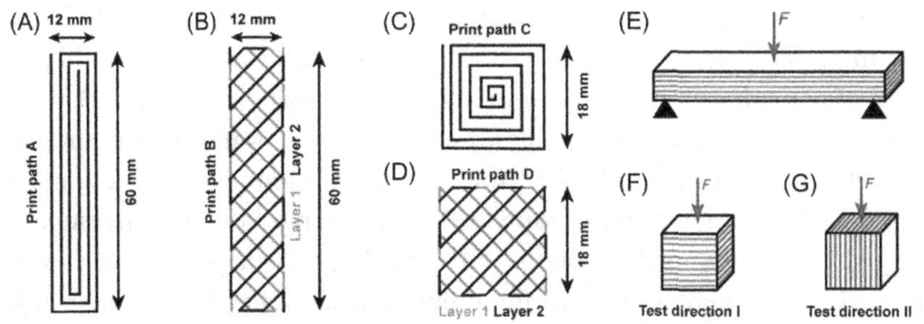

Figure 5.5 Top view of the print paths for specimens used for 3-point bending test. (A) A parallel printed called "print path A." (B) Illustration of "print path B" samples having two different layers (rotated 90 degrees to each other). (C) Top view of print paths for specimens used for uniaxial compressive strength test being parallel printed referred to as "print path C," and (D) being printed in a crosshatch shape called "print path D" (layers rotated 90 degrees to each other were used). Schematic illustration of 3D-printed layers (gray lines) and sample orientation in (E) 3-point bending test and (F) uniaxial compressive strength test for perpendicular layer orientation referred to as "test direction I," and (G) longitudinal layer orientation referred to as "test direction II."

Table 5.3 Flexural strength determined in 3-point bending test for test Beams printed along path A and B

Sample	Print path A (MPa)	Print path B (MPa)
Plain cement paste (without fibers)	10.6 ± 0.7	11.4 ± 0.6
Fiber-reinforced cement paste (1 vol.% carbon fibers)	29.1 ± 1.8	13.9 ± 0.5
Fiber-reinforced cement paste (1 vol.% glass fibers)	12.4 ± 0.8	10.3 ± 0.7
Fiber-reinforced cement paste (1 vol.% basalt fibers)	13.8 ± 0.5	10.2 ± 0.6

5.3.2.1 Flexural Strength of 3D-Printed Reinforced Cementitious Composites

Flexural strength of the tested beams ranges from 10 to 30 MPa, depending on the reinforcement fiber type and print path. All the results are summarized in Table 5.3 for flexural strength and in the supplementary material (Table 5.A1) for deformation modulus in the 3-point bending test. For each testing series at least five specimens were tested. 3D models and photographs of the prepared test specimens are shown in Fig. 5.7A and B.

Samples printed via print path A exhibit a flexural strength of about 10 MPa when plain cement paste is being used and about 13 MPa if glass or basalt fibers are present. The carbon-fiber reinforced test specimen already investigated shows a remarkable increase in strength (30 MPa). The increase in flexural strength, while remarkable, is not quite as high as when the manual nozzle-injection method is used. It is likely that the fibers are broken down by the high pressure used to transport the material through the printing head, thus, the degree of alignment is not quite as high The fiber orientation of samples printed using path A is investigated using thin sections, as described in Chapter 2, Performance-Based Testing of Portland Cement Concrete for Construction-Scale 3D Printing. The resulting histogram (Fig. 5.6) shows that 46% of fibers are within ± 20 degrees of the movement direction of the nozzle, whereas the manual nozzle-injection resulted in 62% of fibers being oriented. The original histograms can be found in the Appendix (Figs. A5.13–A5.14).

The appropriate selection of different fiber types exerts a strong influence on the flexural strength of the test specimens. Stress-deformation diagrams show that plain and carbon fiber-reinforced samples do not show postcracking characteristics (Fig. 5.7C) [34]. In contrast, glass and basalt fiber samples reveal postcracking behavior since stress-deformation curves (provided in the supplementary material [Fig. A5.15]) show residual stress after specimen failure [35,36]. Toughening behavior has been

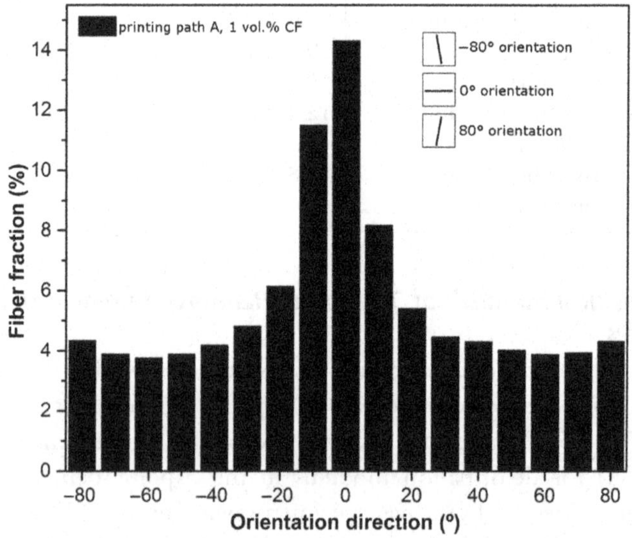

Figure 5.6 Orientation histogram of samples with 1 vol.% fibers printed using path A. *Reproduced from M. Hambach, Hochfeste multifunktionale Verbundwerkstoffe auf Basis von Portlandzement und Kohlenstoffkurzfasern, Dissertation, Augsburg University, Augsburg, 2016 [33], slightly modified.*

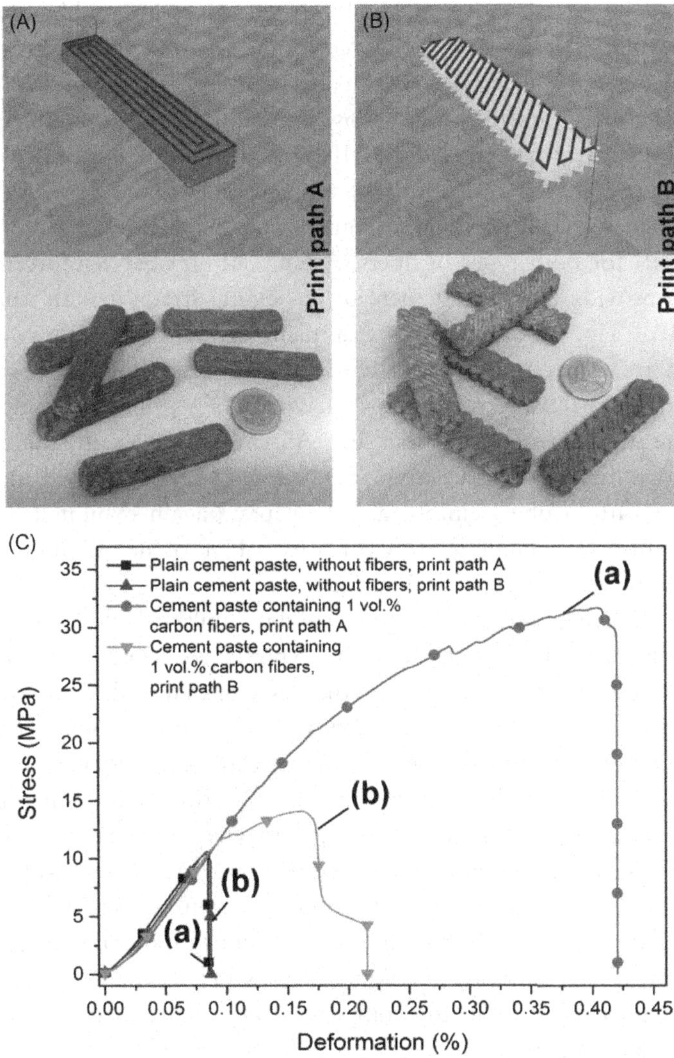

Figure 5.7 3D-printing path models in 3D-printing software and photographs of specimens fabricated via (A) print path A and (B) print path B in the 3-point bending test. (C) Stress-deformation plots for plain cement samples (without fibers) and carbon fiber-reinforced samples in 3-point bending tests proving high flexural strength of print path A samples being reinforced with carbon fibers.

reported for fiber-reinforced concretes [12,37,38] and high-performance biomaterials [39,40] characterized by a remarkable zone of plastic deformation ensuring high flexural strength values. Similar behavior can be observed for carbon fiber-reinforced cement paste processed by print path

A which exhibits the highest flexural strength values in combination with significant plastic deformation, ranging from 0.1% to 0.4% deformation, as shown in Fig. 5.7C. If plain cement paste is being used, samples fabricated with print paths A and B did not show different flexural strength. In both cases (print paths A and B), 10–11 MPa strength can be reported for plain cement paste (Fig. 5.2C). Both basalt and glass fiber reinforcement did not lead to increased flexural strength compared to plain cement paste, yielding 10 MPa for both types of fibers. Again, carbon fiber reinforcement is found to provide the highest increase in flexural strength with values up to 14 MPa. Fig. 5.7C illustrates that plain cement paste shows similar stress-deformation curves irrespective of the chosen print path (A or B). All tested fibers show postcracking behavior when printed via print path B (see the supplementary material, Fig. A5.15). However, glass and basalt fibers show higher residual stress after failure when fabricated via print path A. A carbon fiber-reinforced cement paste again exhibits a zone of plastic deformation ranging from 0.1% to 0.15% of deformation (*orange* (gray in print version) curve in Fig. 5.7C).

Since all fibers used in this study exhibit similar tensile strengths and fiber aspect ratios (Table 5.1), it can be hypothesized that the low flexural strength values observed for cement samples prepared with glass- or basalt fibers might be connected to the low values of Young's modulus reported for these types of fibers. Samples containing glass or basalt fibers exhibit a cracking behavior which is different to the one shown by nonreinforced reference samples, as shown in the supplementary material (Fig. A5.15). A residual stress after failure indicates that the hardened cement paste breaks before the glass or basalt fibers can bear the maximum stress. Consequently, these fiber types will not lead to an enhancement of the maximum flexural strength in the 3D-printing technique in the same way carbon fibers will. However, they still prevent the complete failure of the specimen after the first cracks have formed.

5.3.2.2 Compressive Strength of 3D-Printed Reinforced Cementitious Composites

During compressive strength tests, the cubic test specimens showed minor dependencies on the fiber material and on the exact print path (C or D) being used. In contrast, the direction in which the compressive load was put on the samples has a major influence on the test results. All results are summarized in Table 5.4 for compressive strength and in supplementary material (Table 5.A2) for deformation modulus in compressive strength

Table 5.4 Compressive strength determined in uniaxial compressive strength test

Sample	Print path C, test direction I (MPa)	Print path C, test direction II (MPa)	Print path D, test direction I (MPa)	Print path D, test direction II (MPa)
Plain cement paste, (without fibers)	81.1 ± 10.0	29.6 ± 10.7	77.9 ± 29.3	30.0 ± 10.8
Fiber-reinforced cement paste (1 vol.% carbon fibers)	60.6 ± 9.8	27.4 ± 7.9	82.3 ± 26.0	30.8 ± 14.1
Fiber-reinforced cement paste (1 vol.% glass fibers)	61.0 ± 15.1	20.6 ± 6.8	84.5 ± 22.5	28.1 ± 7.9
Fiber-reinforced cement paste (1 vol.% basalt fibers)	63.0 ± 15.7	33.7 ± 10.7	85.0 ± 31.2	38.6 ± 15.7

test. For each testing series, at least five specimens were tested. The 3D models and photographs of the prepared specimens are shown in Fig. 5.8A and B.

Print path C samples (Fig. 5.8A) tested in test direction I exhibit a compressive strength of 80 MPa for plain cement paste and about 60 MPa for fiber-reinforced test specimens. Different fiber types do not influence compressive strength properties significantly. Stress-deformation diagrams indicate that fiber-reinforced specimens do not show postcracking characteristics [41] (e.g., resulting from fiber pull-out) in a compressive strength test (see the supplementary material, Fig. A5.15). If print path C samples are flipped by 90 degrees in the testing setup (test direction II), compressive strength decreases to around 30 MPa for all tested samples, regardless whether plain or fiber-reinforced cement paste is being used. Reduced compressive strength resulting from separate printing layers is also reported by other authors [13]. Stress-deformation plots show that the first cracks form at around 10 MPa, along the printing layers and—since fiber reinforcement is not effective in test direction II—compressive strength decreases considerably in comparison to test direction I (see the supplementary material, Fig. A5.13). Print path D samples (Fig. 5.8B) tested in test direction I show compressive strength values of about 80–85 MPa,

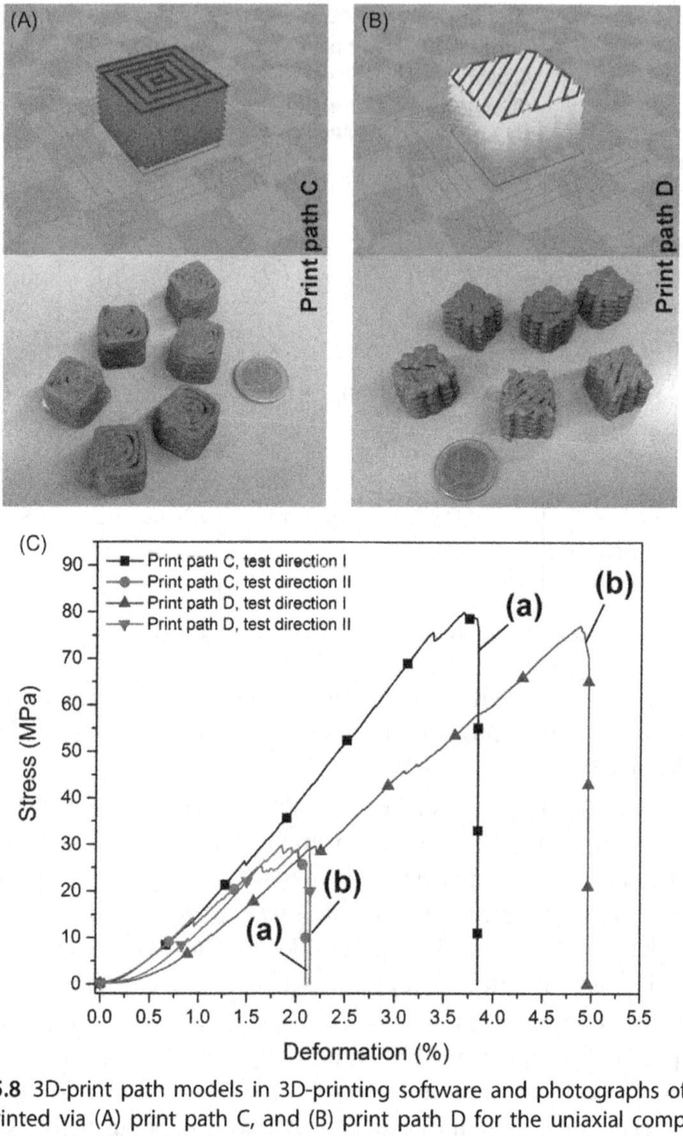

Figure 5.8 3D-print path models in 3D-printing software and photographs of specimens printed via (A) print path C, and (B) print path D for the uniaxial compressive strength test. (C) Stress-deformation plots for plain cement samples (without fibers) in uniaxial compressive strength tests showing high strength for test direction I and low strength for test direction II.

whereas plain cement paste performance is slightly inferior (80 MPa) in comparison to fiber-reinforced samples (85 MPa). Again, the fiber type does not influence compressive strength significantly. Stress-deformation curves do not show postcracking characteristics, as presented in supplementary material Fig. A5.15. Print path D and test direction II result in compressive strength values ranging from 30 to 40 MPa, whereas only basalt fibers show slightly higher strength values of 40 MPa. In plain cement paste, carbon and glass fibers display inferior compressive strength of about 30 MPa. Reduced compressive strength between the printing layers can be reported for print path D, similar to print path C tested in test direction II. Summarizing the results of compressive strength tests, it can be stated that fiber reinforcement, no matter which type of fiber is being used, does not influence the compressive strength of the printed cement paste significantly. However, the test direction (parallel or perpendicular to the layers obtained in 3D-printing) has a major impact on the compressive strength, since direction II results in remarkably lower compressive strength compared to test direction I, no matter which print path is being tested (Fig. 5.8C).

5.3.2.3 Density and Porosity

The photographs of the 3D-printed specimens (Figs. 5.7A, 5.8A and supplementary Fig. A5.16) imply that the samples might exhibit varying porosity, since they were fabricated in layers instead of being casted at once in a mold. Furthermore, no (plate) vibrator has been used for further densification after printing, which could lead to air voids embedded between the layers. The specimen density of each print path sample and a mold-casted reference was measured after drying the specimens in an oven and is given in Table 5.5. Furthermore, the helium porosity of each print path sample and mold-casted cement paste was determined by helium pycnometry, as presented in Table 5.5. Images of polished cross-sections for all print paths (A to D) and the mold-casted specimen are given in the supplementary material (Fig. A5.16).

The results show that cement paste printed by print path A exhibits the highest specimen density and lowest helium porosity (19.7%). Mold

Table 5.5 Specimen density and helium porosity for mold-casted and 3D-printed samples

Sample	Specimen density (g/cm^3)	Helium porosity (%)
Mold-casted cement paste	1.911 ± 0.002	20.2 ± 0.2
3D-printed cement paste, print path A	1.927 ± 0.003	19.7 ± 0.3
3D-printed cement paste, print path B	1.898 ± 0.012	21.3 ± 0.5
3D-printed cement paste, print path C	1.911 ± 0.005	20.3 ± 0.2
3D-printed cement paste, print path D	1.867 ± 0.009	22.1 ± 0.4

casting and print path C result in slightly lower specimen density and consequently higher helium porosity at 20.2% and 20.3%, respectively. In contrast, print path B samples show increased helium porosity at 21.3% and print path D increases helium porosity even further to 22.1%. Since helium gas is used for pycnometry, the pore volume determined contains enclosed air voids (macropores) and micropores which are present in all cement-based binders. All samples were made from the same basic cement paste recipe and were mixed in the same way, consequently the micro pore volume is identical for all samples. Thus, the difference in helium porosity can be ascribed to a varying amount of macropores (trapped air voids) induced through 3D-printing. Although flexural strength values can be enhanced by fiber alignment in 3D-printing, porosity may also increase compared to a mold-casted sample. The highest porosity was measured for samples exhibiting 3D-printed layers being twisted layer by layer (print paths B and D), which leads to an increased number of enclosed pores in the fabricated structures. However, if suitable print path configurations are chosen (print paths A and C), the porosity of the 3D-printed samples is similar, or even lower, than those of the mold-casted references.

5.3.3 3D-Printing of Hierarchical Materials

Although fiber-reinforced cement paste shows high flexural strength values for optimized print paths (print path A), the development of suitable extrusion systems and customized cement paste compositions [42] make largescale 3D-printed structures expensive, especially if very expensive carbon fibers are being admixed. In order to reduce material costs, we also investigated the concept of fabricating 3D-printed hollow formworks that are subsequently filled with common mortar or concrete [43]. The amount of fiber-reinforced infill can be adjusted depending on the strength needed to ensure the construction's stability. Consequently, these structures are made of (expensive) high flexural strength cement paste, containing aligned reinforcement fibers, and standard mortar or concrete acting as a comparably inexpensive filling agent. Filling the voids of the formwork after printing is referred to as *hierarchical structuring*. Similar concepts can be found in natural biomaterials: bones from vertebrates, for instance, display a hierarchical structure consisting of a dense and stiff outer layer, termed cortical bone, and a soft and light core, which is referred to as cancellous bone, thus combining superior mechanical properties and material efficiency in one composite [44–46]. Transferring the principle of hierarchical materials to 3D-printed specimens, a carbon fiber-reinforced cement paste was used for 3D-printing hollow formworks and a simple mortar mixture was used to fill the voids in the printed formworks. Thus, dense specimens were obtained which contain varying amounts of a (rather expensive) fiber-reinforced cement paste and a (rather cheap) mortar filling. Since carbon fibers were found to provide the highest increase in flexural strength, this fiber type was used for fiber reinforcement in the fabricated hierarchical structures.

During 3D-printing preparation software the "shell thickness" (in millimeters) and the amount of material (in vol.%) in the core of a structure (termed "fill density") can be adjusted. The borders of the specimens were printed parallel with a shell thickness of 4 mm to create a confining wall, which basically corresponds to a shell of print path A. The core material was printed in a crosshatch pattern, corresponding to a core of print path B, as illustrated in Fig. 5.9A and B. This combination described

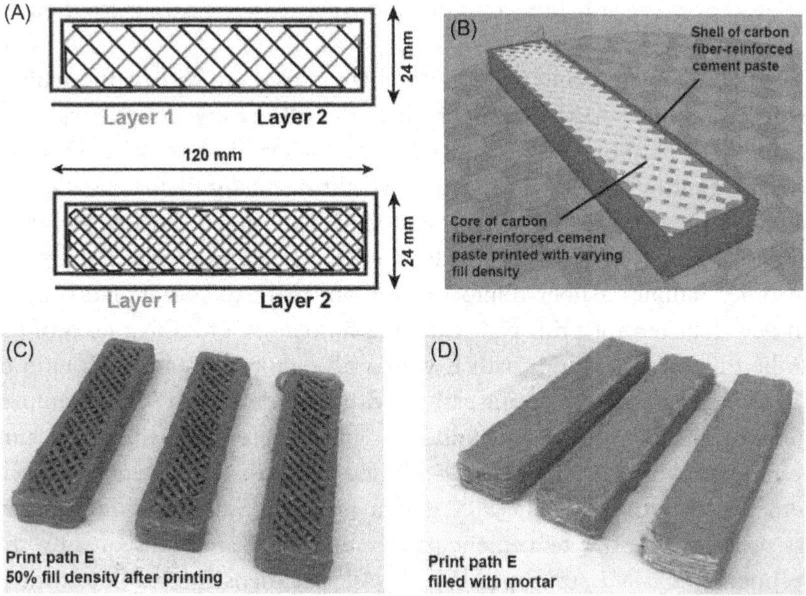

Figure 5.9 3D-printed hierarchical structures. (A) Sketches of "print path E" having a parallel-shaped shell and a crosshatch-shaped core. By controlling the distance between the single lines of the crosshatch pattern, the percentage of fiber-reinforced cement paste in the core area can be adjusted (termed "fill density" in 3D-printing software). (B) 3D illustration of print path E for 50% fill density in the core area. (C) Photograph of specimens obtained after 3D printing (50% fill density), and (D) after loading the structure with mortar.

(of a shell with print path A and a core with print path B) is hereafter termed print path E. The fill density for 3D printing was varied (from 0% to 100% in 25% steps) and the remaining void volume of the specimens was further filled with mortar after printing in order to adjust the fiber-reinforced cement paste to the mortar ratio. For each fill density (0%, 25%, 50%, 75%, and 100%) three specimens (120 mm in length, 24 mm in width, and 12 mm in height) were prepared. Mortar (mixture: 54% by weight of sand, 36% by weight of Portland cement, 10% by weight of water) was poured into the specimens 24 hours after printing to fill the whole structure with mortar and, thus, obtain compact specimens irrespective of the fill density preadjusted for 3D printing (Fig. 5.9C and D). Additional illustrations of the print paths and photographs of the specimens before and after filling with mortar are provided in the supplementary material Fig. A5.17. The fill density adjusted in 3D-printing preparation software, the weight of the specimens after 3D printing, and the weight after filling with mortar are given in Table 5.6. Additionally, the ratio of mortar to fiber-reinforced cement paste was calculated and is used to label the samples.

Hollow molds (print path E not being filled in with carbon fiber-reinforced cement paste during printing, 0% fill density) result in a specimen weight of about 30.7 g and were filled after printing with 55.0 g of mortar, resulting in a total specimen weight of 86.7 g. Thus, the ratio of mortar to fiber-reinforced cement paste is determined to be 1.8, and these samples are referred to as "print path E−1.8." In an analogous fashion, for the other samples a fiber-reinforced cement paste to mortar ratio of 0.8 can be calculated for print path E with a fill density of 25%, a ratio of 0.4 can be calculated for print path E with a fill density of 50% and a ratio of 0.1 can be calculated for print path E with a fill density of 75%. Compact reference specimens (not containing any voids) were obtained by adjusting the fill density to 100% in the 3D-printing software. The weight of specimens was determined to be 90.2 g after printing. Since no filling mortar was used, the mortar to cement paste ratio was 0 and consequently the specimens are called "print path E−0". All specimens exhibit almost identical weight (88.0 ± 1.5 g) after filling with mortar, which proves that most of the (intended) voids could be filled with mortar after printing. Flexural strength tests (3-point bending test) were carried out 28 days after fabrication and the results are given in Table 5.6. Specimen orientation in 3-point bending test was identical to the samples tested and are illustrated in Fig. 5.5E.

Table 5.6 Weight of the hierarchical specimens after printing, after filling with mortar, and the results of the flexural strength measurement in a 3-point bending test

Sample	Amount of "fill density" adjusted in 3D-printing software (%)	Weight of 3D-printed formwork (g)	Weight after filling with mortar (g)	Ratio of mortar to fiber-reinforced cement paste	Flexural strength in 3-point bending test (MPa)
Print path E−1.8	0	30.7 ± 0.3	85.7 ± 0.6	1.8	12.4 ± 0.5
Print path E−0.8	25	48.2 ± 0.7	88.8 ± 1.4	0.8	14.4 ± 0.3
Print path E−0.4	50	65.0 ± 0.3	88.3 ± 0.7	0.4	15.1 ± 0.5
Print path E−0.1	75	77.2 ± 4.4	87.0 ± 6.5	0.1	16.2 ± 0.3
Print path E−0	100	90.2 ± 4.4	90.2 ± 4.4	0	17.5 ± 0.5

Hollow formwork samples infilled completely with mortar (cement paste to mortar ratio of 0.56) were found to exhibit flexural strength of about 12 MPa. By increasing the fill density in the 3D-printing software and, thus, carbon fiber-reinforced cement paste in the core area, flexural strength can be pushed from 12 MPa for "print path E−01.8" (hollow mold infilled with mortar) to 17.5 MPa for "print path E−0" (printed at 100% fill density). Table 5.6 shows that increasing fill density of fiber-reinforced cement paste in 25% steps increases flexural strength in steps of about 1.5 MPa, proving that the fill density in the 3D-printing software can be used to control the desired flexural strength of the printed hierarchical specimens. Additional stress-deformation plots are provided in the supplementary material (Fig. A5.18).

5.4 CONCLUSION

In summary, we have introduced a method that allows for the alignment of fibers in reinforced cement pastes by means of extruding the paste through a nozzle. Using this method between 60% and 70% of fibers in any given sample can be aligned. When using a manual extrusion method through a syringe, flexural strength of samples with 3 vol.% aligned carbon fibers reaches up to 120 MPa.

Furthermore, the application of this method in a fully automated 3D-printing setup is investigated. Composites of cement paste and aligned glass, basalt, or carbon fibers are printed and show effective fiber alignment, leading to a remarkable increase in the flexural strength of the composites. Since the printing process was found to enforce an alignment of the fibers along the print path direction, the build path itself was used to spatially control fiber orientation within the printed structures. Indeed, 30 MPa was found to be the highest flexural strength value achievable by an optimized print path and a content of 1 vol.% carbon fiber. To use the material to its fullest potential and produce printed samples with high fiber content and a flexural strength exceeding 100 MPa, the development of optimized cement pastes and extrusion systems are required. The optimization process will also require a detailed investigation of rheological properties since, for example, viscosity or yield stress are important parameters to ensure a 3D-printing procedure at high fiber volume contents. Other fibers, namely glass or basalt fibers, were not found to increase flexural strength of the composite significantly.

The use of 3D printing provides access to complex geometries and, thus, offers the possibility to print hierarchical structures which have the potential to provide strong and material-efficient structures for future constructions. The tests performed on hierarchical specimens demonstrate that these composite types are viable for fine-tuning mechanical properties and, consequently, could lead to reduced material costs. Further development into more complex 3D-printed structures could include a two-component dispenser system being able to extrude both plain and fiber-reinforced cement pastes at spatially and temporally varying ratios. A two-component extrusion technique could open up the possibility of fabricating novel structures, exhibiting high (flexural) strength and material efficiency in an automated fabrication process without the need to fill structures manually after 3D printing. Of course, the fabrication of large constructions (e.g., entire buildings) seems elusive in the near future since the 3D-printing hardware would first have to be adapted to the largescale processing of Portland cement binders (typically several tons of material per hour).

APPENDIX

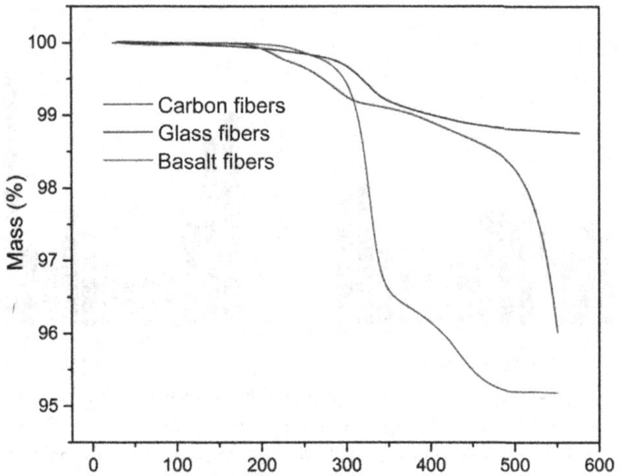

Figure A5.1 TGA curves of carbon, glass, and basalt fibers measured in air. The mass loss, starting at around 200°C, corresponds to the polymer sizing being removed from the fibers. Carbon fiber (surface) oxidation, for enhanced bond strength, starts at around 350°C.

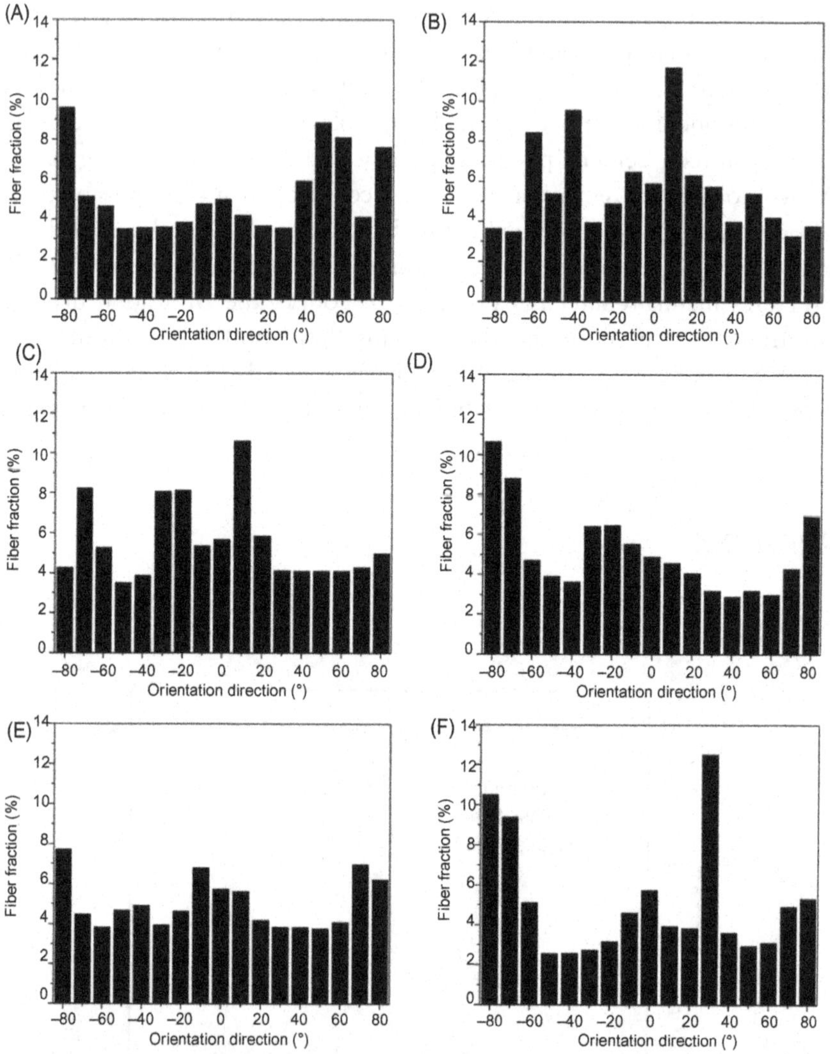

Figure A5.2 First series of orientation analysis in a sample with 1 vol.% randomly dispersed fibers. (A–F) the orientation histogram of the sample. *Reproduced from M. Hambach, H. Möller, T. Neumann, D. Volkmer, Portland cement paste with aligned carbon fibers exhibiting exceptionally high flexural strength (>100 MPa), Cem. Concr. Res. 89 (2016) 80–86, slightly modified.*

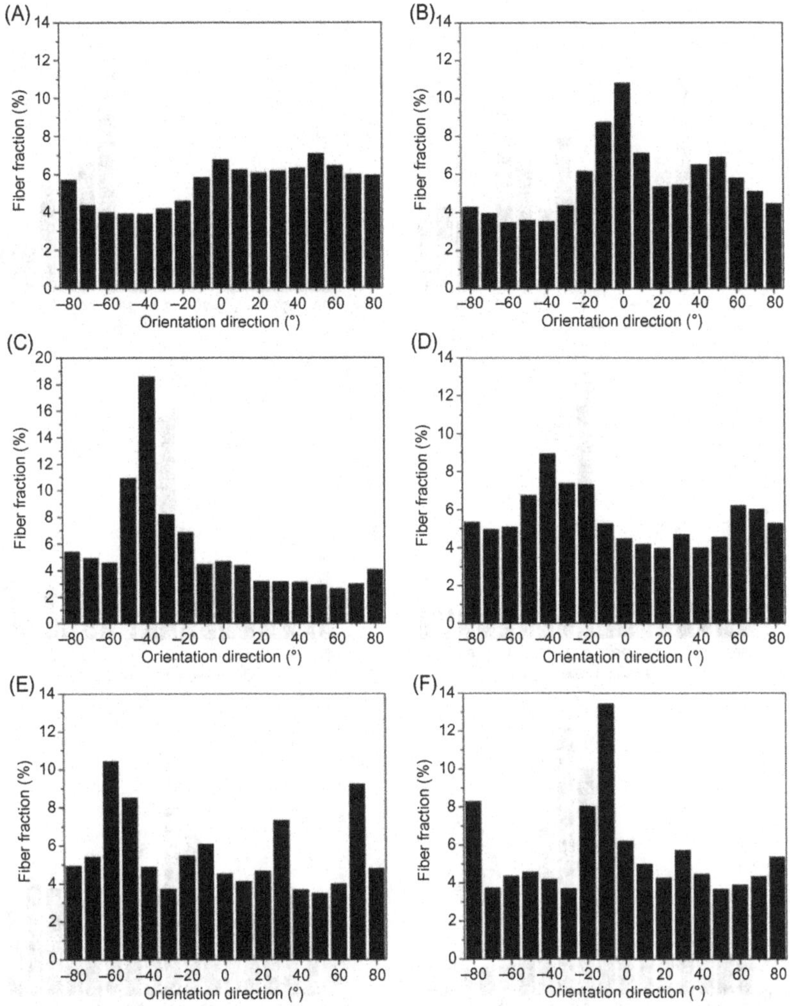

Figure A5.3 Second series of orientation analysis in a sample with 1 vol.% randomly dispersed fibers. (A–F) the orientation histogram of the sample. *Reproduced from M. Hambach, H. Möller, T. Neumann, D. Volkmer, Portland cement paste with aligned carbon fibers exhibiting exceptionally high flexural strength (>100 MPa), Cem. Concr. Res. 89 (2016) 80–86, slightly modified.*

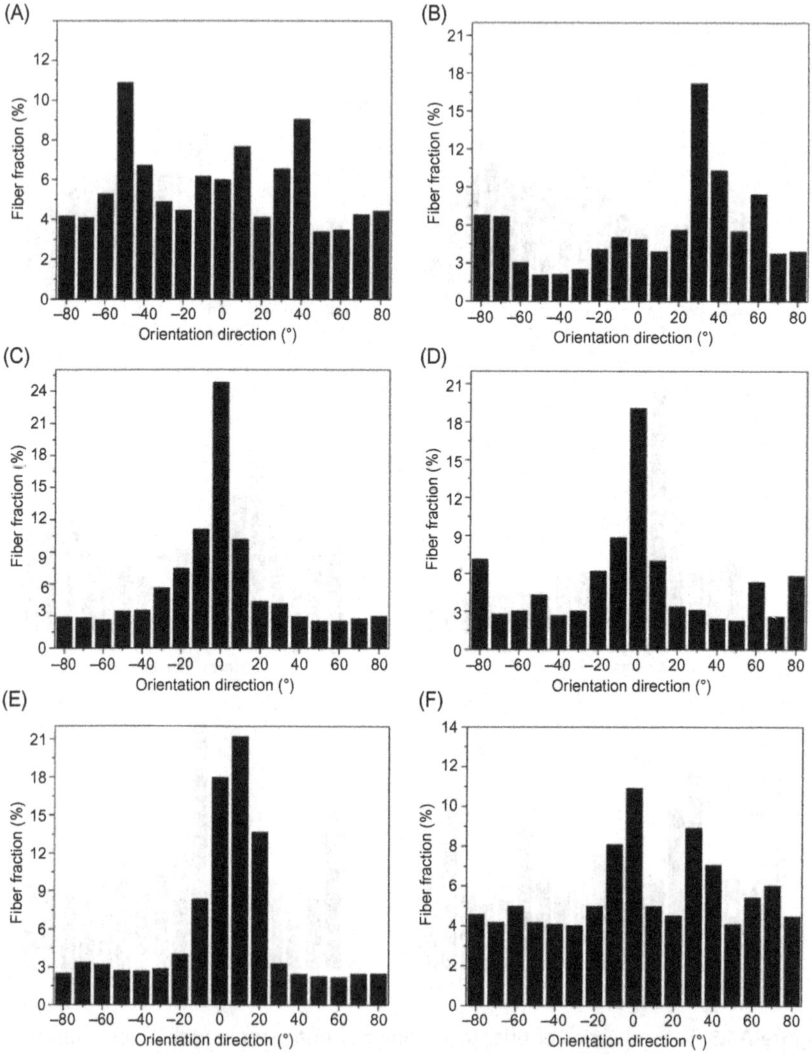

Figure A5.4 Third series of orientation analysis in a sample with 1 vol.% randomly dispersed fibers. (A–F) the orientation histogram of the sample. *Reproduced from M. Hambach, H. Möller, T. Neumann, D. Volkmer, Portland cement paste with aligned carbon fibers exhibiting exceptionally high flexural strength (>100 MPa), Cem. Concr. Res. 89 (2016) 80–86, slightly modified.*

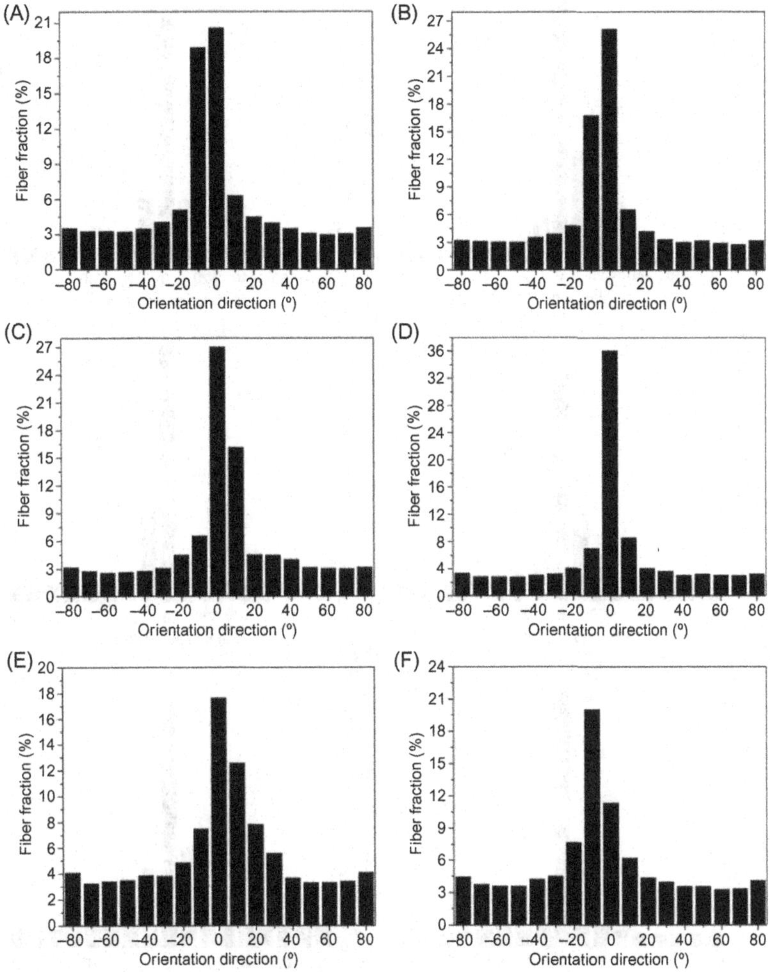

Figure A5.5 First series of orientation analysis in a sample with 1 vol.% aligned fibers. (A–F) show the orientation histogram of the sample. *Reproduced from M. Hambach, H. Möller, T. Neumann, D. Volkmer, Portland cement paste with aligned carbon fibers exhibiting exceptionally high flexural strength (>100 MPa), Cem. Concr. Res. 89 (2016) 80–86, slightly modified.*

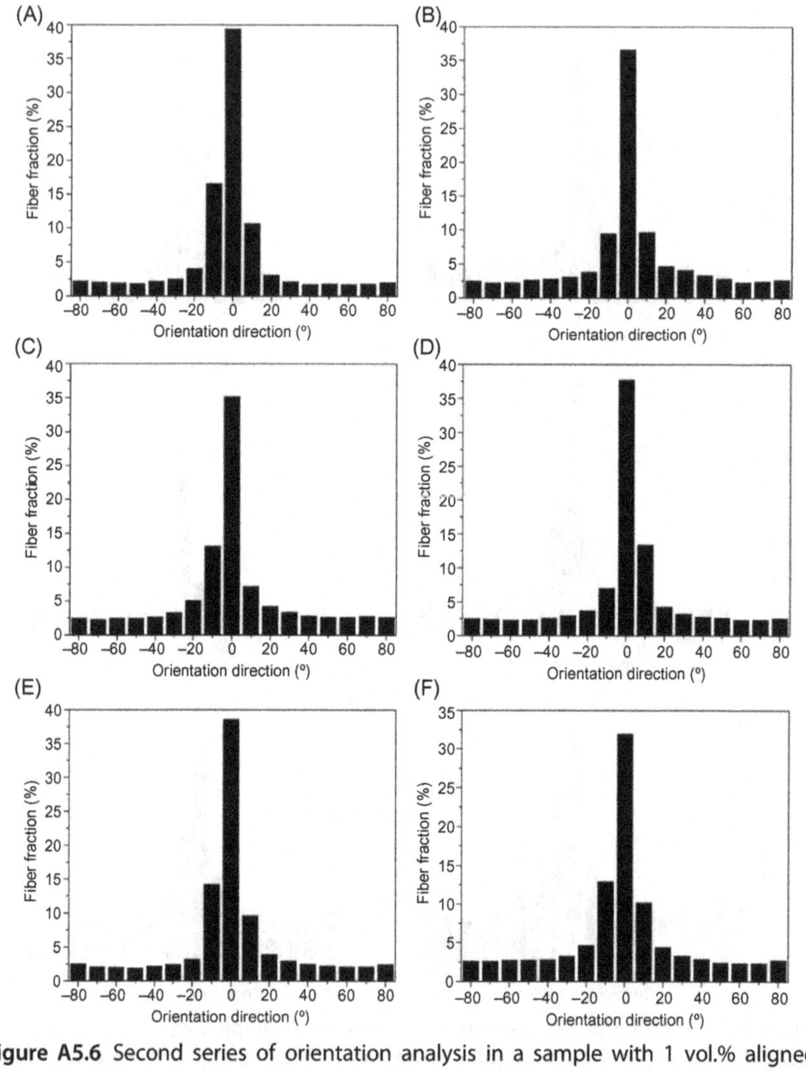

Figure A5.6 Second series of orientation analysis in a sample with 1 vol.% aligned fibers. (A–F) the orientation histogram of the sample. *Reproduced from M. Hambach, H. Möller, T. Neumann, D. Volkmer, Portland cement paste with aligned carbon fibers exhibiting exceptionally high flexural strength (>100 MPa), Cem. Concr. Res. 89 (2016) 80–86, slightly modified.*

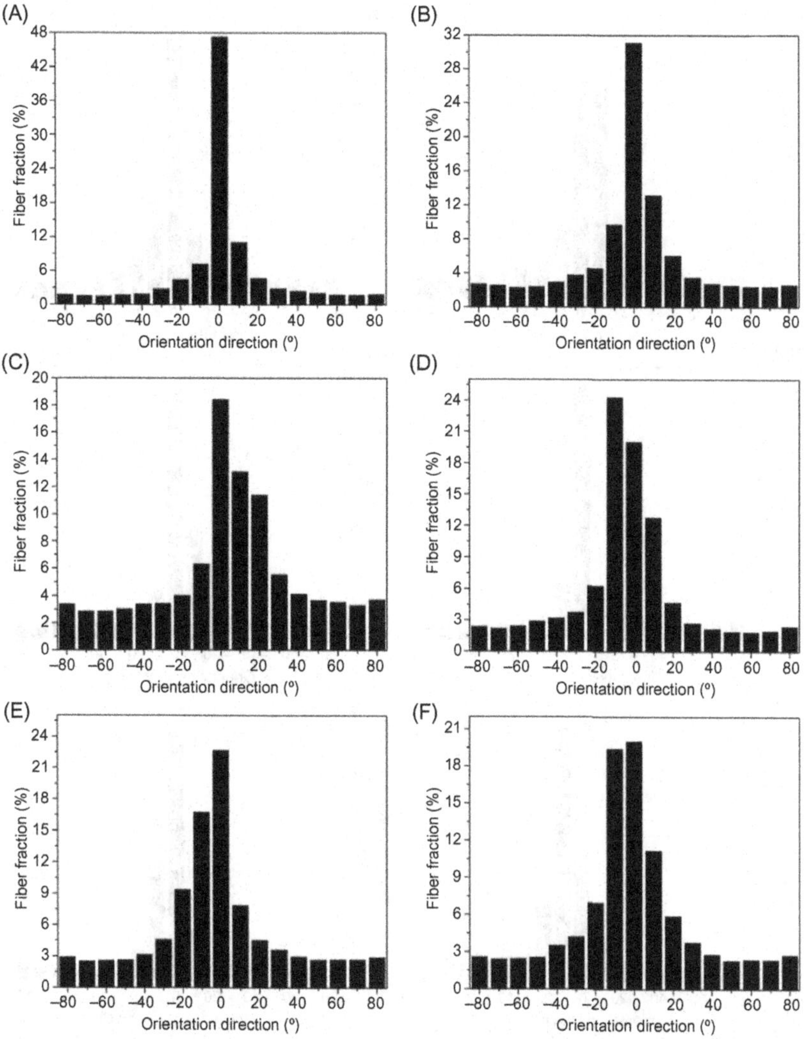

Figure A5.7 Third series of orientation analysis in a sample with 1 vol.% aligned fibers. (A–F) the orientation histogram of the sample. *Reproduced from M. Hambach, H. Möller, T. Neumann, D. Volkmer, Portland cement paste with aligned carbon fibers exhibiting exceptionally high flexural strength (>100 MPa), Cem. Concr. Res. 89 (2016) 80–86, slightly modified.*

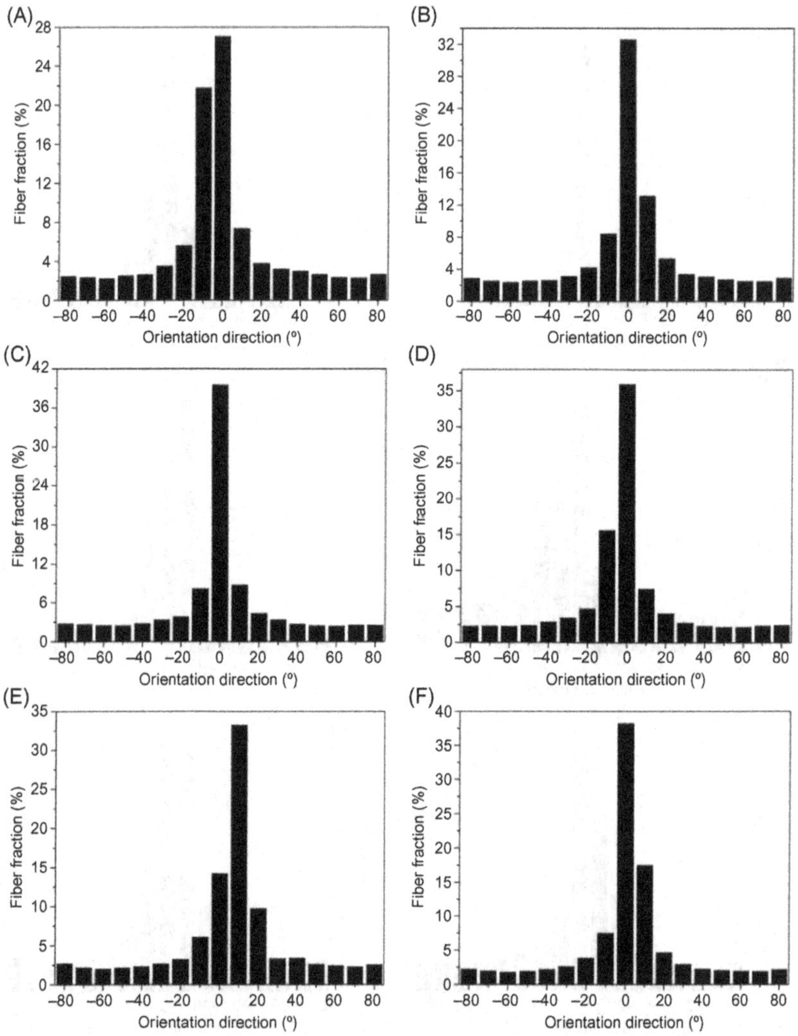

Figure A5.8 First series of orientation analysis in a sample with 3 vol.% aligned fibers. (A–F) show micrographs the orientation histogram of the sample. *Reproduced from M. Hambach, H. Möller, T. Neumann, D. Volkmer, Portland cement paste with aligned carbon fibers exhibiting exceptionally high flexural strength (>100 MPa), Cem. Concr. Res. 89 (2016) 80–86, slightly modified.*

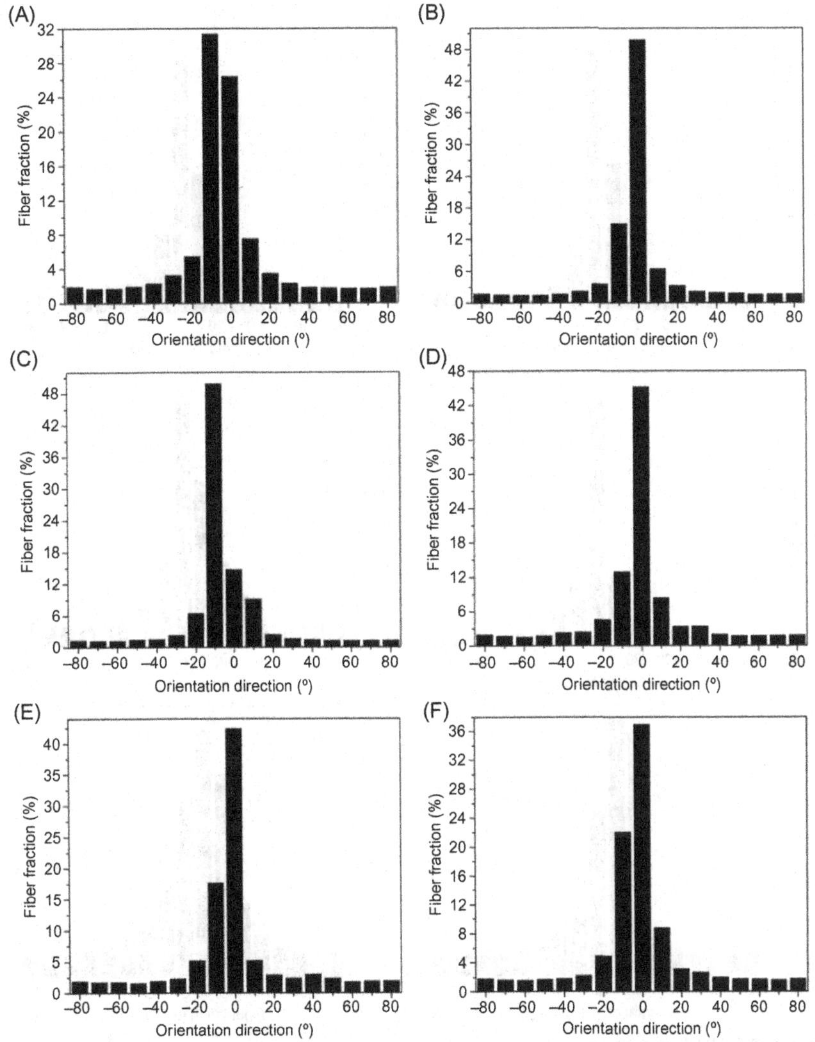

Figure A5.9 Second series of orientation analysis in a sample with 3 vol.% aligned fibers. (A–F) show the orientation histogram of the sample. *Reproduced from M. Hambach, H. Möller, T. Neumann, D. Volkmer, Portland cement paste with aligned carbon fibers exhibiting exceptionally high flexural strength (>100 MPa), Cem. Concr. Res. 89 (2016) 80–86, slightly modified.*

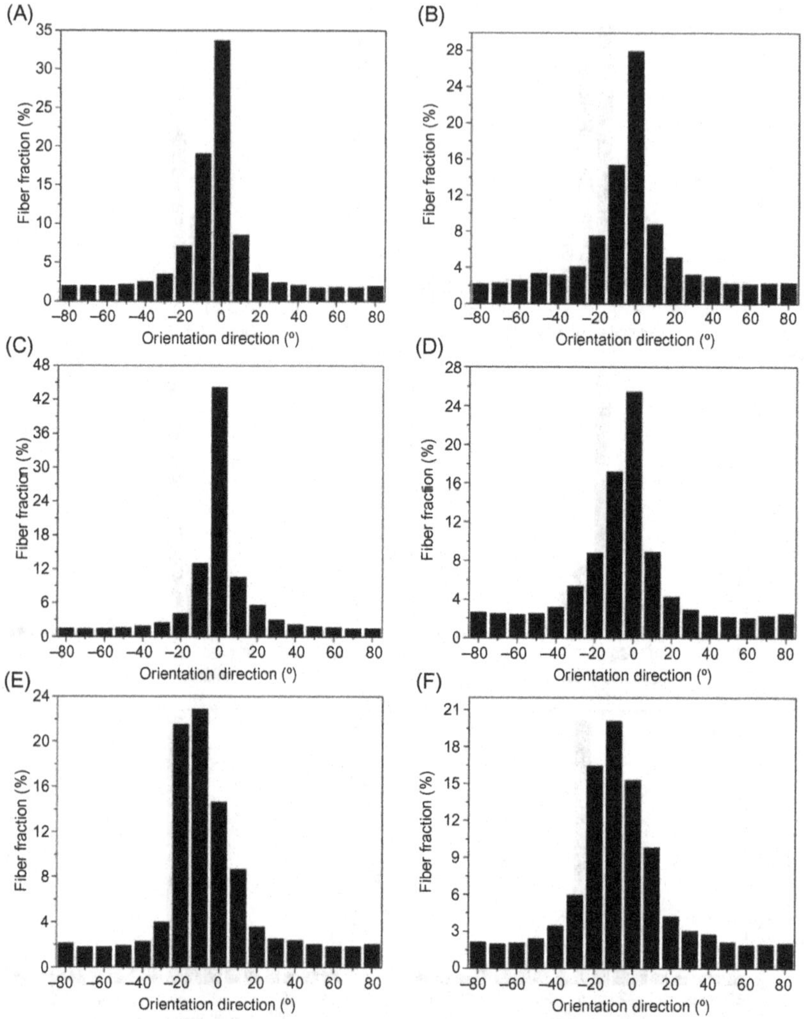

Figure A5.10 Third series of orientation analysis in a sample with 3 vol.% aligned fibers. (A–F) the orientation histogram of the sample. *Reproduced from M. Hambach, H. Möller, T. Neumann, D. Volkmer, Portland cement paste with aligned carbon fibers exhibiting exceptionally high flexural strength (>100 MPa), Cem. Concr. Res. 89 (2016) 80–86, slightly modified.*

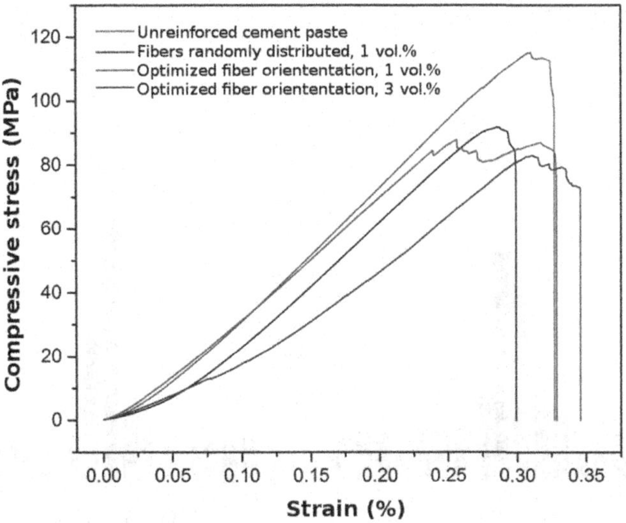

Figure A5.11 Stress–strain diagram for plain cement paste and carbon-fiber composites for the uniaxial compressive test [12]. Plain cement paste shows a high compressive strength around 100 MPa, while carbon fiber-reinforced samples decrease slightly to values at around 80 MPa. *Reproduced from M. Hambach, H. Möller, T. Neumann, D. Volkmer, Portland cement paste with aligned carbon fibers exhibiting exceptionally high flexural strength (>100 MPa), Cem. Concr. Res. 89 (2016) 80–86, slightly modified.*

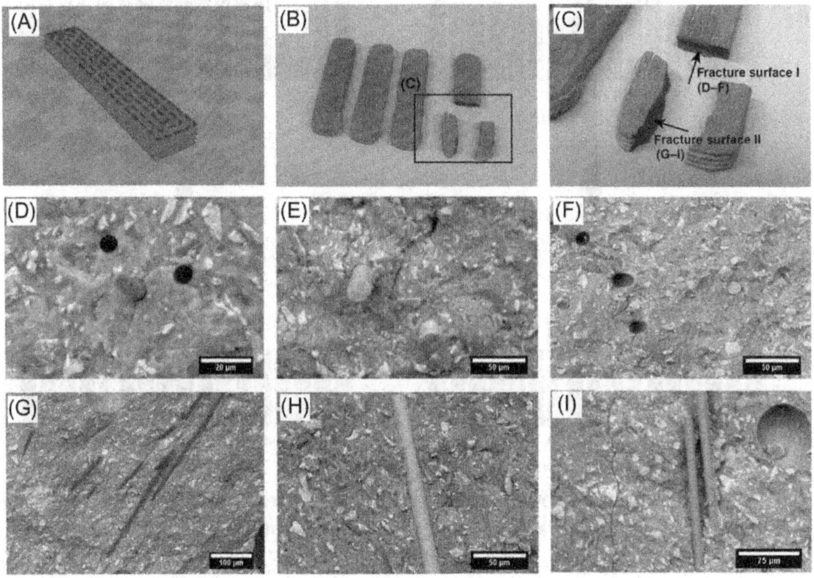

Figure A5.12 (A) 3D model of a print path A with an alignment of fibers within the specimen (fibers sketched as lines along the print path). (B) Photographs of the 3D-printed specimens and preparation of the samples for ESEM micrographs. (C) Detailed view of samples for ESEM micrographs: Fracture surfaces I for perpendicular fiber orientation and fracture surface II for longitudinal fiber orientation. ESEM micrographs of fracture surface I proving perpendicular fiber orientation for (D) carbon fibers, (E) glass fibers, and (F) basalt fibers. ESEM micrographs of fracture surface II proving longitudinal fiber orientation for (G) carbon fibers, (H) glass fibers, and (I) basalt fibers.

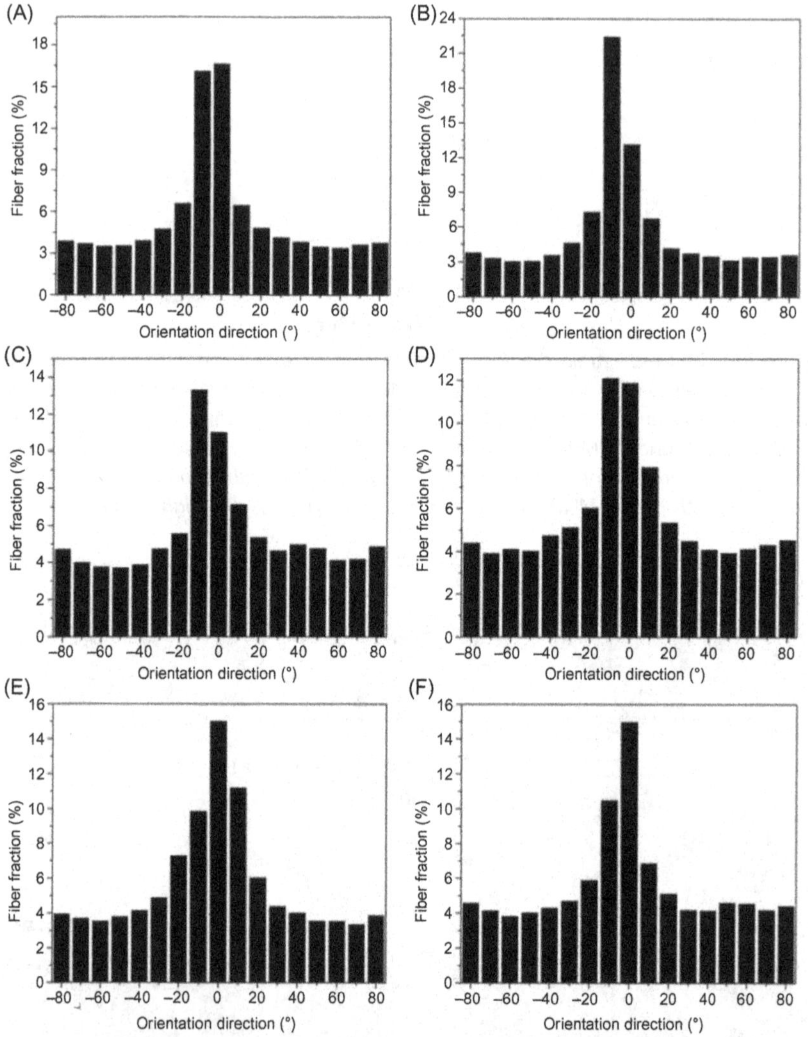

Figure A5.13 First series of orientation histograms of specimens printed with path A and 1 vol.% of fibers. Histograms (A) to (F) each represent measurements of fiber orientation within a single thin section. *Reproduced form M. Hambach, Hochfeste multifunktionale Verbundwerkstoffe auf Basis von Portlandzement und Kohlenstoffkurzfasern, Dissertation, Augsburg University, Augsburg, 2016 [33], slightly modified.*

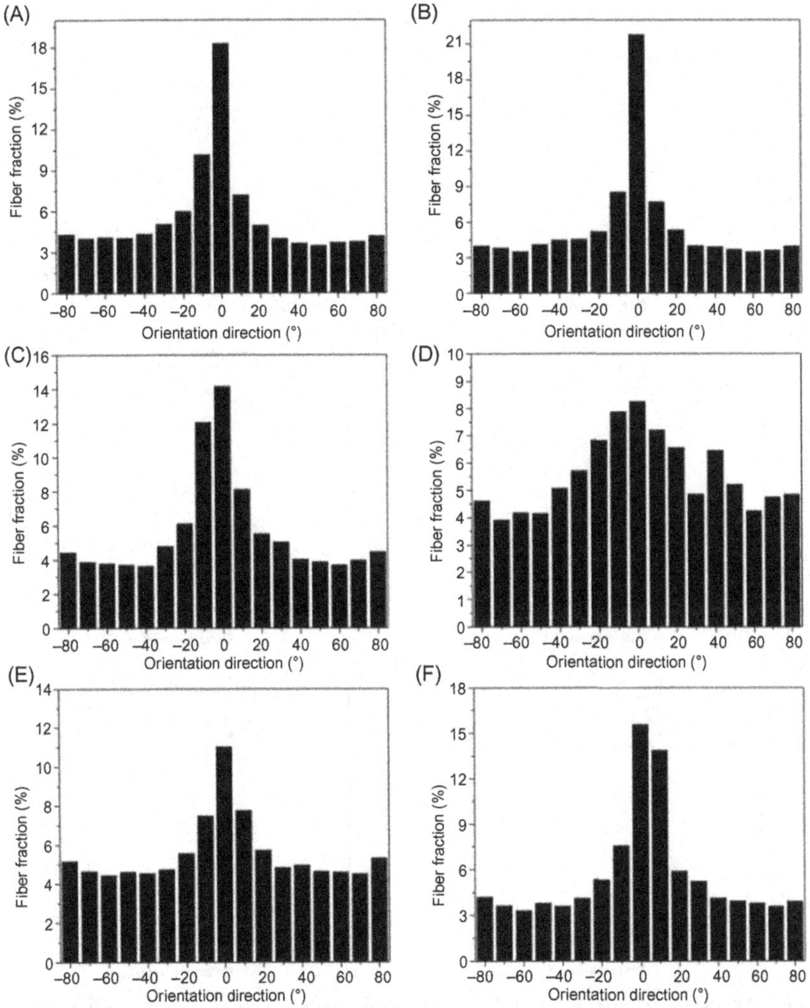

Figure A5.14 Second series of orientation histograms of specimens printed with path A and 1 vol.% of fibers. Histograms (A) to (F) each represent measurements of fiber orientation within a single thin section. *Reproduced form M. Hambach, Hochfeste multifunktionale Verbundwerkstoffe auf Basis von Portlandzement und Kohlenstoffkurzfasern, Dissertation, Augsburg University, Augsburg, 2016 [33], slightly modified.*

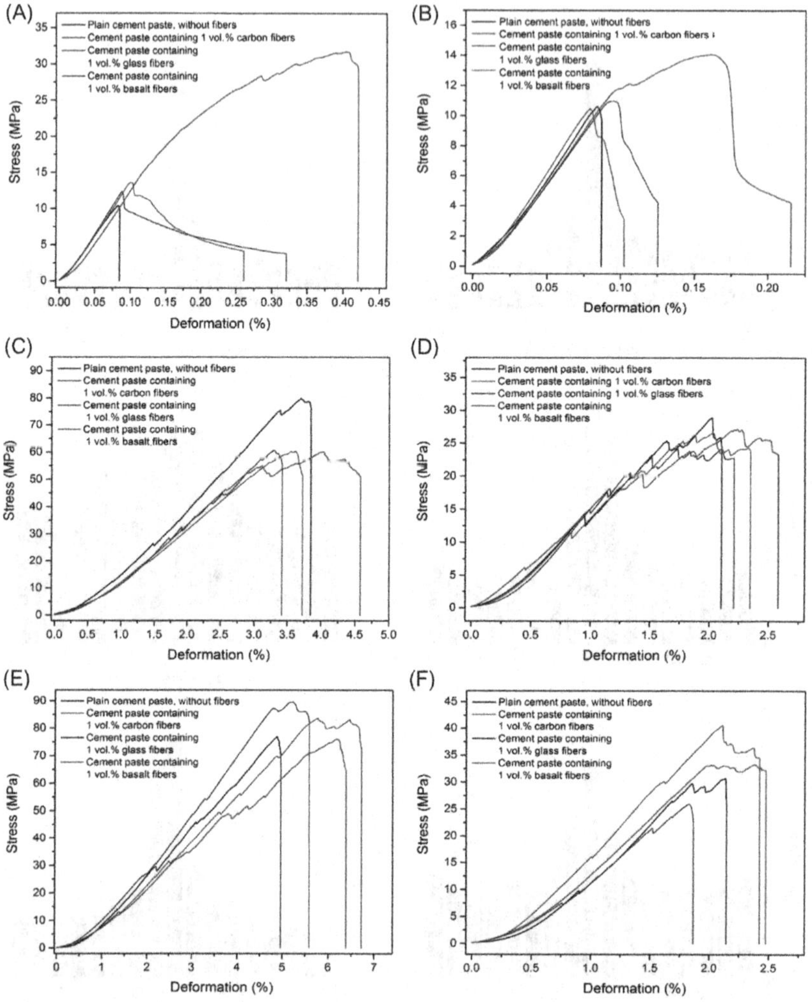

Figure A5.15 Stress-deformation plots: (A) Print path A samples in 3-point bending test. (B) Print path B samples in 3-point bending test. (C) Print path C samples in uniaxial compressive strength test (test direction I). (D) Print path C samples (test direction II) in uniaxial compressive strength test. (E) Print path D samples in uniaxial compressive strength test (test direction I). (F) Print path C samples (test direction II) in uniaxial compressive strength test.

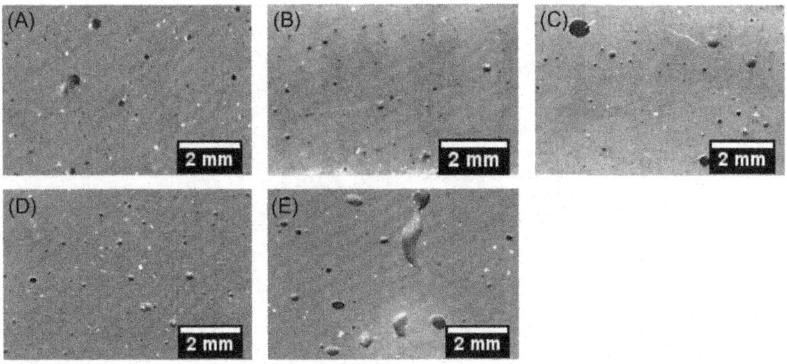

Figure A5.16 Photographs of polished surfaces of (A) mold-casted cement paste, (B) print path A sample, (C) print path B sample, (D) print path C sample, and (E) print path D sample.

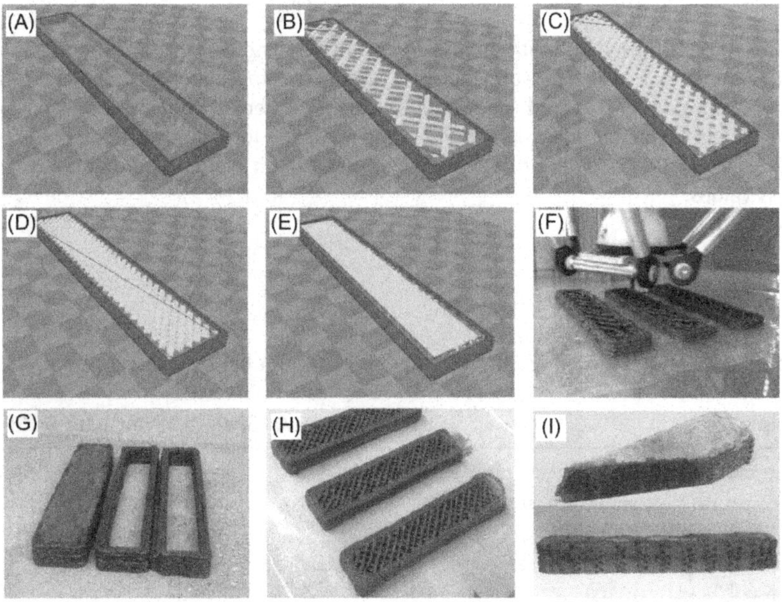

Figure A5.17 3D model of (A) print path E (0% fill density), (B) print path E (25% fill density), (C) print path E (50% fill density), (D) print path E (75% fill density), and (E) print path E (100% fill density) in Cura software revealing varying distances between print path lines (*yellow* (white in print version)) for different fill densities. (F) Photograph of 3D-printing hierarchical specimens. (G) Photograph of print path E (0% fill density) specimens after printing and after filling with mortar. (H) Photograph of print E (50% fill density) specimens after printing. (I) Infill mortar was colored *red* (black in print version) by iron oxide powder to show a transverse and a longitudinal cross-section to illustrate 3D-printed cement paste (*gray*) and mortar filled voids (*red* (black in print version)).

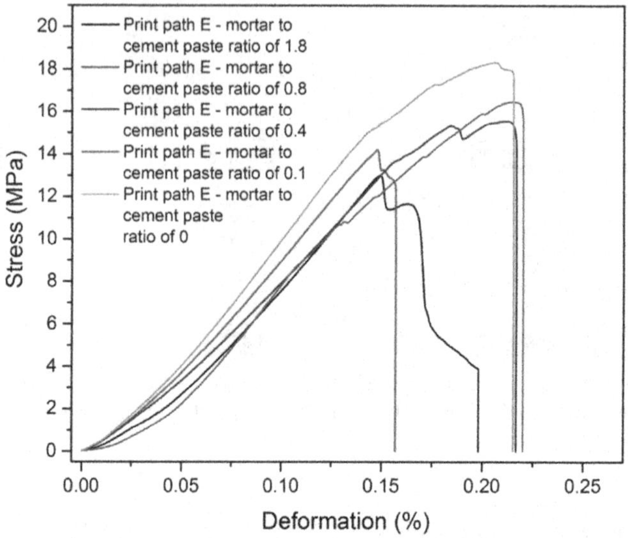

Figure A5.18 Stress-deformation plots in 3-point bending test for specimens printed with print path E and exhibiting different mortar to fiber-reinforced cement paste ratios. "Print path E–1.8" and "print path E–0.8" specimens do not show a significant zone of plastic deformation in 3-point bending test. However, "print path E–0.4", "print path E–0.1," and "print path E–0" samples exhibit plastic deformation at high stress values.

Table A5.1 Deformation modulus determined in 3-point bending test for print path A and B

Sample	Print path A (GPa)	Print path B (GPa)
Plain cement paste, (without fibers)	14.9 ± 0.4	14.1 ± 0.5
Fiber-reinforced cement paste (1 vol.% carbon fibers)	15.8 ± 1.8	13.9 ± 0.5
Fiber-reinforced cement paste (1 vol.% glass fibers)	15.0 ± 2.3	13.3 ± 1.2
Fiber-reinforced cement paste (1 vol.% basalt fibers)	14.5 ± 1.1	14.6 ± 1.2

Table A5.2 Deformation modulus determined in uniaxial compressive strength test for print path C and D measured in test direction I and II

Sample	Print path C, test direction I (GPa)	Print path C, test direction II (GPa)	Print path D, test direction I (GPa)	Print path D, test direction II (GPa)
Plain cement paste, (without fibers)	2.4 ± 0.4	1.9 ± 0.3	2.3 ± 0.5	2.3 ± 0.4
Fiber-reinforced cement paste (1 vol.% carbon fibers)	1.7 ± 0.3	1.9 ± 0.4	2.1 ± 0.4	1.7 ± 0.7
Fiber-reinforced cement paste (1 vol.% glass fibers)	1.8 ± 0.3	1.5 ± 0.2	2.0 ± 0.3	2.0 ± 0.3
Fiber-reinforced cement paste (1 vol.% basalt fibers)	2.0 ± 0.3	1.7 ± 0.4	2.1 ± 0.5	2.1 ± 0.4

REFERENCES

[1] H.-G. Ni, J.-Z. Wang, Prediction of compressive strength of concrete by neural network, Cem. Concr. Res. 30 (2000) 1245–1250.

[2] J.D. Birchall, A.J. Howard, K. Kendall, Flexural strength and porosity of cements, Nature 289 (1981) 388–390.

[3] D.D. Higgins, J.E. Bailey, Fracture measurements on cement paste, J. Mater. Sci. 11 (1976) 1995–2003.

[4] R.F. Zollo, Fiber-reinforced concrete: an overview after 30 years of development, Cem. Concr. Comp. 19 (1997) 107–122.

[5] D.D.L. Chung, Carbon Fiber Composites, Butterworth-Heinemann, Newton, 1994.

[6] N. Banthia, J. Sheng, Micro-reinforced cementitious materials, Mater. Res. Soc. Symp. Proc. 211 (1991) 25–32.

[7] P. Morgan, Carbon Fibers and Their Composites, Taylor & Francis, Boca Raton, 2005.

[8] X.Q. Qian, X.M. Zhou, B. Mu, Z.J. Li, Fiber alignment and property direction dependency of FRC extrudate, Cem. Concr. Res. 33 (2003) 1575–1581.

[9] H. Takashima, K. Miyagai, T. Hashida, V.C. Li, A design approach for the mechanical properties of polypropylene discontinuous fiber reinforced cementitious composites by extrusion molding, Eng. Fract. Mech. 70 (2003) 853–870.

[10] B. Mu, M.F. Cyr, S.P. Shah, Extruded fiber-reinforced composites, Adv. Build. Technol. 1 (2002) 239–246.

[11] B. Shen, M. Hubler, G.H. Paulino, L.J. Struble, Functionally-graded fiber-reinforced cement composite: processing, microstructure, and properties, Cem. Concr. Comp. 30 (2008) 663–673.

[12] M. Hambach, H. Möller, T. Neumann, D. Volkmer, Portland cement paste with aligned carbon fibers exhibiting exceptionally high flexural strength (>100 MPa), Cem. Concr. Res. 89 (2016) 80–86.

[13] T.T. Le, S.A. Austin, S. Lim, R.A. Buswell, R. Law, A.G.F. Gibb, et al., Hardened properties of high-performance printing concrete, Cem. Concr. Res. 42 (2012) 558–566.

[14] A. Perrot, D. Rangeard, A. Pierre, Structural built-up of cement-based materials used for 3D-printing extrusion techniques, Mater. Struct. 49 (2016) 1213–1220.

[15] E. Lloret, A.R. Shahab, M. Linus, R.J. Flatt, F. Gramazio, M. Kohler, et al., Complex concrete structures merging existing casting techniques with digital fabrication, Comput. Aided Des. 60 (2015) 40–49.

[16] G. Villar, A.D. Graham, H. Bayley, A tissue-like printed material, Science 340 (2013) 48–52.

[17] C.B. Highley, C.B. Rodell, J.A. Burdick, Direct 3D printing of shear-thinning hydrogels into self-healing hydrogels, Adv. Mater. 27 (2015) 5075–5079.

[18] B.G. Compton, J.A. Lewis, 3D-printing of lightweight cellular composites, Adv. Mater. 26 (2014) 5930–5935.

[19] J.T. Muth, D.M. Vogt, R.L. Truby, Y. Menguc, D.B. Kolesky, R.J. Wood, et al., Embedded 3D printing of strain sensors within highly stretchable elastomers, Adv. Mater. 26 (2014) 6307–6312.

[20] R. Kruger, J. Groll, Fiber reinforced calcium phosphate cements – on the way to degradable load bearing bone substitutes? Biomaterials 33 (2012) 5887–5900.

[21] S. Bose, S. Vahabzadeh, A. Bandyopadhyay, Bone tissue engineering using 3D printing, Mater. Today 16 (2013) 496–504.

[22] U. Gbureck, T. Hozel, U. Klammert, K. Wurzler, F.A. Muller, J.E. Barralet, Resorbable dicalcium phosphate bone substitutes prepared by 3D powder printing, Adv. Funct. Mater. 17 (2007) 3940–3945.

[23] J.A. Inzana, D. Olvera, S.M. Fuller, J.P. Kelly, O.A. Graeve, E.M. Schwarz, et al., 3D printing of composite calcium phosphate and collagen scaffolds for bone regeneration, Biomaterials 35 (2014) 4026–4034.

[24] S. Christ, M. Schnabel, E. Vorndran, J. Groll, U. Gbureck, Fiber reinforcement during 3D printing, Mater. Lett. 139 (2015) 165–168.

[25] H. Lipson, M. Kurman, Fabricated The New World of 3D Printing, Wiley, Hoboken, 2013.

[26] J. Malda, J. Visser, F.P. Melchels, T. Jungst, W.E. Hennink, W.J.A. Dhert, et al., 25th anniversary article: engineering hydrogels for biofabrication, Adv. Mater. 25 (2013) 5011–5028.

[27] S.V. Murphy, A. Atala, 3D bioprinting of tissues and organs, Nat. Biotechnol. 32 (2014) 773–785.

[28] R. Bogue, 3D printing: the dawn of a new era in manufacturing? Assembly Autom. 33 (2013) 307–311.

[29] D.D.L. Chung, Dispersion of short fibers in cement, J. Mater. Civil Eng. 17 (2005) 379–383.

[30] T. Sugama, L.E. Kukacka, N. Carciello, D. Stathopoulos, Interfacial reactions between oxidized carbon-fibers and cements, Cem. Concr. Res. 19 (1989) 355–365.

[31] X.L. Fu, W.M. Lu, D.D.L. Chung, Improving the tensile properties of carbon fiber reinforced cement by ozone treatment of the fiber, Cem. Concr. Res. 26 (1996) 1485–1488.

[32] J. Tinevez, Directionality plugin for imageJ. <http://fijii.sc/wiki/index.php/Directionality/>, 2015 (accessed 27.05.15).

[33] M. Hambach, Hochfeste multifunktionale Verbundwerkstoffe auf Basis von Portlandzement und Kohlenstoffkurzfasern, Dissertation, Augsburg University, Augsburg, 2016.

[34] M.J. Shannag, R. Brincker, W. Hansen, Pullout behavior of steel fibers from cement-based composites, Cem. Concr. Res. 27 (1997) 925–936.

[35] A. Abrishambaf, V.M.C.F. Cunha, J.A.O. Barros, The influence of fibre orientation on the post-cracking tensile behaviour of steel fibre reinforced self-compacting concrete, Frat. Integrita Struct. (2015) 38–53.
[36] J.A.O. Barros, V.M.C.F. Cunha, A.F. Ribeiro, J.A.B. Antunes, Post-cracking behaviour of steel fibre reinforced concrete, Mater. Struct. 38 (2005) 47–56.
[37] V.C. Li, M. Maalej, Toughening in cement based composites .2. Fiber reinforced cementitious composites, Cem. Concr. Comp. 18 (1996) 239–249.
[38] S.P. Shah, C. Ouyang, Mechanical-behavior of fiber-reinforced cement-based composites, J. Am. Ceram. Soc. 74 (1991) 2727–2753.
[39] R.Z. Wang, H.S. Gupta, Deformation and fracture mechanisms of bone and nacre, Annu. Rev. Mater. Res. 41 (2011) 41–73.
[40] H. Kakisawa, T. Sumitomo, The toughening mechanism of nacre and structural materials inspired by nacre, Sci. Technol. Adv. Mater. 12 (2011).
[41] F. Bencardino, L. Rizzuti, G. Spadea, R.N. Swamy, Implications of test methodology on post-cracking and fracture behaviour of steel fibre reinforced concrete, Compos. Part B-Eng. 46 (2013) 31–38.
[42] T.T. Le, S.A. Austin, S. Lim, R.A. Buswell, A.G.F. Gibb, T. Thorpe, Mix design and fresh properties for high-performance printing concrete, Mater. Struct. 45 (2012) 1221–1232.
[43] N. Oxman, J. Duro-Royo, S. Keating, B. Peters, E. Tsai, Towards robotic swarm printing, Archit. Des. 84 (2014) 108–115.
[44] M.A. Meyers, P.Y. Chen, A.Y.M. Lin, Y. Seki, Biological materials: structure and mechanical properties, Prog. Mater. Sci. 53 (2008) 1–206.
[45] S. Weiner, H.D. Wagner, The material bone: structure mechanical function relations, Annu. Rev. Mater. Sci. 28 (1998) 271–298.
[46] U.G.K. Wegst, H. Bai, E. Saiz, A.P. Tomsia, R.O. Ritchie, Bioinspired structural materials, Nat. Mater. 14 (2015) 23–36.

CHAPTER 6

3D Concrete Printing: Machine Design, Mix Proportioning, and Mix Comparison Between Different Machine Setups

Zeina Malaeb, Fatima AlSakka and Farook Hamzeh
Civil and Environmental Engineering, American University of Beirut, Beirut, Lebanon

6.1 INTRODUCTION

3D concrete printing is an innovative construction method that was recently introduced to the construction industry and has proven to be beneficial in terms of optimizing construction time, cost, design flexibility, and error reduction as well as being environmentally friendly. It involves fabricating a predesigned building element in 2D layers on top of each other, the repetition of which results in a 3D model. The concrete, which is poured out of a printing nozzle, does not require any formwork or subsequent vibration [1].

The technology of 3D printing started in the 1980s [2], but was only introduced to the construction industry in the mid-1990s by researchers at the University of Southern California [3]. The research team has produced extensive material on the topic including developing several machines for research on fabrication using various types of materials. These machines include an XYZ gantry system, a nozzle assembly with three motion control components, and a six-axis coordinated motion control system [4]. The developments in the field remained almost steady until about 2012 [3] when several other parties began investigating different approaches and technologies related to 3D printing. The University of Loughborough has presented a pioneering work in concrete printing [3]. Moreover, the Chinese company, WinSun, displayed its expertise in 3D printing by printing 10 houses within 24 hours, with each house costing a mere $5000 [5]. WinSun then took it to a higher level and printed a five-story building which is currently the highest 3D-printed structure [6].

3D Concrete Printing Technology
DOI: https://doi.org/10.1016/B978-0-12-815481-6.00006-3

Numerous other 3D-printing projects have been executed or are in progress as the interest in advancing and adopting this technology is rapidly growing around the globe.

3D concrete printing promises to improve construction on several fronts: (1) it minimizes the construction process duration by eliminating some time-consuming procedures involved in traditional construction [7]. In a study conducted using value stream mapping, a 60% reduction in construction duration was recorded for 3D concrete printing compared to the conventional methods of construction [8]. (2) Costs incurred on the project are reduced by minimizing overproduction, waste, and labor utilization [9]. (3) 3D concrete printing provides flexibility in constructing structural forms that conventional construction methods are not able to build; and (4) delivers improvements in the overall safety and environmental impact of the structure [10].

One of the greatest challenges in 3D concrete printing consists of designing a proper printing material. The main purpose is to optimize the mix for compatibility with the printer, on the one hand, and with the minimum standards of mechanical behavior on the other. The printable concrete should match several requirements of extrudability, buildability, adhesion, and open time. Great efforts are put into investigating the effect of different materials on enhancing the performance of the concrete. Le et al. studied the significance of using superplasticizers, retarders, and accelerators on workability, strength, and open time [11]. Khoshnevis et al., on the other hand, worked on enhancing the surface finish and geometry resolution of printed concrete by using plaster and clay-like material [12]. Similarly, numerous studies target other properties pertinent to the printing material and extensive research is needed in this field.

This chapter will tackle the topic of 3D concrete printing through studying its impact on several engineering fronts. Chiefly, it concentrates on the construction materials field through developing the appropriate concrete mix that is suitable for 3D printing. The optimal mix for this function is obtained after conducting several tests to verify its properties. On another front, it proposes an appropriate design for the printing machine after investigating the printing methodology required. A structural sample is printed as a proof of concept for the printing technique. Finally, a second study was undertaken using a different printing nozzle and machine is briefly presented to prove that the convenience of a mix is contingent on the machine setup.

This study enhances the understanding of 3D concrete printing as well as its applications in real-life construction projects and contributes to the growing body of scientific knowledge that aims to highlight the significance of this method in improving automation in civil engineering projects, enhancing efficiency in resource management, and showing glimpses of an advanced construction method that promises shorter project durations at lower costs.

6.2 METHODOLOGY

6.2.1 Materials: Mix Design

In designing a concrete mix for 3D concrete printing, it is important to ensure that it meets several essential criteria that are directly related to the methodology of printing the concrete. Therefore, the design of the concrete mix, on one hand, and the printing machine, on the other, must be complementary. The mix that is labeled "optimal" should be able to meet certain target goals, as listed in Table 6.1.

The table presents some goals that seem to be in conflict with each other, and yet all must be targeted. Thus, the challenge is in maintaining an appropriate balance of the different goals. One example is in relation to the compressive strength of the mix, which we aim to maximize. Maximizing the compressive strength in the mix means minimizing the water–cement ratio. However, an appropriate water content must be maintained in the mix to ensure suitable workability of the concrete. In addition, the mix should be flowable throughout the system, yet, upon pouring, the mix must be buildable and each layer should be able to hold itself and subsequent layers. Finally, when poured, the mix should set as fast as possible, but slow enough to ensure appropriate bonding with the subsequent layer.

Table 6.1 Mix goals

Maximize compressive strength		**Maximize workability**
Maximize flowability in the system Maximize speed of concrete setting		Maximize buildability upon pouring Maintain appropriate setting rate so as to ensure bonding with the subsequent layer

To address these goals, we set specific criteria of the mix that are measurable. The five most critical properties of the mix that are studied are extrudability, flowability, buildability, compressive strength, and open time. Both extrudability and flowability relate to the concrete extrusion, flow, and workability, as the aim is to reach a continuous easy-flowing paste from the source to the printing nozzle. Buildability refers to the ability of the concrete layer to hold the subsequent layers above it without collapsing. The concrete must also maintain an appropriate compressive strength. Finally, open time studies the change in the concrete's flowability with time. The goal is to ensure that each printed layer has the capacity to hold itself and harden when poured, and yet stay liquid enough to bond with the layer above it so as to ensure that suitable connection is achieved. On the other hand, the concrete paste must have a certain flowability upon its transfer that must not threaten its ability to stiffen upon pouring [1].

As for the constituents of the concrete mix, aggregates with a maximum size of 2 mm were used since the diameter of the printing nozzle is relatively small (2 cm). Other dry constituents include cement type I and sand. A superplasticizer (Viscocrete) was used with the mix to ultimately increase the workability of the concrete and compensate for the low water–cement ratio. An accelerator is added to the concrete mix allowing it to settle and gain strength at a faster pace when poured. In addition, a retarder is also added to prevent the concrete from settling in the tank.

An appropriate balance of all the constituents has to be reached to ensure proper functioning of the mix. Several experiments were performed in order to determine the exact quantities of materials to be added for the optimal mix. An optimal amount of the different additives had to be found. In fact, quantities of superplasticizers exceeding the found optimum amount might give unwanted results [13].

6.2.2 Mechanics: Machine Design

The design of the 3D printing machine, primarily referring to its printing methodology, is critical to this project's success. An optimal machine design is necessary in order to complement the optimal mix design; only then will project success be achieved. The machine design has to account for several criteria related to both the fresh and printed properties of the concrete mix. The 3D printer is made up of three main components: the concrete tank and pumping mechanism, the printing nozzle, and the motion control system. The concrete is stored in the tank. Then, it is manually pumped to enable it to move to the nozzle for pouring it out. The machine is designed to move on a tri-axial plane (x-y-z) in order to print a 3D element.

6.2.2.1 Motion Along the Axes

For our experiment's purposes, the printer was designed with a specific initial target: to print a specimen composed of a 77 cm × 10 cm structural wall. The wall was to be printed in two parallel lines, each with a length of 77 cm and a width of 2 cm. The lines were separated by a distance of 10 cm. Therefore, the nozzle's route of printing would start by printing one line on the longitudinal axis (x-axis), then moves along the perpendicular axis (y-axis) to get in position to print the next line, parallel to the initial one. The nozzle would then print the other line in the same way it printed the first one, moving back in position along the former line. Finally, the nozzle would move along the z-axis to print layer-upon-layer and complete the 3D design. The machine was designed to move up in 2 cm intervals as the height of each layer was designed to be 2 cm.

A key part of the machine is the vertical component that holds the mobile tank and nozzle. This component functions to ensure the motion along the vertical axis (z-axis) and is designed in a way that allows either hydraulic or manual operation. The element's design allows it to roll on a specific track along the x-axis, along a threaded horizontal bar. This motion is controlled by a rotating drill of adjustable speed and connecting the drill to one end of the threaded horizontal bar. The rotation would then cause the connected vertical component to move along the bar. The speed of motion can be controlled by adjusting the speed of the drill. Finally, the nozzle held by this element is able to move in a direction perpendicular to that of the machine's motion and change positions on the y-axis.

6.2.2.2 Nozzle Design

Another critical component of the machine is the printing nozzle, since its design has a high impact on the nature of the extruded concrete. Attached to the nozzle are two trowels, a side and top trowel, which lag behind it. The trowels have different functions along different axes. The side trowel, which is a vertical trowel located on the outer side, is responsible for straightening the concrete being poured as the nozzle passes by.

The top trowel, a horizontal trowel located at the top, serves to straighten the upper surface of the concrete layer to ensure maximum buildability.

The diameter of the nozzle is directly related to the concrete mix properties, specifically the flowability of the concrete. In short, as the diameter size of the nozzle decreases, the flowability of the mix should be increased to account for it, and vice versa. Therefore, ensuring an optimal nozzle diameter is critical as it affects the design of the concrete mix. In order to determine the ideal nozzle diameter, and before designing the nozzle, the research group experimented with syringes of different opening diameters that ranged from 1 cm to 3 cm. It was found that a diameter size of 2 cm was the most optimal. A diameter less than 2 cm resulted in problems related to the segregation of the concrete mix components while a diameter greater than 2 cm caused problems related to the buildability of the concrete mix as the individual layer was not able to hold itself, let alone other layers.

6.2.2.3 *Tank and Pump Design*

The concrete starts its journey in the tank and is transported to the nozzle to be printed. The pump is the component responsible for transporting the concrete mix to be poured at the nozzle. The pump used should be able to handle the specific concrete mix, taking into account its different properties from the maximum aggregate size to the water–cement ratio. The challenge was in obtaining a pump that would work on such a small scale for the experiment's purpose. No commercial pump was found that was able to account for the concrete pouring mechanism required at this scale. Therefore, a "piston-pump" was designed to carry out this purpose. This pump combines two mechanisms: the pressure application of the syringe, on one hand, and a cement screw pump mechanism, on the other.

The specially designed pump was integrated with the concrete tank, in this case a cylindrical mobile tank. It is mobile since it is connected to the printing machine and moves with it. The printing nozzle is at the bottom of the tank and the piston-pump is at the top. The pump functions by exerting a force on the piston which allows it push the concrete down from the tank and into the nozzle. The most crucial criterion to be managed is the pressure exerted by the pump on the mix. The exerted pressure has a direct relationship with the speed at which the concrete is being poured. The goal is to find an ideal association between the two.

6.2.3 Testing Procedures

In order to ensure that the achieved mix is the optimal one, several tests had to be conducted taking into consideration the target parameters to be achieved. The five parameters (extrudability, flowability, buildability, compressive strength, and open time) together contribute to the success of the printing process. Five tests were designed that targeted these five criteria.

6.2.3.1 Flowability Test

Flowability could be defined as the ease with which concrete flows in a system under given conditions. The concrete mix's flowability is measured by performing the slump flow test [14]. In this test, the mix is poured out of an inverted cone and the time required for the mix to spread to a specific diameter is calculated. Subsequently, the concrete's flow rate can be obtained. The easier the mix expands, the greater its flowability and workability.

6.2.3.2 Extrudability Test

Extrudability is the ability of concrete to be squeezed out of the nozzle in a continuous manner. Knowing the definition of flowability, it is self-evident that higher flowability, up to a certain limit, results in easier extrudability and vice versa. This implies that whatever factors that critically reduce flowability should have a counter effect on extrudability.

As a first step, preliminary trial tests were conducted to test the concrete's extrudability. Note that none of the other properties are of any significance if concrete is not extrudable. Although all the properties need to be fulfilled, there is an order of priority that must be followed. Concrete cannot be extruded if not flowable, cannot be built if not extruded, cannot properly function if not properly built, and, finally, does not have an open time if one of the properties is not met.

Given the fact that extrudability is flowability-dependent, it is influenced by many factors related to the distributions and quantities of the dry constituents in the concrete mix, the properties of different constituents, the water content, the presence of additives, the delivery system, prevailing conditions, and time, etc. Yet the essential aspect to be considered in this study is the mixture proportions. The proportions of the initial mix were set with reference to the concept of slip-form concrete design. This type of concrete, which has a self-compacting property and does not require further consolidation, is an apt starting point as it has similar characteristics to the target mix [15].

Once an extrudable concrete paste is obtained, additional tests were performed to meet the other target requirements. Note that the quantities of additives used play a prominent role in affecting these criteria.

6.2.3.3 Buildability Test

Since buildability refers to the capacity of concrete layers to support themselves in addition to all the subsequent top layers, a viable way to test it relates to the number of buildable layers. The higher the buildability, the higher the number of concrete layers that could be stacked upon each other before the system excessively deforms or collapses.

6.2.3.4 Open Time Test

This criterion is of vital importance considering the printing mechanism of the concrete. The mix's initial and final setting times that are more prominent in traditional concrete pouring, are not relevant in concrete printing [10]. Since the printed concrete isn't poured in one go, as in the traditional method, it is expected that some of its properties change with time from the first printed layer to the last. Therefore, a better representation of the concrete mix's workability change with time is described by measuring the open time. Open time is calculated using the slump flow test to obtain measurements of flowability over specific time intervals.

6.2.3.5 Compressive Strength Test

The concrete's compressive strength is of particular importance due to the fact that the printing mechanism pours the structure in layers rather than in its entirety at once. Since the printing process happens only in a matter of minutes and setting time is assumed to be instantaneous, the targeted strength and strength gain should be high. The target strength of the concrete is determined using BS 1881-116:1983 and 5 cm × 5 cm concrete cubes [16].

6.3 RESULTS AND DISCUSSION

6.3.1 Machine

When it comes to the printing machine's function, a balanced relationship must be maintained between the speed of the machine and the extrusion rate of the concrete mix from the nozzle tip. Otherwise, the printing

Figure 6.1 3D printing machine, experiment 1.

resolution would be greatly impacted; if the linear horizontal speed is larger than the extrusion rate, the extruded filaments would be too thin and might even exhibit some ruptures. On the other hand, if the printing rate is too slow, more material than what is necessary would be deposited resulting in thick concrete layers.

The controlling factor is the extrusion rate of the mix; after the extrusion rate of the optimal mix is calculated, the printer's speed is adjusted to achieve the appropriate criteria for printing. The nozzle extruded the concrete paste at a rate of 0.09 L/s. Therefore, the machine speed that was compatible with this rate was set at 18.76 cm/s. Fig. 6.1 depicts the physical printing machine, designed according to these criteria.

6.3.2 Experimental Results for the Concrete Mix

6.3.2.1 Mix Proportions

A large number of trial mixes had to be designed and tested on the printer before the optimal mixture was achieved. The optimal mix was considered the one that required the lowest water–cement ratio while satisfying all the required criteria of flowability, extrudability, buildability, adhesion, and open time.

Obtaining an optimal combination of different constituents requires designing a control mix. The testing procedures were first based on assessing the extrudability of mixtures prepared by varying the proportions of dry constituents (cement, aggregates, and sand) and water. After all, no progress can be recorded with a mixture that cannot be pumped out of the nozzle. In order to find the optimum diameter of the nozzle, a set of syringes, ranging from 1 to 3 cm in diameter, were used to evaluate the flowability of each mixture. Note, however, that the behavior of concrete varies from one nozzle to another depending on many factors including the nozzle's diameter, length, material type, and surface texture, etc. The syringes help give a relatively close representation of the flowability of the mix. Extrudability was measured by visual inspection; once a mixture has the capability to be pumped out in a continuous manner without any blockage or rupture, the combination is deemed satisfactory.

After conducting many experiments, it was noticed that increasing the quantity of cement while reducing that of sand resulted in a better extrudability. After many tryouts, a convenient combination consisting of a fine aggregate to cement ratio of 1.28 and a fine aggregate to sand ratio of 2 was reached. These proportions of dry constituents were fixed afterwards. The minimum water–cement ratio necessary for proper extrudability was then found to be 0.48. Finally, the optimal nozzle diameter was found to be 2 cm as smaller nozzle sizes led to blockage and segregation.

After reaching the proper proportions of essential constituents, additional experiments were needed to determine the quantities of additives. The aim of this stage was to increase strength by reducing the water–cement ratio without compromising the flowability of the concrete. This was achieved with the help of superplasticizers. The superplasticizer allows reducing the quantity of water while maintaining the flowability of concrete. However, were the superplasticizer excessively added, the mixture would become too workable and lose the ability to support itself and buildability would be lost. Hence, it is integral to achieve a balance between flowability (needed for extrudability and other properties as will be explained) and buildability.

Starting with the control mix, five trials were conducted by progressively reducing the amount of water while increasing the quantity of the superplasticizer until the concrete was no longer buildable. The obtained results are listed in Table 6.2.

It is essential to ensure proper bonding between printed concrete layers. Poor bonds form when concrete starts to set before the next layer

Table 6.2 Testing of superplasticizer

Mix number	Cement (g)	Sand (g)	Fine aggregates (g)	W/C ratio	Superplasticizer (mL)	Retarder (mL)	Accelerator (mL)
1 (Control)	125	80	160	0.48	0	1	0.5
2	125	80	160	0.42	0.5	1	0.5
3	125	80	160	0.39	1	1	0.5
4	125	80	160	0.38	1.1	1	0.5
5	125	80	160	0.36	1.3	1	0.5

is overlaid. Hence, setting time should be delayed providing enough time for the printer to deposit a new layer while the concrete is still flowable. This would allow consecutive concrete layers to intermix and exhibit a sufficient adhesion with each other. For this purpose, a retarder was added in order to delay the setting time of the concrete. At the same time, however, the concrete should be able to harden fast enough to carry the load of the top layers. Hence, an accelerator was added directly before extrusion. The related data presented in Table 6.2 was tentatively selected to conduct the tests and was later optimized.

6.3.2.2 Tests Results

6.3.2.2.1 Compressive Strength Test

Compressive strength tests were undertaken for the five samples. The results are presented in Table 6.3.

As expected, the highest compressive strength corresponds to the lowest water—cement ratio. All the samples exceeded the targeted compressive strength which is set as 40 MPa. Therefore, strength is not considered critical for the purpose of this study. Rather, flowability and buildability were the crucial determinants for the concrete mixture.

6.3.2.2.2 Slump Flow Test

To measure the workability or flowability of the mixture, the slump flow test was conducted on the five mixes. Concrete should have a minimum flowability for ease of delivery from the tank to the nozzle, proper extrudability, and sufficient adhesion. In the slump flow test, workability is measured by recording the time taken for the concrete to reach a certain diameter. The results of flowability measurements are presented in Table 6.4.

The control mixture, designated as mix 1, had an insufficient flowability and was disregarded. As expected, the flowability of concrete increases

Table 6.3 Compressive strengths

Sample number	Compressive strength (MPa)
1	40.6
2	41.5
3	42.3
4	43.5
5	55.4

Table 6.4 Flowability rates

Sample number	Flowability rate (cm/s)
1	–
2	1.1
3	1.13
4	1.2
5	1.4

with the increase of the quantity of superplasticizer. Mix 2 containing 0.5 mL of superplasticizer had the lowest flowability of 1.1 cm/s, while that of mix 3 increased to 1.4 cm/s with an increase of 0.8 mL of superplasticizer.

6.3.2.2.3 Buildability Test

This test was undertaken by counting the number of concrete layers that could be stacked on top each other before a major deformation or collapse occurred. It was aimed at evaluating the capacity of the layers corresponding to each mixture to stand stiff without support while sustaining the load imposed by the top layers. The number of buildable layers was determined for the four mixes. This number decreases as the quantity of superplasticizer is increased. Mixes 2 and 3 showed the highest buildability which greatly decreased for mixes 4 and 5.

6.3.2.2.4 Additional Analysis

The data collected from the slump and buildability tests were plotted together in function of the quantity of superplasticizer added to various mixtures. Fig. 6.2 displays these results.

Note that as the superplasticizer is added, the behavior of the concrete changes slowly until a certain limit (1 mL in this experiment); the slopes of curves corresponding to both flowability and buildability are flat. These

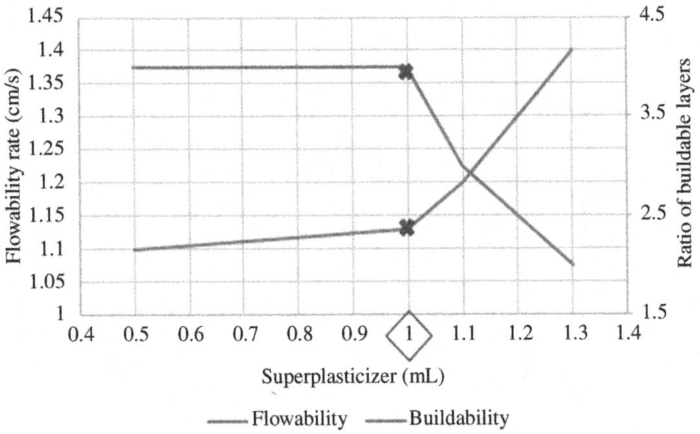

Figure 6.2 Assessment of flowability and buildability of different mixes.

slopes become much steeper as the quantity of superplasticizer is further increased, indicating that flowability becomes excessive and buildability is lost with high proportions of superplasticizer. Hence, meticulous care should be taken when adding additives to concrete.

Based on experimentation, it was found that satisfying the requirements of buildability and extrusion necessitates a flowability rate that is strictly between 1.0 and 1.2 cm/s. A flowability rate that is below 1.0 cm/s resulted in extrudability problems while a rate that is higher than 1.2 cm/s negatively impacted buildability. The corresponding values obtained for mixes 2 and 3 were within the acceptable range of flowability; the mixes are extrudable and provide a sufficient capacity to satisfy buildability. On the other hand, the flowability of mixes 4 and 5 was so high that concrete layers could not be adequately stacked on top of each other.

From Fig. 6.2, it can be concluded that mix 3 (marked in red (black in print version)) is the optimal mix that best confirms to the criteria of buildability and extrudability. Its flowability rate is 1.13 cm/s which is greater than the required rate of 1 cm/s. Moreover, it provides the highest ratio of buildability before it starts steeply declining. Hence, mix 3 was selected to carry out further tests.

6.3.2.3 Retarder and Accelerator Optimization

6.3.2.3.1 Retarder Dosage

Several trial mixes were prepared in order to determine the proper quantity of retarder for concrete. The effect of retarder on the workability of

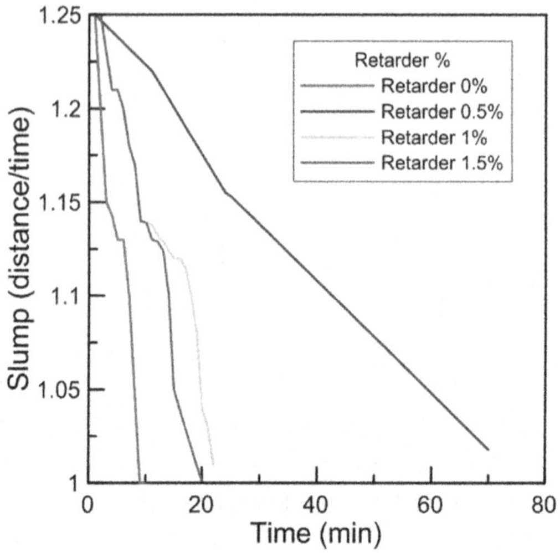

Figure 6.3 Effect of retarder dosage on workability with time.

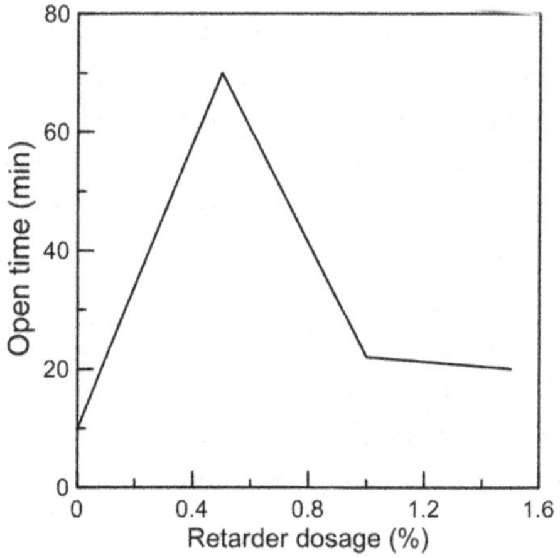

Figure 6.4 Effect of retarder dosage on open time.

the mix with respect to time was examined by varying its dosage from 0% to 1.5%. The results are plotted in Fig. 6.3.

The effect of the retarder on open time was also tested. The obtained results are presented in Fig. 6.4.

Fig. 6.3 reveals that the slowest loss of workability with time corresponds to the mix that contains 0.5% retarder. Retarding the workability loss is favorable for it maintains the minimum flowability needed for extrudability and for proper adhesion between concrete layers. This

Table 6.5 Testing of accelerator

Sample	Cement	Sand	Aggregate	Accelerator (mL)	SP (mL)	W/C
3a	125	80	160	0.5	1	0.39
3b	125	80	160	1	1	0.39
3c	125	80	160	1.5	1	0.39

implies that adding 0.5% is the optimum based on workability measures. This is illustrated in Fig. 6.4 in which the highest open time (around 70 minutes) was also obtained for the mix with 0.5% retarder. Also note that adding more than 0.5% of retarder has a detrimental effect on open time which is steeply reduced to 20 minutes at 1%.

6.3.2.3.2 Accelerator Dosage

Adding an accelerator right before the concrete is extruded from the nozzle serves buildability. The concrete will then have the proper stiffness to settle and set on its own. The optimum quantity of accelerator had to be determined. For this purpose, three trial mixes were prepared and tested. The data is summarized in Table 6.5.

In these tests, it was noticed that an amount of accelerator that exceeds a certain limit (1 mL in this case) negatively affects extrudability as the paste loses flowability becoming too stiff to be squeezed out of the nozzle (Mix 3c). Assuming that extrudability remains possible, another probable effect of a high amount of accelerator, although not tested in this experiment, is the loss of adhesion between layers. On the other hand, a small amount of accelerator did not have a noticeable impact and took a long time to take effect. Thus, an accelerator amount of 1 mL (Mix 3b) was found to deliver the most convenient behavior of concrete.

6.3.2.4 Printed Specimen

After achieving the optimal mix, it was finally tested by printing a wall specimen of 77 cm in length, 10 cm in width, and 10 cm in height, as shown in Fig. 6.5.

6.4 A BRIEF REPRESENTATION OF STUDY TWO

This section briefly presents some relevant parts of a study that followed the first study. This second study was aimed at demonstrating how the term "optimal" is specific to each 3D printing setup. For example, an

Figure 6.5 Machine printing (left) and printed specimen (right).

optimal extrudability for one nozzle might not be extrudable for another. The reasoning behind this will explained next.

6.4.1 Machine Design and Operation

The printer used throughout this study was assembled by integrating a nozzle with Computer Numerical Control (CNC) machinery. The nozzle is 2.5 cm in diameter. The extrusion mechanism of the designed nozzle is based on the rotation of an Archimedes screw connected to a stepper motor. The screw generates pressure that pumps the concrete within the nozzle. A CAD representation of the screw and the nozzle and the fabricated nozzle are displayed in Fig. 6.6.

These were attached to a CNC machine which required G-Codes for operation. To print a given structure, a CAD model was translated into a G-Code which was fed into the CNC machine. The final setup is depicted in Fig. 6.7.

6.4.2 Mix Proportions

The same procedure that was adopted for determining the required quantities of various concrete constituents was followed in this experiment. Numerous trials were conducted in order to find the proper combination of dry constituents. Eventually, the quantities of dry constituents were fixed as follows: 600 g of cement, 650 g of sand, and 250 g of aggregates with a maximum size aggregate (MSA) of 2.36 mm (Sieve #8).

ViscoCrete superplasticizer from Sika was used to increase the flowability of concrete, induce self-compaction properties, and reduce water need. The recommended dosage of ViscoCrete, as instructed by Sika, ranges from 0.2% to 2% by weight of cement [18]. The appropriate

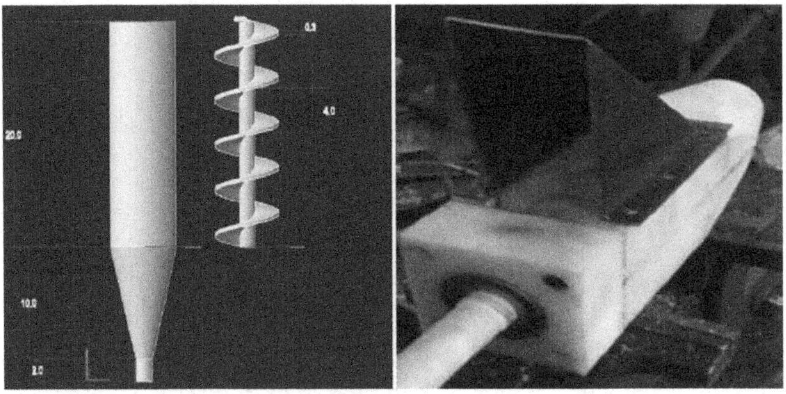

Figure 6.6 Nozzle [17].

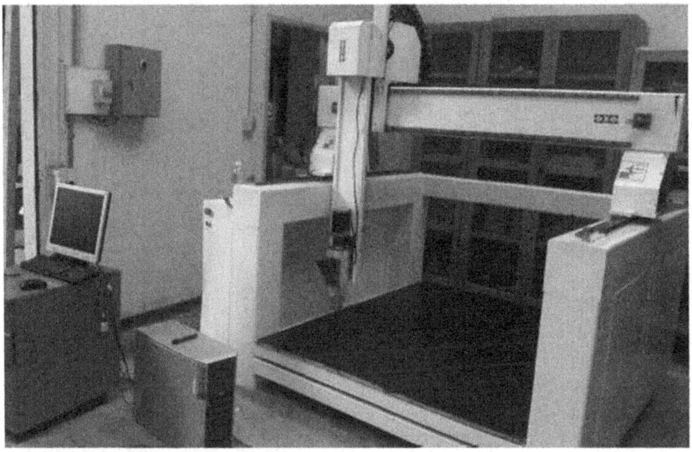
Figure 6.7 Printer setup [17].

amount of superplasticizer was determined to be 0.5% of the cement weight (i.e., 3 g).

Finally, the water content needed for the mix to satisfy the requirements of buildability and extrudability was investigated. To avoid clogging the nozzle, the initial water content added to the mix was high to ensure extrudability. It was then gradually reduced until the concrete became buildable. A portion of the experimental data is shown in Table 6.6. Trial mixes that did not exhibit any buildability capacity were disregarded, and the corresponding reductions of water content were large until the concrete started developing some buildability. Subsequently, the intervals of water reduction were narrowed as concrete became more and more

Table 6.6 Water–cement ratio results

Cement (g)	Sand (g)	Fine aggregates (g)	Superplasticizer (g)	Water (g) (W/C ratio)	Observation
600	650	250	3	243 (0.405)	Concrete is easily extrudable but is too fluid →Reduce w/c to make it buildable
600	650	250	3	241 (0.402)	Concrete is properly extrudable and has a slightly acceptable buildability →Slightly reduce w/c to improve buildability
600	650	250	3	240 (0.400)	Concrete is properly extrudable and buildable →Slightly reduce w/c to check whether buildability could be further optimized
600	650	250	3	239 (0.398)	Extrudability is negatively impacted →Opt for the previous ratio

Figure 6.8 Printed specimen [17].

buildable until the target water–cement ratio (w/c = 0.400) was achieved.

To summarize the results obtained for this printing setup, a proper mix is composed of 600 g of cement, 650 g of sand, 250 g of aggregates, 3 g of Viscocrete, and a w/c of 0.400. Fig. 6.8 displays a concrete sample printed using these proportions.

6.5 COMPARISON OF RESULTS

The results of both experiments are listed in Table 6.7 for comparison. It should be noted that the mix that was deemed optimal in the first experiment was not appropriate in the second one. This emphasizes the notion that what is best for one setup is not necessarily ideal for another. The two mixes are composed of dissimilar proportions of various constituents. Yet each mix performs properly in the printing process (i.e., machine and nozzle setup) it is designed for.

Every element displayed in Table 6.7 differently affects the behavior of the concrete mixture. Printer 2 has a nozzle of a larger diameter than that of printer 1. Accordingly, the maximum size aggregate that was used in designing mix 2 is greater than that of mix 1. It is desirable to increase the maximum size aggregate as the surface area of coarser aggregates is smaller and, hence, requires less water and cement paste. As a result, the concrete becomes stronger and more economical [19]. Hence, it is beneficial to change the maximum size aggregate to match the size of each nozzle making the mixtures more specific to the nozzles they are designed for.

Table 6.7 Summary of results

	MSA (mm)	Cement: sand: aggregate	w/c	Superplasticizer (%)
Experiment 1	2	1: 0.640: 1.28	0.390	0.9
Experiment 2	2.36	1: 1.08: 0.417	0.400	0.5

The extrudability of a mix should decrease if used with a nozzle of a smaller diameter than that it was prepared for.

Regarding the proportions of fine aggregates to coarse aggregates, mix 2 is richer in sand than mix 1. A higher fraction of fine aggregates greatly contributes to flowability, stability, and self-leveling of the concrete [20]. The significance of having a flowable concrete has already been established. As for stability and self-leveling, these are valuable for obtaining a proper adhesion between successive concrete layers. Such a behavior is ensured in the first mix by increasing the amount of superplasticizer. On the other hand, a lower ratio of coarse aggregates, as in the case of the second mix, leads to a reduction in the modulus of elasticity of concrete [18]. In short, mix 1 has better hardened properties of concrete while mix 2 exhibits more favorable fresh properties needed in 3D concrete printing. Recall, however, that both performed well in their own printing setup.

6.6 CONCLUSIONS AND RECOMMENDATIONS

The 3D printer developed by the research team in experiment one moves on all 3 axes to print a specimen composed of a 77 cm wall, and eventually other structural shapes. The printing nozzle, a critical component of the printer, was designed to optimize the concrete extrusion process. The nozzle's diameter is 2 cm and is attached to a side trowel and a top trowel that lag behind it. After thorough experimentation with various mixes, an optimum mix was obtained that was suitable for 3D printing for the given scale. This mix consisted of 125 g of cement, 80 g sand, and 160 g fine aggregates with a w/c ratio of 0.39. In addition, the mix included 1 mL of accelerator and 0.625 mL of retarder. The concrete exhibited a compressive strength equal to 42 MPa. The final mix obtained proves to be of appropriate strength in addition to satisfying the extrudability, flowability, and buildability requirements.

A second study was conducted to validate the conception that the designation of optimum is contingent on the printing setup, conditions, and desired criteria. Another printer was used for experiment two along with the operation mechanism it was based on. The originally designed mix was not compatible with the second printer. An optimal mixture for the second printer was sought and prepared. Finally, the results of both studies were compared, and the tentative conception was confirmed.

The application of the 3D concrete printing prototype developed in this research to real-life cases proves to be fitting, at least for small-scale structures. With the increasing advancements in construction technology comes a promising new method in the form of 3D concrete printing. Through an optimal mix that satisfies all the required criteria, structures will be constructed without the use of formwork. Such technology is appropriate for use in developing countries where the need for an effective, inexpensive, and fast construction method is apparent. Furthermore, it may also be used for building larger structures after taking into account scalability issues and proper logistics. 3D concrete printing is advantageous over traditional construction in that it offers savings in terms of time, overall lifecycle cost, use of labor, environmental harm, and complexity; all significant factors in the construction industry. This method is not only a promising technique for structural design, but is also a potential tool for architectural design. It may be able to offer designers additional design flexibility as the construction is not limited by manual labor or types of formwork.

ACKNOWLEDGMENTS

Research and experiments presented in this paper were supported by the Civil and Environmental Engineering Department at the American University of Beirut. All support is gratefully acknowledged. Any opinions, findings, conclusions, or recommendations expressed in this paper are those of the authors and do not necessarily reflect those of the funding organizations.

REFERENCES

[1] Z. Malaeb, H. Hachem, A. Tourbah, T. Maalouf, N.E. Zarwin, F. Hamzeh, 3D concrete printing: machine and mix design, Int. J. Civil Eng. 6 (6) (2015) 14–22.

[2] I. Pomerantz, J. Cohen-Sabbani, A. Bieber, J. Kamir, M. Katz, M. Nagler, U.S. Patent No. 4,961,154. Washington, DC: U.S. Patent and Trademark Office, 1990.

[3] F. Bos, R. Wolfs, Z. Ahmed, T. Salet, Additive manufacturing of concrete in construction: potentials and challenges of 3D concrete printing, Virtual Phys. Prototyping 11 (3) (2016) 209–225. Available from: https://doi.org/10.1080/17452759.2016.1209867.

[4] B. Khoshnevis, Automated construction by contour crafting—related robotics and information technologies, Autom. Constr. 13 (1) (2004) 5–19.
[5] Kira, Exclusive: WinSun China builds world's first 3D printed villa and tallest 3D printed apartment building, 3D Printer and 3D Printing News, 2015.
[6] I. Hager, A. Golonka, R. Putanowicz, 3D printing of buildings and building components as the future of sustainable construction? Procedia Eng. 151 (2016) 292–299.
[7] C.M. Rouhana, M.S. Aoun, F.S. Faek, M.S. Eljazzar, F.R. Hamzeh, The reduction of construction duration by implementing contour crafting (3D printing), in: Proceedings for the 22nd Annual Conference of the International Group for Lean Construction, 2014, pp. 1031–1042.
[8] F. El Sakka, F. Hamzeh, 3D concrete printing in the service of lean construction, in: 25th Annual Conference of the International Group for Lean Construction, Heraklion, Greece, July 9–12, 2017, pp. 781–788.
[9] B. Khoshnevis, D. Hwang, K.T. Yao, Z. Yeh, Mega-scale fabrication by contour crafting, Int. J. Ind. Syst. Eng. 1 (3) (2006) 301–320.
[10] M. Rahimi, M. Arhami, B. Khoshnevis, Crafting Technologies, 2009.
[11] T.T. Le, S.A. Austin, S. Lim, R.A. Buswell, A.G. Gibb, T. Thorpe, Mix design and fresh properties for high-performance printing concrete, Mater. Struct. 45 (8) (2012) 1221–1232.
[12] B. Khoshnevis, S. Bukkapatnam, H. Kwon, J. Saito, Experimental investigation of contour crafting using ceramics materials, Rapid Prototyping J. 7 (1) (2001) 32–42.
[13] A. Verma, M. Shukla, A.K. Sahu, Use of superplasticizers in concrete and their compatibility with cements, Int. J. Civil Eng. Technol. (2013) 153.
[14] ASTM International, ASTM C1611 / C1611M-14, Standard Test Method for Slump Flow of Self-Consolidating Concrete. West Conshohocken, PA, 2014.
[15] K.T. Fosså, Slipforming of vertical concrete structures. Friction between Concrete and Slipform Panel, 2001.
[16] British Standard Institution, BS 1881-116:1983, Testing concrete, Method for determination of compressive strength of concrete cubes. London, 1983.
[17] H. Ajra, L. Fayed, D. Itani, T. Narciss, N. Skayan, Mix design and automation of concrete 3D printer, In: The Fifteenth FEA Student and Alumni Conference American University of Beirut, April 27–28, 2016, Beirut, Lebanon. Retrieved September 14, 2017, from: http://feaweb.aub.edu.lb/feasac/15/Program/FEASAC15thBook.pdf.
[18] Sika Lebanon, Sika ViscoCrete 20HE, 2011. Retrieved September 9, 2017, from: http://lbn.sika.com/content/near_east/main/en/system/search.html?q = viscocrete& searchtype = document.
[19] S.H. Kosmatka, B. Kerkhoff, W.C. Panarese, Design and control of concrete mixtures. Portland Cement Assoc. 2011.
[20] K. Holschemacher, D. Weiße, Concrete made with sand-rich particle fractions, n.d.

CHAPTER 7

Investigation of Concrete Mixtures for Additive Construction

Todd S. Rushing[1], Peter B. Stynoski[2], Lynette A. Barna[3], Ghassan K. Al-Chaar[2], Jedadiah F. Burroughs[1], Jameson D. Shannon[1], Megan A. Kreiger[2] and Michael P. Case[2]

[1]Geotechnical and Structures Laboratory, US Army Engineer Research and Development Center, Vicksburg, MS, United States
[2]Construction Engineering Research Laboratory, US Army Engineer Research and Development Center, Champaign, IL, United States
[3]Cold Regions Research and Engineering Laboratory, US Army Engineer Research and Development Center, Hanover, NH, United States

7.1 INTRODUCTION

Construction processes, in general, are labor intensive and potentially dangerous, making them ideal candidates for automation. To reduce risk to personnel in the field and overall logistic requirements of construction, the US Army Corps of Engineers has been investigating large-scale, automated, additive manufacturing processes for construction applications. Additive construction is a fabrication technology that uses computer control to exploit the surface-forming capability of troweling to create smooth and accurate planar and free-form surfaces out of extruded materials at a construction scale. This research intended to develop and evaluate the capability to perform construction using an automated, additive process using locally available materials. Thus, the focus of the initial research into additive construction techniques was on full-scale 3D printing using concrete as the printing medium.

Significant advances have been made in the use of cementitious materials for additive construction, but, at the time this project was initiated, the literature regarding large-scale systems capable of 3D printing with concrete was limited. A number of concepts for additive construction were evaluated as described in the literature, such as D-Shape [1], Digital Construction

3D Concrete Printing Technology 2019 Published by Elsevier Inc.
DOI: https://doi.org/10.1016/B978-0-12-815481-6.00007-5

Platform [2], Concrete Printing developed at Loughborough University [3,4], and Contour Crafting [5,6]. Many reported studies were based on nontraditional concrete mixtures or mortars. For instance, the work done at Loughborough University investigated the use of mixtures that contained no coarse aggregate which, strictly speaking, is a mortar [7]. Some of the work presented by researchers at the University of Southern California investigated concrete containing synthetic aggregates, such as glass beads [8].

Since the ultimate goal of this project was to employ an automated construction system in a variety of environments and locations across the world, it was desirable to use concrete mixtures that comprised aggregates, binders, and other materials that are commonly found worldwide and that can be tailored to available raw materials. A major differentiating requirement for this research was the use of coarse aggregates in the printable concrete mixture. Also, a variety of commercial materials are available worldwide that can be added to concrete to impart the desired fresh and hardened properties. Since the present application required a combination of fresh and hardened properties that were different from those needed for traditional concrete, several means of concrete modification were considered.

This chapter highlights several stages of concrete mixture design and testing that were conducted at the US Army Engineer Research and Development Center (ERDC) as part of the Automated Construction for Expeditionary Structures (ACES) program, a three-year effort that began in the fall of 2014. Generally, the materials were investigated in three phases. The initial phase was a scoping study that was intended to develop a printable concrete mixture containing natural coarse and fine aggregates as a proof of concept. It further examined the effects of various additives. The second phase was an extension of the first in that it examined still more additives, while it also tested the concept that the printable concrete mixture could be adapted with minor modifications to use different natural additives, that is, that locally sourced materials were suitable for concrete printing even if the naturally occurring materials were quite different. The third phase of materials research, some of which was ongoing at the time this manuscript was produced, was an expanded laboratory study aimed at a more systematic approach to determining the effects of combinations of additives on the fresh rheological properties of a printable concrete mixture. For this last phase, a processed masonry sand was used rather than a natural sand to reduce aggregate variability so that the mixture properties could be better related to controlled parameters.

Beyond the phases of materials research, the majority of the work within the ACES program was dedicated to designing, building, and implementing the mechanical and control systems for a large-scale 3D printing system, performing structural, architectural, and energy modeling and design for the structures to be printed in execution of the program objectives, and, finally, printing many prototype structural elements and a full-scale structure suitable for human occupation. A summarized description of the products and results of the overall program are provided in Section 7.6.

7.2 PHASE I MATERIALS RESEARCH

7.2.1 Objectives

The primary objective of the initial materials research conducted at the ERDC's Geotechnical and Structures Laboratory (GSL) was to qualify a near-conventional concrete mixture for large-scale material extrusion in additive construction, or 3D printing with concrete, on a structural scale. A conventional concrete mixture is defined here as one that uses naturally occurring coarse and fine aggregates, standard Portland cement, admixtures, and conventional mixing techniques.

A trial traditional mixture was determined using the "weight-method." It was necessary for the trial mixture to display the following properties:

Flow in the fresh state must be sufficient to allow pumping and extrusion through a narrow square nozzle (nominally 45 mm × 45 mm).

High static yield strength is needed to provide shape stability after extrusion.

Plastic viscosity must be sufficient to prevent aggregate segregation.

The coarse aggregate size must be small to minimize the possibility of clogging during extrusion (9.5 mm was chosen).

Early properties, notably setting time, must be controlled to allow each layer of construction to support subsequent layers.

7.2.2 Experimental Approach

Starting with a typical concrete formulation, several additives to the concrete including chemical admixtures and supplementary cementitious materials (SCMs) were investigated based on their ability to provide the desired properties. Fly ash and silica fume were investigated as SCMs because of their influence on the rheology of concrete due to their small particle size and round particle morphology [9,10]. It was further expected that fly ash and silica fume would aid in controlling compressive strength development, depending on the percentage of cement substituted [11,12]. In light of the challenges posed by conventional concrete reinforcing bars in the additive construction process, discrete fiber reinforcement was investigated due to its potential to mitigate shrinkage cracking and increase tensile and flexural strength [13,14]. Relatively short fibers, nominally 12.5 mm long, were chosen to minimize flow problems. Bentonite was investigated for inclusion at low addition rates due to its potential for improving fresh-shape stability, as noted in reports on bentonite's use in slip forming [15]. A polycarboxylate-based superplasticizer was also investigated to improve the fresh concrete workability at a low water content without decreasing concrete strength.

A series of tests was performed on various concrete mixtures to explore their applicability toward the additive construction process. To that end, the mixtures were tested for flow, setting time, early compressive strength, and flexural strength.

7.2.3 Materials

7.2.3.1 Conventional Mixture Designs

A baseline trial mixture, herein labeled A0, was proportioned using traditional concrete ingredients and was selected as a starting point for testing. The mixture included a conventional or typical cement to fine aggregate to coarse aggregate ratio of about 1:2:3. A type I/II Portland cement was used in this study. The coarse aggregate was a 9.5 mm crushed limestone and the fine aggregate was manufactured limestone sand, both sourced from Calera, Alabama. A rheology-controlling admixture (RCA), MasterMatrix 33, supplied by BASF, was included in the baseline mixture at a high dosage rate of about 1300 mL/100 kg of cement in order to provide the increased workability needed for extrusion. Additionally, the following three modifications were made to the baseline mixture for testing:

A0: Baseline concrete mixture

A1: Increase in cement content to provide more paste

A2: Addition of bentonite clay to improve wet-shape stability

A3: Addition of a superplasticizer to increase flow

The specific materials and quantities listed in Table 7.1 were calculated assuming a density of 2370 kg/m^3 based on the use of limestone

Table 7.1 Conventional mixture proportions, kg/m^3

Mixture	Water	Type I/II cement	Limestone sand	Limestone 3/8-in.	RCA	Additive type	Additive amount
A0	223	358	790	994	4.5	None	–
A1	208	441	736	979	5.6	None	–
A2	211	447	746	939	5.7	Bentonite	22
A3	213	451	752	946	5.7	SP	2.9

aggregates in concrete. Material quantities are reported as the amount in kilograms required to produce one cubic meter of concrete.

The A0 mixture was prepared and loaded into the extruder apparatus, but was too stiff to press through the nozzle. Modifications A1–A3 were performed in an attempt to increase the flow, but none of these measures were sufficient to get the desired extrudability because of the high content of 9.5 mm limestone aggregate.

7.2.3.2 Nonconventional Mixture Designs

After the conventional mixture design failed to produce sufficient flow characteristics for extrusion, the amount of coarse aggregate was significantly reduced while the fine aggregate was increased. These adjustments led to a nonconventional extrusion mixture. Where the conventional mixture had a cement to fine aggregate to coarse aggregate ratio of about 1:2:3, the extrusion mixture was chosen to have a ratio of 1:3:1 and a water to cement ratio (w/c) of about 0.5. Like the A0 mixture, the second control mix, labeled B0, additionally contained the Navitas 33 RCA at a dosage of about 1300 mL/100 kg of cement. Furthermore, this B0 mixture was varied six times with one additional additive in each iteration. Metal fiber was obtained from Baumbach, nylon fiber was from Nycon, the superplasticizer was ADVA 190 from W.R. Grace, the Class C fly ash was from Redfield, Arkansas, the bentonite clay was Aquagel from Baroid, and the silica fume was Force 10,000 D from W.R. Grace. These mixture proportions are listed in Table 7.2 as amounts in kilograms required to produce one cubic meter of concrete on a basis of 2370 kg/m^3. The fiber contents in mixtures B1 and B2 were not included in the volumetric calculation; the reported equivalent amounts of fiber were added to the freshly mixed concrete.

B0: Control extrusion mixture

B1: B0 with ½-in.-long metal fiber reinforcement added

B2: B0 with ½-in.-long nylon fiber reinforcement added

Table 7.2 Nonconventional extrusion mixture proportions, kg/m^3

Mixture	Water	Type I/II cement	Limestone sand	Limestone 9.5 mm	RCA	Additive type	Additive amount
B0	215	430	1290	430	5.4	None	–
B1	215	430	1290	430	5.4	12.5 mm metal fiber	39
B2	234	426	1278	426	5.4	12.5 mm nylon fiber	5.3
B3	215	429	1287	429	5.4	Superplasticizer	4.3
B4	215	344	1290	430	5.4	Fly ash	86
B5	232	422	1267	422	5.3	Bentonite	21
B6	215	387	1290	430	5.4	Silica fume	43

B3: B0 with superplasticizer added
B4: B0 with fly ash replacing 20% mass of cement
B5: B0 with bentonite clay added
B6: B0 with silica fume replacing 10% mass of cement

7.2.4 Methods

To be applicable for printing, concrete must be extrudable through a nozzle. To mimic the extrusion conditions, we chose to examine how the mixtures would pass through a common clay extruder. A Bailey Ceramics Standard nine clay extruder assembly was selected for qualifying the test materials based on extrudability. This apparatus was chosen because the rheology of clay and fresh concrete is relatively similar and the manual extrusion process is simple and cost-effective. The clay extruder assembly was modified by welding a tapered nozzle that reduced to a 45 mm × 45 mm opening at the end so as to mimic the opening for the additive construction extrusion nozzle for the large-scale 3D printer that was being designed concurrently with the materials study. Although there was no existing standard test method for this material/process, this qualitative approach provided an initial assessment of how the material flowed and an opportunity to observe whether the material might create an obstruction. It also allowed assessment of the shape stability of the material once it exited the nozzle. A simplified schematic of the extruder is shown in Fig. 7.1.

Each fresh concrete mixture was loaded into the extruder and pressed through the outlet by manually pulling the arm, thus, applying a load to the plunger and forcing the test material through the funnel. This test method, depicted in Fig. 7.2, provided a qualitative and comparative indicator of the suitability of each material for a simple extrusion process. The

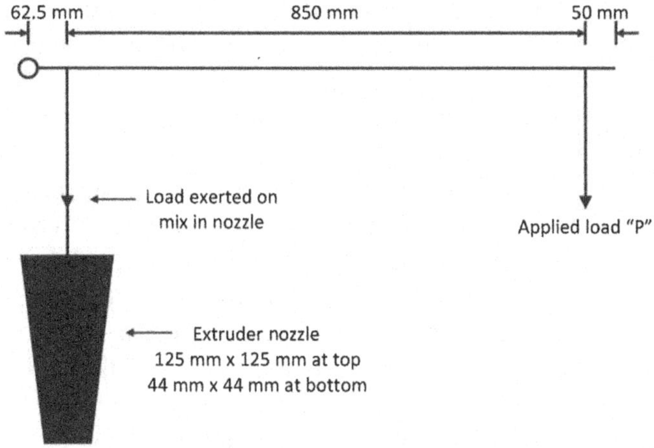

Figure 7.1 Schematic of extruder assembly.

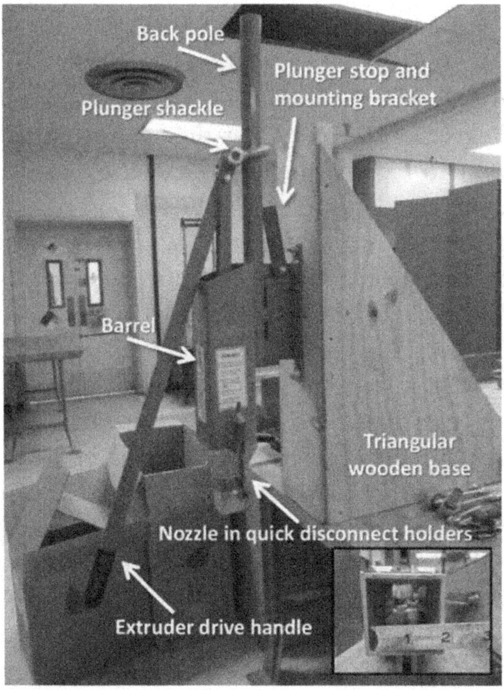

Figure 7.2 Extruder apparatus for qualifying mixtures for printability.

apparatus did not specifically quantify the load applied to the mixture and, therefore, did not measure the pressure required to extrude each concrete mixture.

After preparing each batch of material and qualifying for extrudability, materials that were deemed suitable were put through a series of standard empirical tests. A drop table test was performed for each mixture in accordance with ASTM C1437, with the table conforming to ASTM C230, to measure the relative flow characteristics. Initial and final times of setting were measured using the Vicat method according to ASTM C191. Sets of 50-mm cubes were prepared from each material, and the specimens were cured at room temperature in a 100% humidity environment. Unconfined compression testing was performed on the cured cubes in accordance with ASTM C109, with six cubes tested for each mixture. Finally, to quantify the flexural strength of each mixture, a flexural strength test was performed in accordance with ASTM C78.

7.2.5 Phase I Results

7.2.5.1 Extruder Testing

All four of the conventional concrete mixtures listed in Table 7.1 were qualitatively evaluated using the extruder assembly. A0 was prepared having a w/c of 0.6, but the mix locked up in the extruder funnel and excessive bleeding was observed under the applied pressure. An attempt was made with A1 to increase the paste content by adding cement while alleviating bleeding by reducing w/c to 0.47; however, the mix would not pass through the extruder apparatus. A2 and A3 included bentonite and superplasticizer, respectively, to test for improvements in extrudability. While the modified mixtures behaved somewhat more favorably compared with A0, neither offered a reasonable degree of extrudability in terms of a large-scale material extrusion for additive construction. It was deemed by inspection that the coarse aggregate content was too high, causing interlocking of the aggregate and clogging in the funnel section of the device. Additionally, an unacceptable degree of bleeding of the concrete was observed when pressure was applied via the plunger mechanism. As such, no further tests were performed on the conventional mixture designs (A0, A1, A2, and A3).

The nonconventional concrete mixtures listed in Table 7.2 were also qualitatively evaluated using the extruder assembly and visually inspected for suitable flow characteristics. As a result of changing the cement to fine aggregate to coarse aggregate ratio to 1:3:1, all of these extrusion mixtures

Table 7.3 Rating of mixture extrudability (1 = worst, 5 = best)

Mixture	B0	B1	B2	B3	B4	B5	B6
Rating	1	4	2	5	4	1	3

could be pressed through the extruder assembly with reasonable effort. An estimate of the amount of pressure required to extrude the concrete was also made. Without apparatus to quantify the load applied to the fresh concrete and, thus, the amount of pressure required to induce extrusion, a qualitative rating of 1 to 5 was assigned to each material based on the relative ease with which it was extruded. A rating of 1 indicated that the mixture was very difficult to extrude and a rating of 5 indicated the mixture was very easy to extrude. These ratings are listed in Table 7.3.

Although the B3 mixture containing the superplasticizer was very easy to extrude, it was much too fluid and exhibited almost no shape stability; therefore, B3 was not tested further. The water content of the B3 mixture should be reduced if it is to be considered for additive construction. The addition of steel fibers seemed to increase the flowability of the mixture unexpectedly, as the B1 material extruded well compared to the control material, B0. The B2 material containing nylon fibers flowed better than the control, but not nearly as well as B1 with steel fibers. Note that the water content was increased in B2 to obtain suitable flow because of the affinity of nylon for water. In each material containing fibers, the shape stability of the material after extrusion was improved over the B0 material. The replacement of a fraction of the Portland cement by fly ash (B4) and, separately, silica fume (B6) improved extrudability relative to the control material. The water content was increased in mixture B5 because the bentonite absorbed a significant amount of water, thereby drying the concrete. Surprisingly, even with the extra water, the addition of bentonite did not appear to help the overall flow. The bentonite did, however, aid in the shape stability of the extruded material, which is an important consideration for an extrudable material for additive construction.

7.2.5.2 Drop Table Test

The nonconventional extrusion mixtures, excluding B3, were evaluated for flow using the drop table test. After forming the standard cone shape and imparting 25 drops on the table apparatus, the spread of each material was measured in four locations according to the standard method at angles of 0, 45, 90, and 135 degrees. The flow number was determined as a

Table 7.4 Flow numbers from drop table tests

Mixture	B0	B1	B2	B4	B5	B6
Average spread	80%	104%	104%	124%	64%	56%

Table 7.5 Vicat time of setting (measured in min)

Mixture	B0	B1	B2	B4	B5	B6
Initial set	128	108	115	113	145	144
Final set	192	183	205	215	235	230

percent of the original cone base diameter by calculating the four measurements for each material. The flow number results are listed in Table 7.4.

Most of the extrusion mixtures exhibited medium to high flow (spread) in the drop table test. Test B4, with fly ash, flowed the most, followed by the two samples containing short fiber reinforcements, B1 and B2. Tests B5 and B6 exhibited lower flow, corresponding to good shape stability, meaning, they spread out very little under multiple impacts. It was expected that fly ash would increase flow, while silica fume would reduce flow [16,17]. It is of interest to note that there seems to be little correlation between the qualitative extrudability rating reported above and the flow measure from the drop table. The key difference in these two flow evaluation methods is that the extruder method applied a steady stress, while the drop table imparted multiple impulse stresses.

7.2.5.3 Time of Setting

The time of setting by the Vicat method was performed for each of the nonconventional extrusion mixtures with the exception of B3. For each test, the fresh concrete was sieved through a No. 4 sieve to remove coarse particles in accordance with ASTM C191. Samples of B1 and B2 were collected for Vicat testing prior to the addition of fibers; therefore, B0, B1, and B2 samples were essentially the same, and the results provide an indication of the test variability. Initial and final set results in minutes are reported in Table 7.5. While most of the setting times were about the same for the mixtures, the initial set was delayed somewhat in B5 containing bentonite and in B6 containing silica fume. The delay in B5 could be due to the increased water included to achieve consistent flow, while the delay in B6 may be attributed to the reduced cement content since silica fume was added as a cement replacement. No delay in setting was observed in B4 which included fly ash, which was consistent with expectations [16,17]. Generally, time of setting is an important consideration for large-scale concrete extrusion in an additive construction application because it indicates workability time as well as the time required for the deposited layer to begin gaining structural capacity to support subsequent layers placed on top of it. Although not studied in this phase, setting time can be increased or reduced according to the needs of the full-scale process by the use of chemical accelerators or retarders.

7.2.5.4 Compressive Strength Testing

Unconfined compression testing was performed on each nonconventional extrusion mixture with the exception of B3. Furthermore, 50 mm cube samples were used to evaluate B0, B4, B5, and B6, while B1 and B2 (which contained fibers) were evaluated by preparing 75 mm × 150 mm cylinders. The unconfined compressive strength tests of the materials were performed after 1 and 7 days. The values charted in Fig. 7.3 are the averages of six compression test results at each condition, that is, six specimens were tested instead of the typical three samples per test condition.

Generally, for the structural purpose of supporting layers in the case of additive construction, all of the mixtures were sufficiently strong and there was not a large variation in strength. The lowest strength was exhibited by B2 containing nylon fibers. This could be due to the additional water that was added to this mixture to get a workable material. However, a

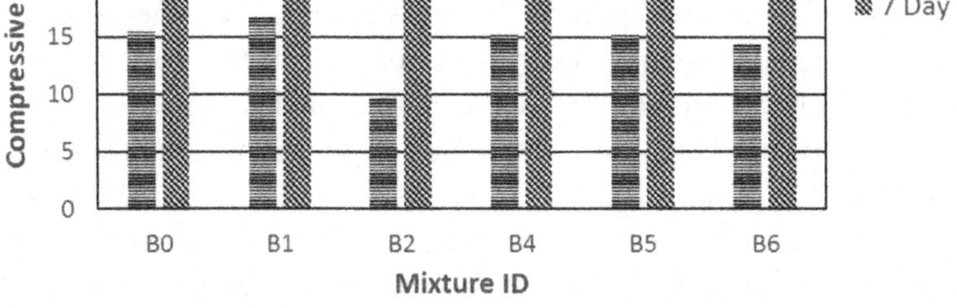

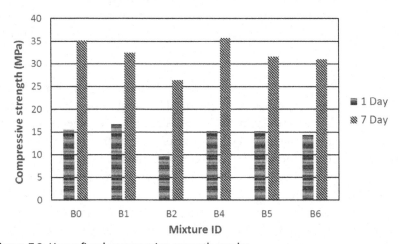

Figure 7.3 Unconfined compressive strength results.

similar amount of additional water was included in B5 containing bentonite without the corresponding decrease in compressive strength. Another rationale for the strength decreases with nylon fibers is that the fibers themselves are low modulus (as opposed to the steel fibers in B1) and act as defects in the concrete matrix under compression. The reinforcing fibers in B1 and B2 did not increase the compressive strength compared with the control B0. This is because the test reports the peak strength; the reinforcing fibers are expected to engage after crack formation resulting in improved tensile softening behavior. The inclusion, independently, of fly ash (B4), bentonite (B5), and silica fume (B6) did not notably alter the compressive strength compared with the control after 1 and 7 days. The favorable observation from Fig. 7.3 is that using the additives as a means to improve the flow and shape stability of the fresh concrete was not significantly detrimental to the concrete's structural properties. This tolerance is important for large-scale concrete extrusion in additive construction because it allows for the addition of a variety of flow aids to enable the inclusion of coarse aggregates while designing for the desired workability properties.

7.2.5.5 Flexural Strength Tests

Flexural tests were performed on 1 beam sample from each of the nonconventional extrusion mixtures, excluding B3, after 7 days. Beams were $76 \times 76 \times 305$ mm and were tested in a three-point bending configuration with a 250 mm span and the load applied at mid-span. The flexural testing load versus displacement data are plotted in Fig. 7.4.

The peak strength, f, was calculated for each beam using the equation $f = 3PL/2bd^2$, where P is the peak load, L is the span, b is the specimen width, and d is the specimen depth. Peak flexural strengths are listed in Table 7.6.

All the unreinforced materials (B0, B4, B5, and B6) resulted in peak flexural strength values that were similar. The samples that included silica fume, B6, achieved the greatest flexural strength. Like the compressive strength results, the flexural test observations imply that the additives do not significantly deteriorate the concrete's structural capacity.

The observations are more interesting regarding the reinforced beam samples. The inclusion of metal fibers (B1) or nylon fibers (B2) was not expected to have much effect on the peak flexural strength. However, both of the reinforced beams were weaker in flexural testing than the unreinforced beams. The reason for this decrease in peak capacity was

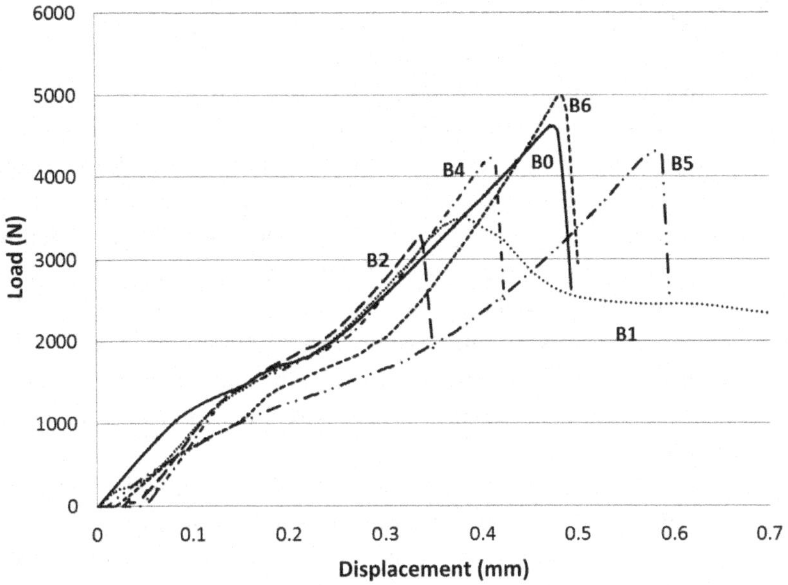

Figure 7.4 Flexural strength comparison.

Table 7.6 Peak flexural strength, MPa

B0	B1	B2	B4	B5	B6
3.98	3.04	2.83	3.64	3.72	4.30

probably due to the effective decrease in the cross-sectional area of the concrete matrix with the inclusion of fibers. The two types of fiber reinforcements were expected to provide some postpeak load-bearing capacity, resulting in a softening behavior in the load-displacement curve. Residual capacity, after crack formation, was observed in the sample containing metal fibers. The load-displacement test data for the sample containing nylon fibers did not show postpeak softening, but the reason is that a threshold setting in the load program was exceeded after the first crack formation, and the software ended the test rather than continuing to collect data during the softening behavior. The postpeak response indicated that the metal fibers provided some reinforcement through a crack-bridging mechanism. Crack-bridging was visually observed in the samples containing nylon fibers, but the response was not measured because the test was terminated.

7.2.6 Phase I Conclusions

A conventional mixture design containing a typical proportion of coarse aggregate was not adequate for additive construction applications because of insufficient flow through a nozzle apparatus. To allow for adequate flow, the concrete material must include a large proportion of fine materials such as sand, but it can still incorporate a significant content of coarse (9.5 mm) aggregate which is expected to reduce shrinkage. Chemical admixtures and additives appeared to have varying degrees of success in aiding flow and shape stability. The addition of fly ash provided the best improvement in flow, while the addition of bentonite provided the best shape stability. The use of superplasticizer significantly increased fluidity of the mixture, but its use should be accounted for by reducing the water content of the mixture.

Experiments showed that the addition of short reinforcing fibers did not reduce flow and, in fact, appeared to aid flow in most cases. The inclusion of fibers improved the shape stability of the fresh materials to some extent; thus, fibers are expected to be beneficial in additive construction since they offer some degree of reinforcement in both the fresh and hardened states. Considering that nylon fibers are hygroscopic and tend to absorb water from the mixture, it was of interest to investigate hydrophobic fibers such as polypropylene in further experiments.

Little correlation was observed between the results of the empirical drop table test and the qualitative extruder test. Although both the extruder and the drop table tests were used to determine flow characteristics of each material under some externally applied stress condition, there were two notable differences in these tests: (1) the extruder test provided a degree of confinement due to the presence of the funnel, as opposed to the unconfined condition experienced in the drop table test; and (2) the extruder test relied on a steadily applied pressure to induce flow, whereas the drop table produced a repeated acute stress, like an impulse load. It was expected that the dynamic impact associated with the drop table played a key role in the difference between the two tests. The use of the drop table and the manual extruder apparatus together was intended to provide an easily attainable indication of how a test material would flow in an additive construction process that involves large-scale material extrusion of concrete through a nozzle.

Broadly, this phase demonstrated the ability to include coarse aggregates in a mixture for 3D concrete printing. Although the effects of the additives were evaluated one by one, an optimized mixture was thought likely to include a combination of the testing additives, for example, the base 1:3:1 concrete mixture, plus silica fume, plus superplasticizer, plus fibers. Material combinations, as well as pumping through a real nozzle configuration, were studied later in the program.

7.3 PHASE II MATERIALS RESEARCH

7.3.1 Experimental Summary

After the conclusion of the initial materials scoping study, a second phase of materials investigation was performed at a different laboratory, the ERDC's Cold Regions Research and Engineering Laboratory (CRREL). The second phase began with the control extrusion mixture proportion, B0, from the first phase, but the aggregates native to the second laboratory location were used. Due to the differences in the aggregate characteristics, the B0 proportion of 1:3:1 no longer yielded a mixture with suitable workability for extrusion and, thus, 3D printing. After a few bench scale iterations, trial and error led to a new baseline mixture, specific to the CRREL aggregates, with cement to sand to coarse aggregate ratios of 1:2.3:1, a w/c of 0.47, and rheology-controlling admixture (RCA) at a dosage of 1300 mL/100 kg of cement. This new baseline mixture served as the control mixture for comparing the effects of several SCMs, admixtures, fibers, and other additives one by one.

Table 7.7 presents a selection of the additives and their amounts studied in the second phase materials investigation. In the cases where admixtures were used, the water content was accounted for and the mix water was reduced accordingly; however, due to excessive fluidity, the w/c was reduced slightly in the cases of the accelerator and shrinkage reducing admixtures.

The mixtures were tested for fresh and hardened properties, as well as for extrudability using the same method as for the first phase investigation. A selection of the measured properties included flow using the flow table method (ASTM C1437), time of setting using the penetration resistance method (ASTM C403), compressive strength (ASTM C109) using 50-mm cubes, and flexural strength (ASTM C78) using $76 \times 76 \times 300$ mm prisms. Note that the time of setting method was different from the Vicat method used in the first phase study due to the availability of test equipment, so the second phase setting time data cannot be directly compared with the first phase results. The phase II test results are listed in Table 7.7.

Table 7.7 Phase II concrete mixture properties

Material added to phase 2 baseline mixture	Amount added	Flow (%)	Time of setting (min)		Compressive strength (MPa)		Flexural strength (MPa)
			Initial	Final	1-day	7-day	7-day
Control	Nothing	105	199	317	23.9	37.2	5.01
Polypropylene fibers	5.6 kg/m^3	77	221	347	20.5	33.2	5.21
Accelerator	3.9 L/100 kg cement	101	146	223	24.3	37.1	5.43
Shrinkage reducer	2.5 L/m^3	88	251	360	23.0	36.7	5.24
Full range water reducer	325 mL/100 kg cement	73	240	310	25.1	40.5	5.74

7.3.2 Phase II Conclusions

This research demonstrated that the extrudable concrete mixture could easily be adapted to a new location using different natural aggregates. It investigated several admixtures that were not included in the first phase, specifically, accelerator, shrinkage reducer, and a different water reducer. The accelerator shortened setting time as expected. Many other additives were studied and further property tests were performed in the course of the second phase study, but only a selection of the experiments and results have been highlighted here. For example, slump, splitting tensile strength, drying shrinkage, and early-age strength development were measured. Furthermore, most of the additives used in the first phase were tested again in the second phase. However, because some of the observations were duplicated and others had little bearing on the formulation chosen for subsequent 3D printing work, much of the information has been omitted from this chapter. Full details of the second phase materials work are being prepared for publication elsewhere.

7.4 CONCRETE MIXTURE FOR 3D PRINTING

The materials investigation described thus far was performed during the first year of the ACES program. The purpose of the first and second phase studies was to understand the constituents and their effects in order produce a first draft of a concrete mixture containing coarse aggregate that would qualify for concrete printing. Concurrently with the concrete

Table 7.8 Concrete mixture proportions for 3D printing, kg/m^3

Water	Type I/II cement	Fine agg.	Coarse agg. 9.5 mm	Bentonite	Silica fume	Fly ash	RCA	HRWR
204	419	1064	462	14.0	23.3	23.3	5.8	3.2

study, a gantry style robotic system for 3D printing was being designed and constructed at a third US Army ERDC Laboratory, the Construction Engineering Research Laboratory (CERL) in Champaign, Illinois, which was the location where the remainder of the program tasks were executed.

Local natural aggregates from the CERL location were selected, and, based on the observations and knowledge gained up, a version-one printable concrete mixture was formulated. The selected coarse and fine aggregates were more similar to those used in the second phase materials study; therefore, the printable mixture had proportions, by weight, of cement to fine aggregate to coarse aggregate equal to 1:2.3:1, like the second phase baseline mixture. The mixture further included silica fume, fly ash, bentonite, superplasticizer, and RCA. The effects of combinations and interactions of the additives had not been investigated at this point in the project, but a printable mixture was required as a practical matter to continue with testing and optimization of the printer apparatus. Although it was understood that the large number of mixture components resulted in a material that was overly complex, and that further materials research would be needed to establish a simpler and more user-friendly material, this array of additives was selected due their individually observed benefits. The printable mixture, described in Table 7.8, was successfully and repeatedly batch mixed and pumped through a nozzle, allowing the team's focus to shift to other program requirements, such as completing the concrete printing machine and modeling the structural and other requirements of a full-scale 3D-printed building.

7.5 PHASE III MATERIALS RESEARCH

7.5.1 Experimental Summary

As the mixture reported in Table 7.8 was being used to finalize the printer development and to print prototype structural elements and structures, research continued on the concrete mixture to generate

Table 7.9 Base concrete and mortar mixtures for phase III, kg/m^3

	Water	Cement	#100 Masonry sand	9.5 mm pea gravel
Mortar	265	625	1405	0
Concrete	210	485	1095	485

increased understanding of how to control and ultimately optimize its formulation and properties. This third phase studied a concrete mixture with a cement to fine aggregate to coarse aggregate proportion equal to 1:2.3:1, as well as a mortar mixture with a cement to fine aggregate ratio equal to 1:2.3. In both cases the w/c was 0.43. For the fine aggregate, a masonry sand was used instead of a natural sand in an effort to reduce experimental variability. Mortar batches were 1.8 L, and concrete batches were 18 L. The starting mortar and concrete mixture proportions are listed in Table 7.9.

The additives tested in the third phase included silica fume, fly ash, bentonite, superplasticizer, shrinkage reducer, RCA, accelerator, and fibers. Various addition rates or dosages for each additive were investigated. In addition to the individual additives, the effects of the following additive combinations were studied: RCA plus accelerator, superplasticizer plus bentonite, superplasticizer pus accelerator, RCA plus superplasticizer, RCA plus bentonite, and fly ash plus silica fume. Property measurements included flow using the drop table test, time of setting, compressive strength, and drying shrinkage. Also, an ICAR rheometer was used to measure static yield strength as well as to construct a flow curve from mean torque readings at various rotation rates.

The details and results of the third phase study are being prepared for publication elsewhere; however, we can report on some of the interesting preliminary observations. Extensive shrinkage cracking was observed in printed elements, and controlling shrinkage is considered to be crucial in this application. The observed plastic drying shrinkage was not controllable by a shrinkage reducing admixture. Shrinkage cracking, however, was very nearly eliminated when polypropylene fibers were added at only 0.025% on a volume basis. The rheological measurements confirmed expectations that static yield strength increased with the addition of bentonite and decreased with the addition of SP. The RCA had little to no effect on static yield strength. Some general rules-of-thumb derived from experimental observations are reported in Table 7.10.

Table 7.10 Trends Observed in Additive Testing

Admixture	Set time	Early strength	Static yield strength
Accelerator	↓↓	↑↑	Not tested
RCA	↓	↓	O
Superplasticizer	↑↑	↓	↓↓
Shrinkage reducer	O	↓	Not tested
Bentonite	↓↓	O	↑↑
Class C fly ash	↑	↓	↓
Silica fume	↓	↓↓	↑↑

7.5.2 Phase III Conclusions

Although not complete at the time this manuscript was written, this materials research suggested that a much simpler printing mixture, compared with the mixture used in the ACES program, is a probable best solution for 3D concrete printing. The researchers speculated that it should be possible to design a simple mixture of water, cement, and aggregates to achieve a printable mix with static and dynamic rheological properties similar to an empirically good printing mix, particularly if aggregate gradation is controlled. In realistic field conditions including the absence of desired aggregate gradations and limited control over moisture content, the admixtures studied here should allow the mix designer to produce a printable mix by following the rules of thumb in Table 7.10. A further consideration was that the careful use of superplasticizer together with silica fume or bentonite should provide enough control over the fresh properties so that much higher print rates could be feasible.

7.6 3-D PRINTING WITH CONCRETE

A significant effort in research, development, and demonstration was conducted over the course of the ACES program, and a few of the program highlights are presented in this chapter. Three generations of 3D printers were designed, built, and implemented. The first and smallest prototype, ACES-1, was developed as a low-cost system that could be easily fixed in the field, scaled up to print a full-sized building, and adjusted to a desired build envelope (minimum 1.2 × 1.2 × 1.2-m). This system utilized command and control software typically used in computer numerical control (CNC) applications to allow for additional axes than are traditionally used in 3D printing. Fig. 7.5 depicts ACES-1 printing beam samples. Despite

Figure 7.5 ACES-1 printing the first layer of two beam elements with sinusoidal infill.

the low-cost nature of ACES-1, its fabrication and operation generated a wealth of experience within the build team as they prepared to design and construct the second, larger prototype, dubbed ACES-2.

The second prototype printer, ACES-2, was designed and constructed based on the ACES-1 design with additional improvements. This system was also designed as a low-cost system with expandable, low-maintenance rails and utilized components that were easy to obtain, for example, I-beams. Despite the increased print envelope (adjustable $5.2 \times 10.3 \times 2.75$-m), ACES-2 was designed to fit into a single ISO container for transport. This printer produced numerous elements in the laboratory while the build team modified and improved the hardware and software. ACES-2 would eventually print a full-scale barracks hut (B-hut), a 48 m^2 structure commonly used on Army installations for habitation and operation management. While printing of the B-hut was performed intermittently over several days due to personnel requirements, the total print time for the B-hut was just 21.5 hours, a significant improvement over the production time for conventional B-huts. Fig. 7.6 is a photo of the finished B-hut, with ACES-2 partially visible behind the building.

A third-generation printer, ACES-3, is under development in cooperation with NASA to provide increased mobility of the print envelope. The printer, shown in Fig. 7.7, is designed for nearby transport as a trailer to reduce printer delays between buildings, or it can be broken down to fit into an ISO or Army pallet for longer hauls. It is meant for outdoor use, can be easily leveled, and enables printing over rough and uneven terrain. At the time of writing, some of the hardware elements of ACES-3

Figure 7.6 ACES-2 printer gantry (background, right) and 3D printed barracks hut, including conventional roofing materials and visible anchoring reinforcement (internal vertical and horizontal reinforcement not shown).

Figure 7.7 ACES-3 prototype printer preparing to move and print over a concrete platform.

were still under development, so information about prints using this, the most rugged of the three prototypes, was not available.

7.7 CONCLUSIONS

The ACES program explored the use of many potential concrete mixture constituents for extrusion-based additive construction, or 3D concrete printing. Naturally occurring aggregates including coarse aggregates were included as a program requirement. Experiments with multiple natural aggregates showed that viable mixtures contained equal parts of cement and coarse aggregates, with a larger proportion of fine aggregate that varied depending on the aggregate characteristics or source. The observations suggest that a printable mixture can be developed via a few experimental trials using the raw materials that are available in most locales that support conventional concrete construction.

Controlling the fresh properties of the concrete mixture was found to be far more important and challenging than designing for hardened properties. The additives that seemed to provide useful means of adjusting the fresh properties were bentonite, silica fume, fly ash, superplasticizer, and RCA; although, interestingly, the latest round of experiments did not find a quantifiable effect of the RCA on the rheological properties measured. A notable concern with the hardened material was the appearance of extensive shrinkage cracking. Reinforcing fibers included at low addition rates were found to be much more effective at controlling this cracking than the use of shrinkage reducing admixtures, and the fibers did not appear to negatively impact fresh properties. Sufficient compressive strength for structural design was obtained in all of the mixtures tested.

A full-scale structure was successfully printed using concrete that was based on natural coarse and fine aggregates, thus, demonstrating the feasibility of additive construction using locally sourced materials. Further refinement and optimization of the printing mixture continues with the goal of developing a simple and tunable printing formulation that can be adapted for expedient construction in a variety of environments.

The interested reader can find many additional sources of information on the ACES program. Videos of several printing operations are available online, multiple journal articles have been published or are in press, and several ERDC Technical Reports are being published to document the program in detail. The program was also spotlighted in feature technology articles of the Society of American Military Engineers' *The Military Engineer* magazine [18] and the American Society of Civil Engineers' *Civil Engineering* magazine [19].

ACKNOWLEDGMENTS

This research was funded by the US Army. The authors would also like to acknowledge the work of NASA Marshall Space Flight Center and NASA Kennedy Space Center under an Interagency Agreement as well as Caterpillar, Inc. under a Cooperative Research and Development Agreement. ACES team members include: William Brown, Brandy Diggs-McGee, Jim Miller, Kurt Kinnevan, Patrick Keane, Eric Kreiger, Bruce MacAllister, Russ Northrup, Michael Pace, John Vavrin, Shawn Waddell, Rich Weichsler, Jacob Wagner, Tanner Wood, Justine Yu, Jade Woodard, and Charles Smith, Jr., among numerous others.

REFERENCES

[1] G. Cesaretti, E. Dini, X. De Kestelier, V. Colla, L. Pambaguian, Building components for an outpost on the lunar soil by means of a novel 3-D printing technology, Acta Astronaut. 93 (2013) 430–450. Available from: https://doi.org/10.1016/j.actaastro.2013.07.034.

[2] S. Keating, N.A. Spielberg, J. Klein, N. Oxman, A compound arm approach to digital construction, Robotic Fabrication Architecture, Art and Design (2014) 99–110. Available from: https://doi.org/10.1007/978-3-319-04663-1_7.

[3] S. Lim, R.A. Buswell, T.T. Le, S.A. Austin, A.G.F. Gibb, A. Thorpe, Development in construction-scale additive manufacturing processes, Autom. Constr. 21 (1) (2012) 262–268.

[4] S. Lim, R.A. Buswell, T.T. Le, R. Wackrow, S.A. Austin, A.G.F. Gibb, A. Thorpe, Development of a viable concrete printing process, in: The 28th International Symposium on Automation and Robotics in Construction (ISARC2011), 29 June–2 July, 2011.

[5] B. Khoshnevis, D. Hwang, K.-T. Yao, Z. Yeh, Mega-scale fabrication by contour crafting, Int. J. Ind. Syst. Eng. 1 (3) (2006) 301–320.

[6] B. Khoshnevis, Contour crafting: state of development, Solid Freeform Fabr. Proc. (1999) 743–750.

[7] T.T. Le, S.A. Austin, S. Lim, R.A. Buswell, Mix design and fresh properties for high-performance printing concrete, Mater. Struct. 45 (2012) 1221–1232. Available from: https://doi.org/10.1617/s11527-012-9828-z.

[8] T. Di Carlo, Experimental and numerical techniques to characterize structural properties of fresh concrete relevant to contour crafting (Ph.D. dissertation), University of Southern California, 2012.

[9] Y. Li, S. Zhou, J. Yin, Y. Gao, Effects of fly ash on the fluidity of cement paste, mortar, and concrete, International Workshop on Sustainable Development and Concrete Technology, Central South University, PRC, 2009, pp. 339–345.

[10] A. Peled, M. Cyr, S. Shah, High content of fly ash (Class F) in extruded cementitious composites, ACI Mater. J. 97 (5) (2000) 509–517. Available from: https://doi.org/10.14359/9283.

[11] H.A. Mohamed, Effect of fly ash and silica fume on compressive strength of self-compacting concrete under different curing conditions, Ain Shams Eng. 2 (2011) 79–86.

[12] D. Raharjo, A. Subakti, Tavio, Mixed concrete optimization using fly ash, silica fume and iron slag on the SCC's compressive strength, Procedia Eng. 54 (2013) 827–839. Available from: https://doi.org/10.1016/j.proeng.2013.03.076.

[13] R.S. Olivito, F.A. Zuccarello, An experimental study on the tensile strength of steel fiber reinforced concrete, Composites 41 (2010) 246–255. Available from: https://doi.org/10.1016/j.compositesb2009.12.003.

[14] S. Shah, Do fibers increase the tensile strength of cement-based matrices?, ACI Mater. J. 88 (06) (1991) 595–602. Available from: https://doi.org/10.14359/1195.

[15] N. Tregger, T. Voigt, S. Shah, Improving the slipform process via material manipulation, Advances in Construction Materials 2007, Grosse CU, Springer, Berlin Heidelberg, 2007, pp. 539–546.
[16] Integrated materials and construction practices for concrete pavement: a state-of-the-practice manual, in: P.C. Taylor, S.H. Kosmatka, G.F. Voigt, M. Brink (Eds.), FHWA Publication No. HIF-07-004, US Department of Transportation, 2007.
[17] Caltrans, State of California Department of Transportation, Guidelines for the Design & Inspection of Concrete, 2010.
[18] L. Link, Printing an expeditionary building, The Military Engineer 109 (711) (2017) 42–43.
[19] C. Cardno, Army corps successfully prints full-size concrete building, ASCE Civil Eng. (2017) 38–39. November 2017.

FURTHER READING

ASTM C78-10, Standard Test Method for Flexural Strength of Concrete (Using Simple Beam With Third-Point Loading), ASTM International, West Conshohocken, PA, 2010, https://doi.org/10.1520/C0078-10e1.
ASTM C109-13, Standard Test Method for Compressive Strength of Hydraulic Cement Mortars (Using 2-in or [50-mm] Cube Specimens), ASTM International, West Conshohocken, PA, 2013, https://doi.org/10.1520/C0109-13.
ASTM C191-13, Standard Test Methods for Time of Setting of Hydraulic Cement by Vicat Needle, ASTM International, West Conshohocken, PA, 2013, https://doi.org/10.1520/C0191-13.
ASTM C230-14, Standard Specification for Flow Table for Use in Tests of Hydraulic Cement, ASTM International, West Conshohocken, PA, 2014, https://doi.org/10.1520/C0230-14.
ASTM C403-16, Standard Test Method for Time of Setting of Concrete Mixtures by Penetration Resistance, ASTM International, West Conshohocken, PA, 2016, DOI: https://doi.org/10.1520/C0403-16.
ASTM C1437-13, Standard Test Method for Flow of Hydraulic Cement Mortar, ASTM International, West Conshohocken, PA, 2013, https://doi.org/10.1520/C1437-13.

CHAPTER 8

Method for the Enhancement of Buildability and Bending Resistance of Three-Dimensional-Printable Tailing Mortar

Zhijian Li[1,2], Li Wang[2] and Guowei Ma[2,3]
[1]College of Architecture and Civil Engineering, Beijing University of Technology, Beijing, P.R. China
[2]School of Civil Engineering and Transportation, Hebei University of Technology, Tianjin, P.R. China
[3]School of Civil, Environmental and Mining Engineering, The University of Western Australia, Crawley, WA, Australia

8.1 INTRODUCTION

In recent few years, significant progress has been made in developing various 3D printing techniques to accommodate the need for construction-scale 3D printing. Many attempts have been conducted to explore the potential of 3D printing in the building and construction industries, such as D-shape, contour crafting, and concrete printing [1–6]. Such techniques are well-suited to the production of one-off, complex structures that would often be difficult to produce using traditional manufacturing methods. Cementitious materials that are compatible with 3D printing promote rapid application of this innovative technique in the construction field with the added advantages of low cost, high efficiency, design flexibility, and being environmentally friendly [6–10]. It is critical to ensure a complementary connection between the designs of the printable mix and printing machine. Currently, various 3D-printable mixtures have been continuously developed, such as high-performance composites [11,12] and fiber-reinforced mixtures [13], among others. A number of specific implementation practices have been presented, for example, a five-story apartment 3D printed by WinSun [14], the Big Delta project [15], a castle printed in situ [16], and architectural elements [17], etc. All of these projects demonstrated the great potential and feasibility of 3D printing in constructing large-scale building components.

3D Concrete Printing Technology
DOI: https://doi.org/10.1016/B978-0-12-815481-6.00008-7

The printable property of fresh cementitious materials and the mechanical behavior of the hardened structures are of great concern for current 3D printing technologies [11,18–20]. One of the important printable characteristics is the buildability, which refers to the ability of cement paste to retain its extruded shape under self-weight and the resistance to the pressure from upper layers [11,21,22]. Buildability can be considered as the early-stage stiffness. Good buildability is a basic requirement for 3D-printable mixtures. A feasible approach for improving the buildability is to appropriately extend the paste age of the mixture [23–25]. A longer time allows further cement hydration and, therefore, contributing to the mix acquiring a certain stiffness.

However, this approach would minimize the fluidity and adhesive behavior of the fresh paste, which may result in poor hardened structural capability and integrity. In the printing process, the concrete components are created by bonding the extruded filaments together to form each layer without using extra formworks [26–28]. The rheology and flowability of the fresh material must allow its fluent extrusion to form small filaments. Concrete paste of low fluidity and adhesion is likely to form voids between filaments and weak bonding interface between adjacent filament and layers, which are negative for the product's overall mechanical performance [2]. Therefore, it is of great significance for the cement paste to optimize and coordinate the buildability and mechanical properties.

Proper treatment can be applied to improve the bonding between the layers and to decrease the voids formed by the filaments. Viscosity modifying agents (VMA) are water-soluble polymers that control the flow characteristics and rheological performance of concrete mortars [29–31]. VMA is also a good material to ensure extrudability as it reduces the cement paste's permeability and, therefore, reduces the risk of water drainage [32–35]. Research investigations have indicated that adding VMA in cement pastes decreases the flowability at a constant water content; however, it can increase the corresponding yield stress and plastic viscosity [20,30,31,36]. The results of these studies show that the addition of VMA enhanced water retention in the concrete and thereby reduced bleeding and increased the degree of hydration [2]. However, it also reduced the surface tension which is beneficial to the paste spreading on the interface between two adjacent layers [3]. It is, therefore, feasible and promising to improve the bonding effect of extruded layers by incorporating the proper dosage of VMA. The 3D concrete printing process differs from the conventional fabrication process. This innovative technique is appropriate for prefabricated structural components. Proper curing methods can be employed to the hardened structures to improve the degree of cement hydration and the compactness of microstructures in the matrix. However, there has been little investigation regarding the influence of VMA and postcuring methods on the viscous and bonding properties of cement motors used for 3D printing.

The objective of this study is to optimize the structural integrity and mechanical performance of the components printed at a favorable buildability situation. To this end, this chapter firstly proposes a 3D-printable cementitious material that is suitable for the extrusion-based printing process. The influence of paste age on the buildability of fresh motors was evaluated to reach a desirable buildability. Meanwhile, the bending resistance of printed prism specimens fabricated at different paste ages was tested. Thereafter, various amounts of VMA and different curing methods were applied to improve the structural capacity by increasing the bonding force between the adjacent extruded layers. In particular, X-ray CT scanning was implemented to characterize the weak bonding interfaces formed in the layered extrusion processes.

8.2 MATERIAL AND METHOD

8.2.1 Material Preparation

Rapid hardening Portland cement P. O 42.5 R, fly ash, and silica fume were used as binding materials. Local river sand with a specific surface area of 0.101 m^2/g and copper tailings with a specific surface area of 0.141 m^2/g served as the fine aggregates. Highly efficient polycarboxylate-based superplasticizer with a water reducing rate of more than 30% and a solid content fraction of 37.2 % were adopted to achieve the required flowability for the mixture. Flowability should be controlled to ensure the fresh paste is smoothly and continuously transported from the storage system to the nozzle without blockage and disruption, therefore, realizing the compatibility between workability and printing process. Additionally, low shrinkage is essential as the free-form components are built without formwork. A small number of polypropylene fibers were employed to reduce the cracking produced by water evaporation. Table 8.1 shows the mixture proportions of the raw materials used for material preparation. The chemical compositions of tailing determined by X-ray fluorescence (XRF) analysis are listed in Table 8.2 and the particle size distribution parameter of tailings is presented in Table 8.3. After a

Table 8.1 Mixture proportions of the raw materials used for 3D printable cementitious material

Mix no.	Natural sand	Tailings	Replacement (%)	Cement	Fly ash	Silica fume	Water	Super-plasticizer (%)	Polypropylene fiber (kg/m^3)
R30	0.72	0.48	40	0.7	0.2	0.1	0.27	1.083	1.2

Table 8.2 Chemical composition of tailings by mass ratio

SiO_2	Al_2O_3	Fe_2O_3	CaO	MgO	SO_3	Na_2O	P_2O_5	K_2O	MnO
39.77	4.61	20.16	22.29	7.17	3.05	1.32	0.26	0.44	0.23

Table 8.3 Particle size distribution parameters for copper tailings

	d(0.1) (μm)	*d*(0.5) (μm)	*d*(0.9) (μm)	Average (μm)
Tailing	38.03	123.75	375.10	24.60

Note: The *d*(0.1), *d*(0.5), and *d*(0.9) denote the sizes of particles that smaller than or equal to the 10%, 50%, and 90% of the total particle mass.

series of attempts and trials, it was found that the most suitable mix for 3D concrete printing is comprised of a tailing to sand mass ratio of 2:3, so that 40% natural sand was replaced by mining tailings [23].

In the preparation process, polypropylene fibers and the dry powders (i.e., cement, fly ash, silica fume, natural sand, and tailings) are firstly blended for three minutes to obtain a uniform mixture. Then, one half of the total amount of water along with the superplasticizer was added and stirred for two minutes. Subsequently, the second half of the total amount of water together with superplasticizer is poured in and stirred for another 2 minutes.

8.2.2 Prism Specimen Manufacture

After blending, the fresh paste is delivered to the material storage tank equipped in our self-designed printing system to manufacture a 40-layer structure [23]. The printed structure is illustrated in Fig. 8.1A. It is built up by vertically stacking extruded filaments with a length of 250 mm and width of 30 mm without collapse. The opening of the printing nozzle is 8 mm × 30 mm. After a series of trials, the optimal operational parameters were determined for smooth printing process. The extrusion rate, V_e, of the cementitious material is designed as 5.4 L/min and the printing speed, V_p, is controlled at 450 mm/min. The extrusion rate of fresh mortar is

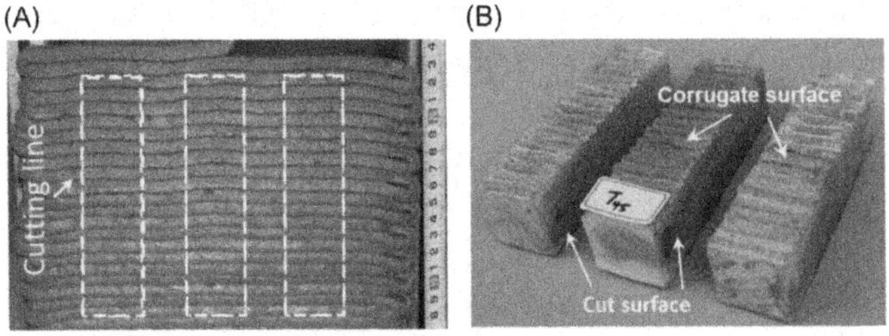

Figure 8.1 (A) Forty-layer structure manufactured through an extrusion-based printing system. (B) Prism specimens with corrugated surface sawed off from the printed structure.

Table 8.4 Testing procedures designed for the evaluation of bending resistance

Series no.	Curing method	Paste age (min)	Viscosity modify agent (%)
T0	Standard curing	Casting	0
T1	Standard curing	15	0
T2	Standard curing	30	0
T3	Standard curing	45	0
T4	Standard curing	45	1.0
T5	Standard curing	45	1.5
T6	Standard curing	45	2.0
T7	Water curing	45	0
T8	90 steam curing	45	0

managed by the rotation of a mixing blade firmly connected to a drive motor. V_p accounts for the moving speed of the printing nozzle, which is also controlled by a drive motor. For more detailed information, readers can refer to our previous research [23].

Thereafter, prism specimens with sizes of 30 mm × 30 mm × 120 mm were sawed off from the 3D-printed structure. As Fig. 8.1B shows, the prism samples have corrugated surfaces and the layers are perpendicular to the printing direction. Then the printed specimens are smoothed to eliminate the influence of the corrugated surface on the fracture behavior, since cracks are prone to be initiated from the transition zones between two layers. Specimens manufactured from a mold-cast were taken as the control reference.

8.2.3 Testing Procedure

Table 8.4 presents testing procedures designed to find optimal solutions for 3D-printable mortar to reach a favorable buildability and structural capacity. Flexural behaviors are investigated in both mold-cast and printed specimens. Herein, three factors are considered, that is, paste age, VMA dosage, and curing method.

Paste age refers to the duration from when the raw materials have been blended to starting printing. The liquid and viscous property of fresh pastes are crucial for the bonding performance between layers, which greatly depends on the paste age. It is expected that the shorter the paste age, the higher the bonding strength between layers. Three paste ages ranging from 15 to 45 minutes and three VMA contents in the range of 0.5%–1.5% were taken into account. Longer paste ages were not considered due to the fresh mortar becoming stiff and could not be printed. Once the specimens are demolded after 24 hour-curing at room temperature, they were cured by three different methods to monitor the strength development with time. The water curing method is designed to directly immerse the specimens into water at room temperature (approximately 20°C) for 7 days. The steam-curing method is used to cure the samples with steam at 90°C for 72 hours. The standard curing method is to place the samples in a moist cabinet for proper curing with an ambient temperature of 20 ± 1°C and relative humidity of 95 ± 5% for 7 days.

By the testing procedures given in Table 8.4, series No. T1–T3 were designed to evaluate the influence of paste age on the flexural performance of printed specimens; No. T4–T6 were applied to measure the impact of VMA on the flexural behaviors; and T3 and T7–T8 were used to assess the effect of the curing method on flexural performances. The flexural properties of both printed and casted specimens were tested according to ASTM C348-14 [37]. As shown in Fig. 8.2, a three-point bending jig was mounted on a digital servo-control universal testing machine with a loading capacity of 100 kN to measure the flexural tensile strength. A symmetrical three-point loading setup with a beam span of 70 mm was used for the flexural tests. The loading rate was set to 48 N/s. The load-displacement curves for all specimens were automatically recorded until specimen failure. At least three identical specimens for each scenario were tested.

8.2.4 X-Ray Computed Tomography Characterization

Nondestructive X-ray computed tomography (CT) has been widely adopted to provide accurate identification of the meso/microscopic structure of engineering materials [38–40]. In this section, we employ

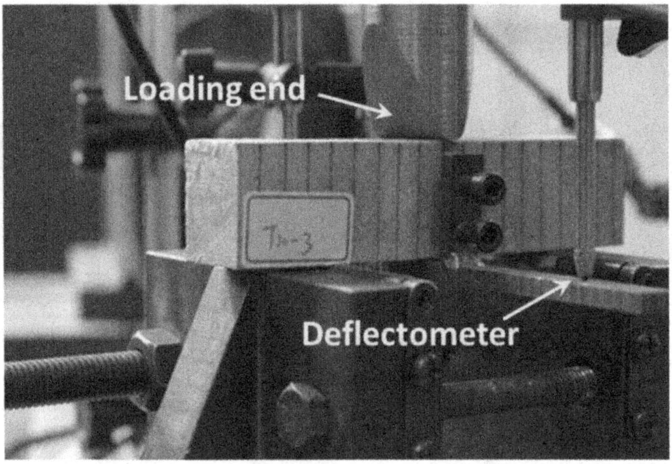

Figure 8.2 Set up for three-point bending test.

advanced CT technology to detect the voids and weak bonding interfaces formed by the filaments as well as the relative position between the fracture path and interfaces. For scanning, a high-resolution CT with a maximum spatial resolution of 10 μm was adopted which satisfied the needs of reconstructing meso-level structures. Due to the specimen being beam-shaped, the scanning was focused on the zones in the vicinity of the fracture surfaces rather than the whole specimen, aiming to improve the clarity of the detecting images. Moreover, the two separated parts of the prism samples after the bending test were assembled together to facilitate the CT characterization.

8.3 RESULTS AND DISCUSSION

8.3.1 Effect of Paste Age on Buildability

To evaluate the influence of paste age on the buildability of fresh mortar paste, 20-layer structures were built up by the proposed tailing mortar at paste ages of 15, 30, and 45 minutes. Fig. 8.3 presents the constructed structures designed with dimensions of 30 mm (W) × 250 mm (L) × 160 mm (H). The printed structures illustrate that the fresh mortar material can be stably stacked in the vertical direction without collapse, indicating an acceptable buildability. However, apparent interfaces may form between adjacent layers due to the inherent layer structure, which is negative to the structural integrity of the 3D-printed models.

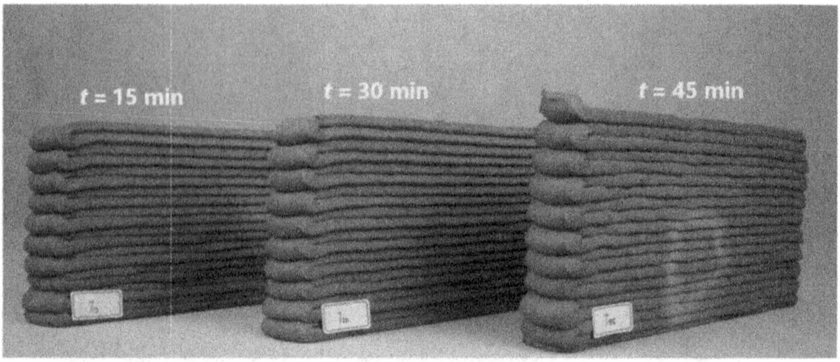

Figure 8.3 Structures constructed by the proposed tailing concrete at different paste ages.

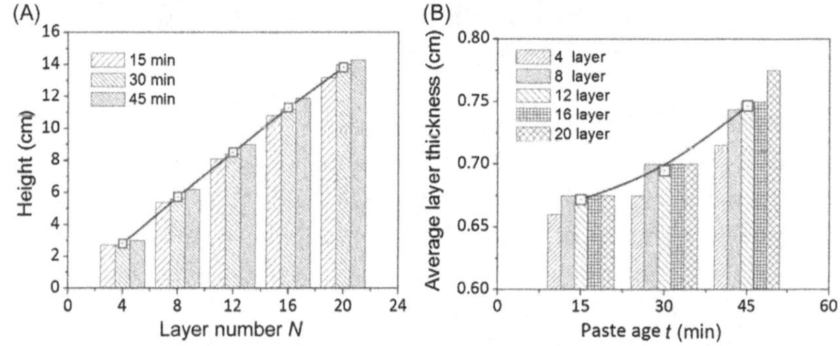

Figure 8.4 Buildability evaluation of printed structures. (A) Overall height of structures stacked with different numbers of layers. (B) Average layer thickness changes with the paste age.

A longer paste age of the cement-based materials promotes the hydration degree of the binding materials and facilitates the fresh paste to turn from a fluidity state into a plastic state [41–43]. From the measured data presented in Fig. 8.4, the height of the structures stacked by a certain number of layers of filaments increases with paste age. Similarly, the average layer thickness also increases with the paste age. Therefore, increased stiffness contributes to the development of buildability, that is, the ability of the paste to retain its extruded shape and sustain the weight of the subsequent layers.

At the paste age of 45 minutes, the measured average thickness of layers under the pressure of self-weight is 7.5 mm, accounting for 93.8% of the nozzle diameter of 8.0 mm. The optimal value is the thickness of the printed filament with no slump. In most cases, it equals to the size of the nozzle. Measured results indicate that the material at the paste age of 45 minutes can perform favorable loading capacity. Good buildability is featured by a sufficient stiffness or unobvious deformation. It is feasible to improve the buildability by increasing the paste age. However, prolonging the paste age will reduce the surface chemical activity to a large extent, produce more voids between two adjacent filaments, and form relative weak interfaces between layers; therefore, producing negative influences of the mechanical integrity of the printed structures. In most cases, proper deformation of the filament is expected to fill the voids to improve the mechanical capacity of the printed structures through enhancing the contact and bonding of adjacent filaments [44]. Therefore, there is a balancing relationship between the buildability (stiffness) and void filling (mechanical capacity).

8.3.2 Effect of Paste Age on the Bending Performance

Increasing the buildability through prolonging the paste age may minimize the bonding force between adjacent layers. The bending behavior of prism specimens printed at different paste ages were evaluated. Fig. 8.5A shows the relationship between load (P) and deflection (δ) (displacement in the middle span) of both casted and printed prism specimens in the three-point bending process. From the P-δ curves, all specimens under load proceeded quickly to failure with the rapid extension of cracks and instant fractures. The P-δ curves displayed no postpeak deformation, indicating obvious brittle failure. Cracking took place in the middle section of the specimens, leading to rough fracture surfaces. The variations in the 7-day flexural strength (f_{flx}) and ultimate deflection (δ_{ult}) in the span centers of the prism specimens printed at different paste ages are presented in Fig. 8.5B and C, respectively. The flexural strength, f_{flx}, of specimens printed at paste ages of 15, 30, and 45 minutes was 31.4%, 33.4%, and 46.1% lower than the casted samples, respectively. Meanwhile, the midspan defection, δ_{ult}, of specimens printed at paste ages of 15, 30, and 45 minutes was 49.5%, 40.1%, and 36.2% lower than the mold-casted samples, respectively. The longer the paste age, the lower the bending resistance. Additionally, the fracture energy G_f was calculated based on the load-deflection curves. G_f quantifies the energy necessary to create a unit area of the crack surface projected onto the plane parallel to the crack direction [45]. The weakly bonded interfaces of printed specimens are prone to produce cracks and result in low fracture energy. As the data in

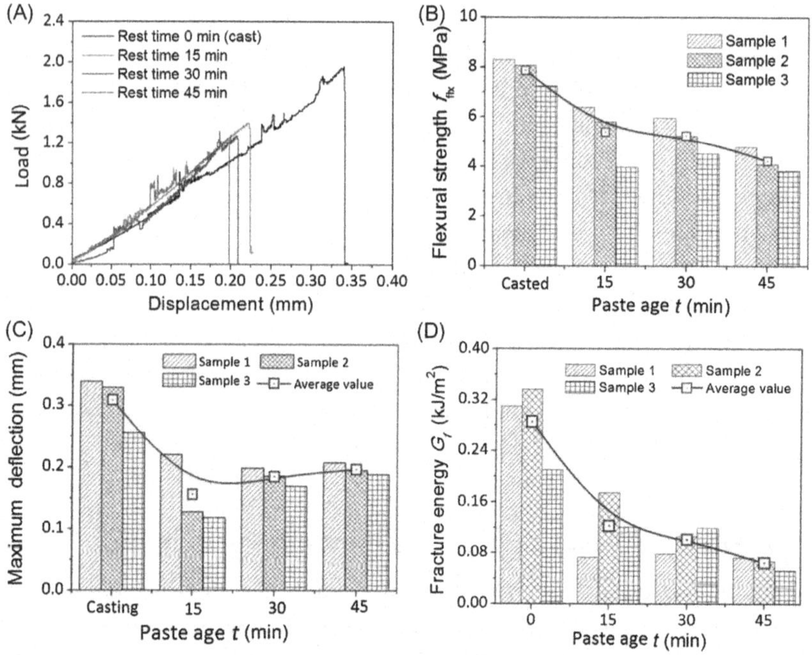

Figure 8.5 Evaluation of the bending behaviors. (A) Load-deflection curves; (B) flexural strength; (C) peak-load deflection; and (D) fracture energy of casted and printed prism specimens.

Fig. 8.5D shows, the printed structures perform lower crack resistance to the applied bending load. From these test results, in general, the printed specimens are of lower bending/fracture resistance relative to the casted ones, which were mainly derived from the bonding strength between adjacent layers being weaker than that of the matrix mass.

Rough fracture surfaces will form once the specimen fails in the bending loads. The two separated parts are assembled together, as shown in Fig. 8.6, to facilitate the CT scanning. Fig. 8.7 illustrates the X-ray CT characterization of microscale structures of prism specimens in vicinity of the bending fractures. As the CT images show in Fig. 8.7A, there are no obvious voids other than the intrinsic pores formed during the hydration process. The paste age of the 15 minutes specimen produced a trivial impact on the mortar matrix. However, the layered structure and significant porosity in the micron order can be seen clearly when the paste age prolonged to 30 and 45 minutes, which was characterized by a series of microscale voids/pores, as shown in Figs. 8.7B and C. These voids possess

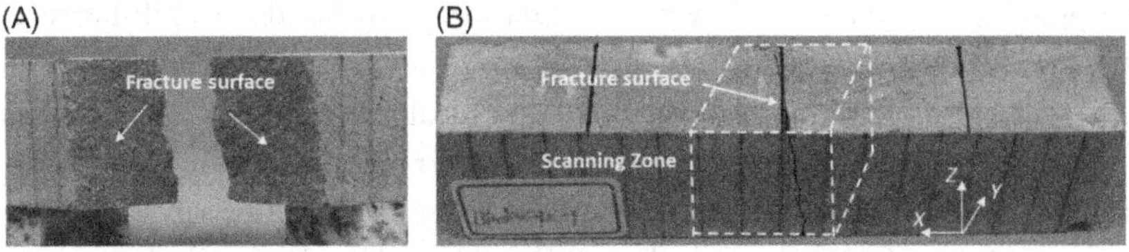

Figure 8.6 (A) Rough surfaces formed after the failure of a prism specimen during the bending test. (B) Illustration of the computed tomography scanning zone.

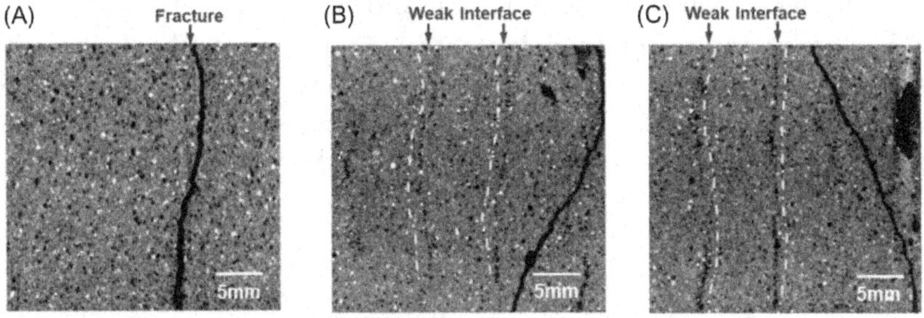

Figure 8.7 Microscale characterization of structures in the vicinity of the bending fractures through X-ray computed tomography scanning. Arbitrary XY cross section of specimens fabricated at paste ages of: (A) 15 min; (B) 30 min; and (C) 45 min.

comparative size to the intrinsic pores in the cement matrix. A distinct feature is that they are discontinuously distributed along the boundary of the extruded filaments, as shown by the yellow (white in print version) line marked in the CT images. From a mesoscopic point of view, the curved boundary may be induced by the instability of the material's extrusion through the small nozzle. The weak interface becomes more noticeable as the paste age increases. The voids located at the boundary of the layers demonstrate the interfaces. Due to the reduction of chemical activity and rheology of the extruded paste, there must be a certain discontinuity between the interlayers. However, all the poorly bonded interfaces will not be certainly displayed by a line of voids. When the printing processes (the speed of deposition and extrusion) are well controlled, the surfaces are flat contacted and there are not distinct microvoids produced in the interstices between the filaments. The hardened properties of mortar are likely to be affected by the layered structure and sometimes may result in certain mechanical anisotropy.

8.3.3 Effect of VMA on Flowability

The 3D concrete printing process differs significantly from traditional fabrication processes. The failure modes, or the path of the fracture surface, is strongly influenced by the layer delamination. The bond strength between stacking layers relies on certain specific treatment methods. It is expected that VMA is a good agent to modify the bonding properties between two adjacent layers due to its potentiality in water retention and surface tension reduction. However, incorporating a certain amount of VMA into the cementitious mixture may lower the flowability of the fresh material. Fig. 8.8 depicts the fluidity of the proposed tailing mortar with different VMA contents. Fluidity is characterized by the spreading diameter of fresh paste through a vibrate table test in accordance with ASTM C1437-07 [46]. The measured results indicate that, as predicted, the higher the VMA content, the more viscous and the lower the flowability of the cement paste. Based on our previous measurement, the mortar material that keeps a spreading diameter ranging in 17.4–21 cm performs

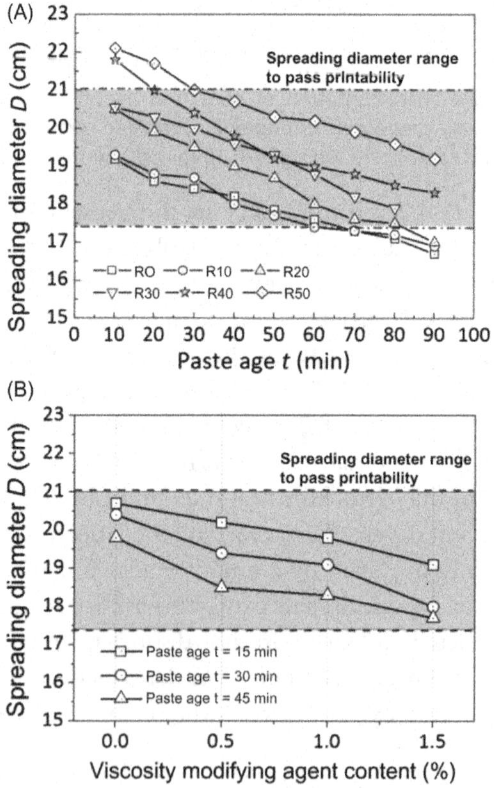

Figure 8.8 (A) Spreading diameter range for acceptable printability. (B) Relationship between fluidity and the viscosity modifying agent content of cement pastes at different paste ages.

acceptable printability [23]. As the data presented in Fig. 8.8 shows, the cement paste with a designed content of VMA (0%−1.5%) retain a spreading diameter within the recommended range and, therefore, meets the requirement for printing. However, further addition of VMA is not applicable for the fresh mixtures to acquire acceptable printability due to mortar with low flowability possibly blocking the material transition system and so cannot be printed.

8.3.4 Effect of VMA on Bending Behavior

It is expected that the bonding properties of the extruded layers will be improved by using a certain amount of VMA. To verify this expectation, the bending behavior of prism specimens fabricated with different VMA contents were investigated. The paste age for printing is designed at 45 minutes because the mixture can acquire favorable buildability at this time. The load-deflection curves of printed specimens with different VMA contents are shown in Fig. 8.9A. The results indicate that the prism specimens all perform obvious brittleness. The VMA does not produce an obvious influence on the failure patterns of printed structures. As the data presented in Fig. 8.9B show, the flexural strength increases with the addition of VMA and the strength of samples with 1.5% VMA is approximately 26% higher than those without VMA. The results demonstrate the positive contribution of VMA to the mechanical property of printed structures. Meanwhile, the addition of VMA greatly enhanced fracture resistance. As illustrated in Fig. 8.9C, G_f was improved by 17.6%, 42.6%, and 54.5%, respectively, when 0.5%, 1.0%, and 1.5% VMA were employed. However, 3D-printed beams show no obvious variation in maximum deflection with the increasing amount of VMA, as shown in Fig. 8.9D. Generally, the bonding properties between adjacently extruded layers were effectively improved by incorporating a certain amount of VMA.

To further probe the influence of VMA on the bonding property between filaments, CT scanning technique was implemented to characterize the microscale interfaces between adjacent filaments and the fractured zones of tested samples. The detection method is similar to that illustrated in Fig. 8.6. From the scanning results, when 0.5% VMA is added, the fracture induced by the applied bending loads propagates along the weak bonding face, as the CT image shows in Fig. 8.10A, which proves the negative impact of the bonding interface on the mechanical properties of the laminated structure. When the addition of VMA is increased, the

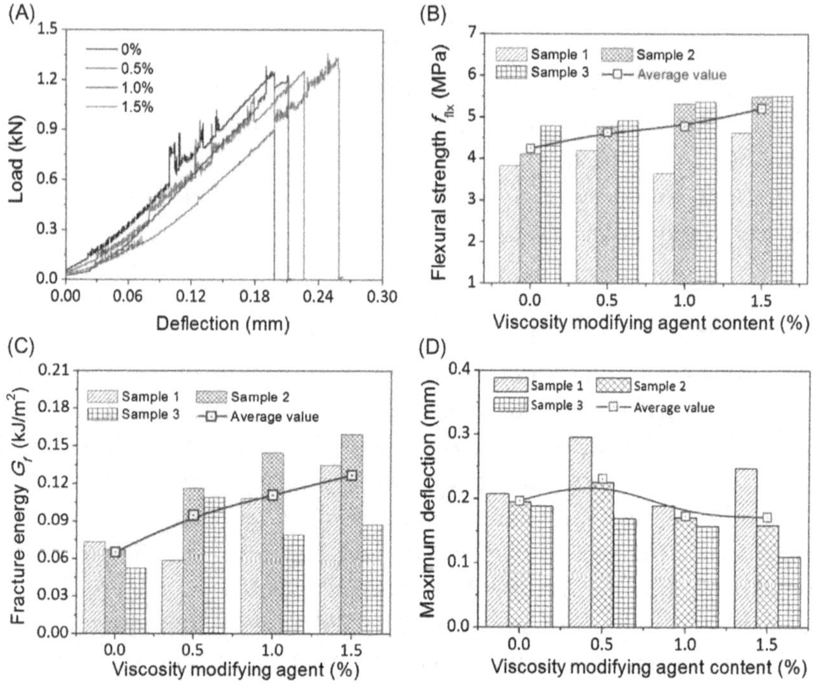

Figure 8.9 Evaluation of bending behavior: (A) load-deflection curves; (B) flexural strength; (C) fracture energy; and (D) maximum deflection of printed prism specimens.

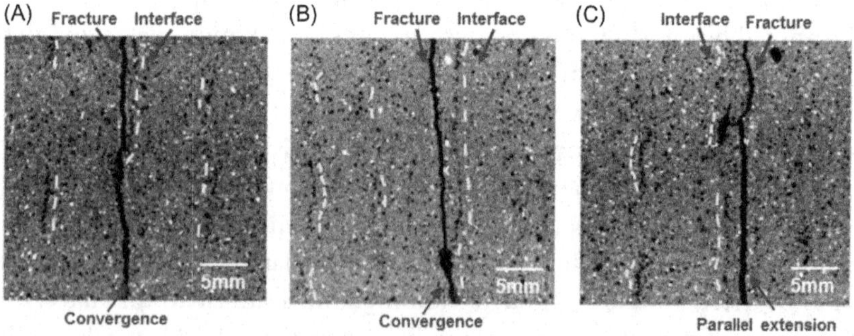

Figure 8.10 Computed tomography images characterizing the microscale structures in the vicinity of the bending fractures of specimens incorporating a viscosity modifying agent content of: (A) 0.5%; (B) 1.0%; and (C) 1.5%.

induced cracks are prone to propagate, in parallel or crossed, along the interface due to the adhesive property between filaments being improved to a certain degree. As the detected result illustrated in Fig. 8.10B, only a small part of crack is converged with the weak interface when 1.0 wt% of VMA was applied. It is, therefore, expected that the interfaces will have a trivial influence on the structural capacity when the adjacent layers are strongly bonded. The CT image presented in Fig. 8.10C proves this expectation. Cracks extend parallel to the interface instead of converging with it. The addition of VMA eliminated the influence of the interlayer delamination on the fracture path to a large extent. The fractures can be normal, parallel or cross-layer structures. From these measured results, it is feasible and promising to employ proper VMAs for the material preparation to reach desirable structural performances.

8.3.5 Effect of the Curing Method on the Bending Behavior

Due to its layered build-up character, 3D printing of cementitious material is appropriate for prefabricated structural components. Proper curing can promote the strength development of the concrete structures. Fig. 8.11 shows the evaluation of the bending behavior of prism specimens cured through the standard curing method, water curing method, and steam-curing method. From these test results, the distinct brittleness of mortar material has not been modified through the different curing methods. The beam specimens instantly fractured once the peak load was reached (Fig. 8.11A). As shown by the measured results given in Fig. 8.11B, the bending strength of standard cured specimens were close to the water-cured ones. However, the 90°C steam-curing methods significantly enhanced strength development. The flexural strength of specimens with steam curing was about four times higher than that achieved by standard curing. Steam curing improves the maximum mid-span deflection of specimens by about 50% relative to standard-cured samples. These findings are mainly derived from in fact that silica fume reacts with calcium hydroxide ($Ca(OH)_2$) during the hydration of cement and promotes the formation of calcium silicate hydrates (C−S−H). This phase links the various components together allowing the creation of a dense and compact cementitious matrix and results in high mechanical properties [47].

From the test results, steam curing significantly contributes to the improvement of the microstructures of the cement matrix as well as the

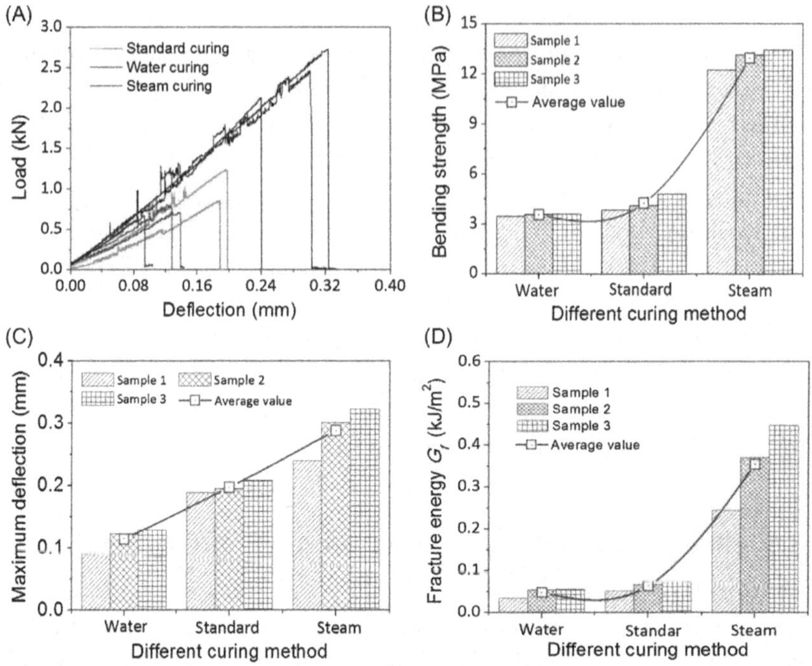

Figure 8.11 Evaluation of the flexural behavior of prism specimens cured by using different methods: (A) Load-deflection curves; (B) flexural strength; (C) maximum deflection; and (D) fracture energy of prism specimens.

elimination of the interlayer delamination, therefore, resulting in structural capacity enhancement of the printed structures. While this is sufficient for some construction applications, further improvements to strength are necessary for many construction applications.

8.4 CONCLUSIONS

This chapter investigated the structural capacity of components printed at favorable buildability situations. The paste age, VMA content, and curing method are considered to optimize the buildability of fresh paste and the mechanical strength of the hardened material to meet the structural capacity requirements and demands of the printed stricture. The following conclusions can be drawn from this study:

1. The buildability of a proposed tailing material can be controlled by adjusting the paste age. The longer the paste age, the better the buildability. At a paste age of 45 minutes, the average layer thickness is

75 mm, accounting for 93.8% of the optimal designed value. A low-slump characteristic represents a well buildability. It is feasible to improve the buildability by adjusting the paste age.

2. The flexural strength of specimens printed at a paste age of 45 minutes accounts for 46.1% of the mold-casted samples. From the CT identification, the weak-bonding interfaces are characterized by discontinuously distributed small voids along the boundary of the extruded filaments. The weak interface becomes more noticeable as the paste age increases. The inherent nature of layer delamination negatively influences the structural integrity and capacity of the printed models.
3. Incorporating 1.5% viscous modifying agent can increase the flexural strength and fracture energy by 25% and 54.5%, respectively. The addition of VMA eliminates the influence of the interlayer delamination on the fracture path to a large extent. The flexural strength of material with 1.5% VMA measures 67% of the mold-casted ones. The flowability of fresh paste must be taken into account to meet the basic requirements of a desirable printability when a certain amount of VMA is adopted.
4. The steam-curing method introduced in this study increased the strengths approximately four times from the original strength. Flexural strengths of 12.93 MPa can be achieved with this postprocessing method. The inherent nature of the layered structure becomes less distinct as components are cured. Heat curing at 90°C may not be an applicable means for the rapid manufacturing processing; however, it is a promising post-treatment method for enhancing mechanical performances.

This chapter investigated the structural integrity and bending resistance of 3D printed structures. It is applicable and beneficial to enhance the 3D-printed structures using the proposed methods. However, there are still certain mechanical mismatches between the printed and mold-cast specimens. The next step for research is to investigate how to reduce the weakening impact of layer delamination on the structural performance of components. Additionally, the mechanical anisotropy of the printed laminar structure needs further study as, currently, 3D-printed objects are either unreinforced, or reinforcement is applied manually; fiber-reinforced cement mixtures or fiber-reinforced polymers that show great potential to increase the ductility of the printed mortar should be developed. Further research will also be devoted to explore the frontiers of 3D printing and promote its effective application in real-life construction scenarios.

ACKNOWLEDGMENT

The authors are grateful to the financial support by the National Major Research Instrument Development Project of the National Natural Science Foundation of China (Grant No. 51627812) and the opening project of State Key Laboratory of Explosion Science and Technology (Beijing Institute of Technology, Grant No. KFJJ13-11M).

REFERENCES

[1] A. Kazemian, et al., Cementitious materials for construction-scale 3D printing: laboratory testing of fresh printing mixture, Constr. Build. Mater. 145 (2017) 639–647.
[2] T.T. Le, et al., Hardened properties of high-performance printing concrete, Cem. Concr. Res. 42 (3) (2012) 558–566.
[3] G.J. Gibbons, et al., 3D printing of cement composites, Adv. Appl. Ceram. 109 (5) (2010) 287–290.
[4] G. Cesaretti, et al., Building components for an outpost on the Lunar soil by means of a novel 3D printing technology, Acta Astronaut. 93 (2014) 430–450.
[5] R.A. Buswell, et al., Freeform construction: mega-scale rapid manufacturing for construction, Autom. Constr. 16 (2) (2007) 224–231.
[6] B. Khoshnevis, Automated construction by contour crafting—related robotics and information technologies, Autom. Constr. 13 (1) (2004) 5–19.
[7] N. Labonnote, et al., Additive construction: state-of-the-art, challenges and opportunities, Autom. Constr. 72 (2016) 347–366.
[8] M. Attaran, The rise of 3-D printing: the advantages of additive manufacturing over traditional manufacturing, Bus. Horiz. (2017).
[9] J. Zhang, B. Khoshnevis, Optimal machine operation planning for construction by contour crafting, Autom. Constr. 29 (2013) 50–67.
[10] R.A. Buswell, et al., Design, data and process issues for mega-scale rapid manufacturing machines used for construction, Autom. Constr. 17 (8) (2008) 923–929.
[11] T.T. Le, et al., Mix design and fresh properties for high-performance printing concrete, Mater. Struct. 45 (8) (2012) 1221–1232.
[12] C. Gosselin, et al., Large-scale 3D printing of ultra-high performance concrete–a new processing route for architects and builders, Mater. Des. 100 (2016) 102–109.
[13] M. Hambach, D. Volkmer, Properties of 3D-printed fiber-reinforced Portland cement paste, Cem. Concr. Comp. 79 (2017) 62–70.
[14] Winsun, WinSun China builds world's first 3D printed villa and tallest 3D printed apartment building. http://www.3ders.org/articles/20150118-winsun-builds-world-first-3d-printed-villa-and-tallest-3d-printed-building-in-china.html, 2015.
[15] WASP, The first adobe building. http://www.wasproject.it/w/en/3d-printers-projects/, 2016.
[16] A. Rudenko, 3D concrete house printer. http://www.designboom.com/technology/3d-printed-concrete-castle-minnesota-andrey-rudenko-08-28-2014/, 2015.
[17] C. Gosselin, et al., Large-scale 3D printing of ultra-high performance concrete – a new processing route for architects and builders, Mater. Des. 100 (2016) 102–109.
[18] A. Perrot, D. Rangeard, A. Pierre, Structural built-up of cement-based materials used for 3D-printing extrusion techniques, Mater. Struct. 49 (4) (2016) 1213–1220.
[19] P. Feng, et al., Mechanical properties of structures 3D printed with cementitious powders, Constr. Build. Mater. 93 (2015) 486–497.
[20] G. Ma, L. Wang, A critical review of preparation design and workability measurement of concrete material for largescale 3D printing, Front. Struct. Civil Eng. (2017). Accept for publication.

[21] S. Lim, et al., Developments in construction-scale additive manufacturing processes, Autom. Constr. 21 (2012) 262–268.
[22] A. Perrot, D. Rangeard, A. Pierre, Structural built-up of cement-based materials used for 3D-printing extrusion techniques, Mater. Struct. 49 (4) (2016) 1–8.
[23] G. Ma, Z. Li, L. Wang, Printable properties of cementitious material containing copper tailings for extrusion based 3D printing, Constr. Build. Mater. (2017). In revision.
[24] H.W. Reinhardt, C.U. Grosse, Continuous monitoring of setting and hardening of mortar and concrete, Constr. Build. Mater. 18 (3) (2004) 145–154.
[25] T. Voigt, T. Malonn, S.P. Shah, Green and early age compressive strength of extruded cement mortar monitored with compression tests and ultrasonic techniques, Cem. Concr. Res. 36 (5) (2006) 858–867.
[26] J. Pegna, Exploratory investigation of solid freeform construction, Autom. Constr. 5 (5) (1997) 427–437.
[27] Anon, Innovative rapid prototyping process makes large sized, smooth surfaced complex shapes in a wide variety of materials, Mater. Technol. 13 (2) (1998) 53–56.
[28] L. Feng, Y. Liang, Study on the status quo and problems of 3D printed buildings in China, Glob. J. Hum. Soc. Sci. Res. 14 (2014) 7–10.
[29] M. Lachemi, et al., Performance of new viscosity modifying admixtures in enhancing the rheological properties of cement paste, Cem. Concr. Res. 34 (2) (2004) 185–193.
[30] A. Leemann, F. Winnefeld, The effect of viscosity modifying agents on mortar and concrete, Cem. Concr. Comp. 29 (5) (2007) 341–349.
[31] M. Benaicha, et al., Influence of silica fume and viscosity modifying agent on the mechanical and rheological behavior of self compacting concrete, Constr. Build. Mater. 84 (2015) 103–110.
[32] A. Perrot, D. Rangeard, Y. Melinge, Prediction of the ram extrusion force of cement-based materials, Appl. Rheol. 24 (5) (2015) 53320.
[33] A. Perrot, et al., Ram extrusion force for a frictional plastic material: model prediction and application to cement paste, Rheol. Acta 45 (4) (2006) 457–467.
[34] A. Perrot, et al., Modeling the ram extrusion force of a frictional plastic material. 2006.
[35] A. Perrot, et al., Use of ram extruder as a combined rheo-tribometer to study the behaviour of high yield stress fluids at low strain rate, Rheol. Acta 51 (8) (2012) 743–754.
[36] G. Ma, L. Wang, Y. Ju, State-of-the-art of 3D printing technology of cementitious material — an emerging technique for construction, Sci. China Technol. Sci. (2017). Accept for pubilication.
[37] ASTM C348-14, Standard Test Method for Flexural Strength of Hydraulic-Cement Mortars, ASTM International, West Conshohocken, PA, 2014, www.astm.org.
[38] S. Lu, E.N. Landis, D.T. Keane, X-ray microtomographic studies of pore structure and permeability in portland cement concrete, Mater. Struct. 39 (290) (2006) 611–620.
[39] E. Gallucci, et al., 3D experimental investigation of the microstructure of cement pastes using synchrotron X-ray microtomography (μCT), Cem. Concr. Res. 37 (3) (2007) 360–368.
[40] M. Zhang, et al., Computational investigation on mass diffusivity in Portland cement paste based on X-ray computed microtomography (μCT) image, Constr. Build. Mater. 27 (1) (2012) 472–481.
[41] A. Boumiz, C. Vernet, F.C. Tenoudji, Mechanical properties of cement pastes and mortars at early ages: evolution with time and degree of hydration, Adv. Cem. Based Mater. 3 (6Part18) (1996). 3599.

[42] T. Voigt, Y. Akkaya, S.P. Shah, Determination of early age mortar and concrete strength by ultrasonic wave reflections, J. Mater. Civil Eng. 15 (3) (2003) 247–254.
[43] G. Trtnik, et al., Possibilities of using the ultrasonic wave transmission method to estimate initial setting time of cement paste, Cem. Concr. Res. 38 (11) (2008) 1336–1342.
[44] T.T. Le, et al., Hardened properties of high-performance printing concrete, Cem. Concr. Res. 42 (3) (2012) 558–566.
[45] Z. Zhao, S.H. Kwon, S.P. Shah, Effect of specimen size on fracture energy and softening curve of concrete: Part I. Experiments and fracture energy, Cem. Concr. Res. 38 (8) (2008) 1049–1060.
[46] Chinese National Testing Standard, GB/T 2419-2005, Test method for fluidity of cement mortar.
[47] J. Yang, et al., An investigation on micro pore structures and the vapor pressure mechanism of explosive spalling of RPC exposed to high temperature, Sci. China Technol. Sci. 56 (2) (2013) 458–470.

FURTHER READING

ASTM C1437-07, Standard Test Method for Flow of Hydraulic Cement Mortar, ASTM International, West Conshohocken, PA, 2007, www.astm.org.

CHAPTER 9

Mechanical Properties of Structures 3D-Printed With Cementitious Powders

Peng Feng[1], Xinmiao Meng[2], Jian-Fei Chen[3] and Lieping Ye[1]
[1]Department of Civil Engineering, Tsinghua University, Beijing, P.R. China
[2]Department of Civil Engineering, Beijing Forestry University, Beijing, P.R. China
[3]School of Planning, Architecture and Civil Engineering, Queen's University Belfast, Northern Ireland, United Kingdom

9.1 INTRODUCTION

Recently, rapid prototyping (RP) technologies, especially three-dimensional printing (3DP), have been successfully used in many areas [1,2] such as manufacturing industry [3], medical applications [4], and food preparation [5]. It is believed that RP technologies may change the whole field of production in the future [1,6]. The 3DP technique can satisfy the requirements of diversification, industrialization, and informatization for engineered construction, making it possible to build a structure by printing. An important issue in achieving wide application of large structures is the understanding of the mechanical behavior of 3D printed products so that they can be designed to be printed in an optimal way and behave as designed.

A typical 3DP technique is an advanced RP technique using a specialized digital geometric model to construct 3D objects layer by layer using binding powder materials [1,7]. There are many other forms of RP techniques similar to 3DP [2], including selective laser sintering [8], fused deposition modeling [9], digital light processing [10], and stereo lithography [11], etc. Fig. 9.1 shows the relationships between these techniques. These techniques are all generally called three-dimensional printing due to their abilities to produce 3D objects directly, but this chapter focuses on the 3DP technique using powder materials.

The 3D printing process used in this research is as follows [1,7,12]:

1. 3D digital model is first built using computer aided design (CAD) software.

3D Concrete Printing Technology
DOI: https://doi.org/10.1016/B978-0-12-815481-6.00009-9

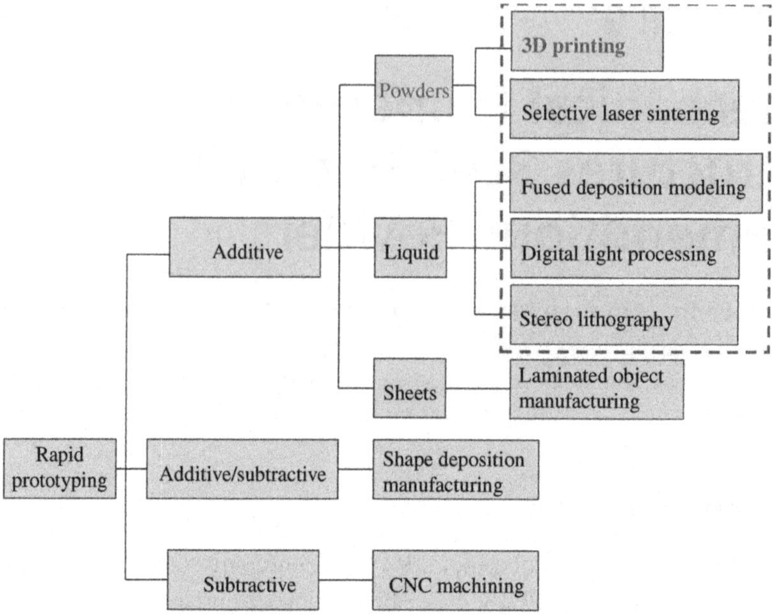

Figure 9.1 Classification of rapid prototyping technologies.

2. The CAD model is then converted into standard triangulation language (STL) format.
3. The STL file is sliced into many thin, digital layers.
4. Each layer, containing the geometric information, is transmitted to the 3D printer in sequence.
5. The printer constructs each layer atop another according to the received data.

During printing, the nozzle of the 3DP printer sprays an adhesive agent (glue) along the predetermined paths (strips), one by one, in each layer. The adhesive agent binds the powder together to form a hardened material. When all the layers are printed one atop another, a complete 3D object is constructed.

Currently, the 3DP technique can print products using powder materials such as sand, plaster, cement, metal, and ceramic [13–15]. It is, thus, suitable to use liquid–particle mixed cementitious materials such as cementitious concrete composed of cement and aggregate to produce civil engineering projects.

The application of 3DP in engineering manufacture may shorten the development cycle, reduce production cost, and improve productivity.

When applied to civil engineering, it has the potential of reducing the number of site workers, speeding up the construction process, and reducing risks during construction. A 3DP model can also be very conveniently linked to a building information model [16], making the whole process from design, construction, management, maintenance, and even decommission, digital.

The past few years have experienced a rapid development of 3D printers. More and more materials have been studied and used in 3DP processes. The size of 3D printers has also increased rapidly, making it likely that in the near future it will be possible to build large and complex-shaped structures by printing. In 2010, Italian inventor Dini developed a large 3D printer named D-shape [17] which made it possible to print buildings with irregular shapes. Dini is currently cooperating with Dutch architect Ruijssenaars to print a Mobius strip-like building. Construction using D-shape is four times faster than the traditional method and costs are halved. In 2012, Novikov and colleagues, architects at the Institute for Advanced Architecture of Catalonia in Spain, invented a robot named Stone Spray which can use organic materials as the base material [18]. This robot constructs architectural shapes by depositing a mix of soil and eco-friendly binder with the help of a jet spray system. Professor Khoshnevis at the University of Southern California in the United States conducted a research project called contour crafting over several years [19,20]. He designed a large printer to construct concrete buildings through extruding cementitious concrete layer by layer [21], expecting to print a 760 m^2 building within 20 hours. Currently, he is researching how to apply such a technique to build extraterrestrial settlement infrastructures [22]. Since 2005, Buswell and colleagues at Loughborough University in the United Kingdom have been conducting a program called "3D Concrete Printing" aimed at automation in the construction sector [23,24]. 3D concrete printing is also based on the extrusion of cement mortar, but has a smaller resolution of deposition [16,25].

Although limited research on constructing large buildings using 3D printers has been conducted, little research on the mechanical properties of 3D-printed structures is available [26–28]. Since 3D-printed layered cementitious materials have different microscopic structures from traditional structural materials such as concrete and steel, their mechanical behaviors may also be different. The mechanical properties of 3D printed structures are the main focus of this chapter.

This chapter presents a study of 3D-printed products made from a powder material that has the potential for wide applications in future engineering structures. The microstructure of the material was analyzed and a series of tests were conducted for its mechanical properties. The test data provide the basis for establishing a mechanical model which is essential for advanced analysis of the behaviors of 3D-printed structures.

9.2 3D PRINTED SPECIMENS

9.2.1 Specimens

9.2.1.1 Material Composition

In this study, a cementitious material was adopted to build 3D objects. The material was a mixture of plaster powder ZP150 and binder ZB60, with a volume fraction of 21.8% for the binder. The main ingredients of ZP150 were plaster, vinyl polymer, and carbohydrate. The main components of the ZB60 binder were humectant and water.

9.2.1.2 Specimen Preparation by Printing

The specimens were prepared using a Spectrum Z510 3D printer produced by Z corporation using a HP 4810 A 11 nozzle. The printer can print products up to 356 mm (length) × 254 mm (width) × 203 mm (height).

The 3D printer consists of a feed bin and a build bin (Fig. 9.2) [29]. The feed bin is filled with plaster powder before printing. At the start of the 3D printing process, a roller mounted together with the print head on the gantry spreads the powder to form a base layer 3.18 mm thick covering the base of the build bin. The print nozzle then applies the binder solution at predetermined locations based on the digital geometric information, strip by strip, until one layer is constructed. The feed bin is then raised by the thickness of one layer, while the build bin lowers by the same distance to allow the next layer to be constructed. Once the powder bed is prepared again and the nozzle is cleaned, construction of the next layer begins. The above steps are repeated until all the layers are printed one atop another to complete the object. Fig. 9.2 provides a schematic view of the printing process.

9.2.1.3 Coordinate System

To study the detailed structure of 3D-printed components, a coordinate system is defined in reference to the actual printing procedure (Fig. 9.3).

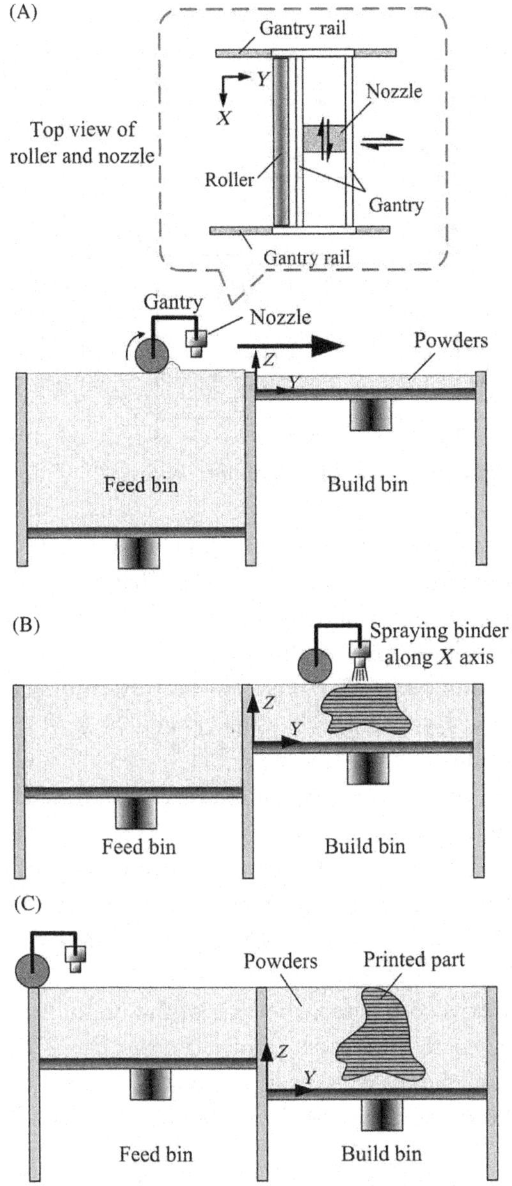

Figure 9.2 Schematic 3D printing process: (A) Preparation for printing; (B) during printing; and (C) end of printing.

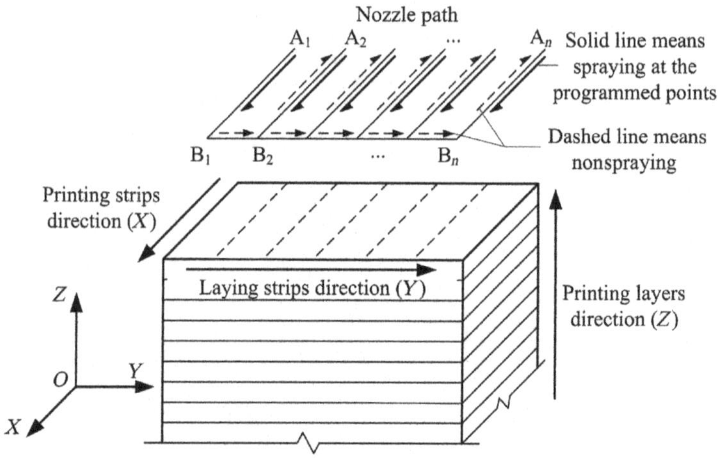

Figure 9.3 Directional description of 3D printing process.

The X axis is defined as the strip direction in which the nozzle moves when it sprays the binder along the gantry, so that all the strips are parallel to the X axis. The Y axis is perpendicular to the strip (X) direction in the plane of a printing layer, that is, the direction in which the nozzle moves from one strip to the next. The Z axis is the vertical direction perpendicular to the printing layer (Fig. 9.3). The process of 3DP can be described using the coordinate system:

1. The nozzle moves along a predetermined line parallel to the X axis and sprays the binder at a controlled rate. The travel distance and area of spray are determined by the digital geometric data of the object.
2. After completing one strip, the nozzle moves a distance equal to the width of a strip in the Y direction and then starts to print the next strip.
3. Once one layer is completed, the base of the build bin is lowered by a distance equal to the thickness of one layer in the Z direction. While the powder of the next layer is spread by the roller the nozzle is cleaned at the same time and then the printer starts to print the next layer.

Fig. 9.3 shows the process of printing a cubic specimen. The nozzle sprays binder from A_1 to B_1, and then moves to B_2 and further to A_2 without spraying. It sprays again when it moves from A_2 to B_2 to complete the second strip. This process is repeated until the nozzle sprays from A_n to B_n to finish one layer. The layering process is repeated until the cube is printed.

When printing cubic specimens specific to the following tests, the nozzle moved in the *X* direction at a high speed of about 460 mm/s when spraying, which was much faster than that in *Y* direction. Clearly, the time between printing two adjacent strips was much shorter than that between adjacent layers. This difference in time intervals between printing adjacent strips and layers may have a significant influence on the mechanical properties of 3D printed specimens.

Once printing was finished, the base of the build bin was raised to remove the printed products from the powder bed. The extra powder that still covered the objects was first brushed off and then vacuumed up. Finally, to accelerate solidification and accordingly increase strength, the freshly printed objects were placed in an oven at 60°C for 3 hours to become completely dry. The density of the hardened specimens was about 1.34 g/cm^3. Fig. 9.4 shows the actual procedure of printing specimens.

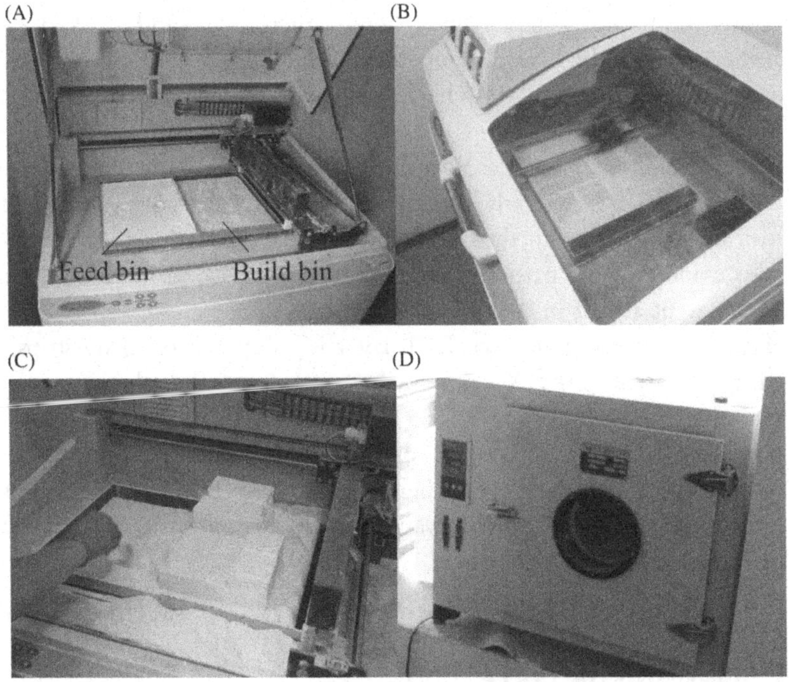

Figure 9.4 Printing procedure of 3D printer Spectrum Z510: (A) Preparation of powder bed; (B) during printing; (C) removal of printed components; and (D) curing in oven.

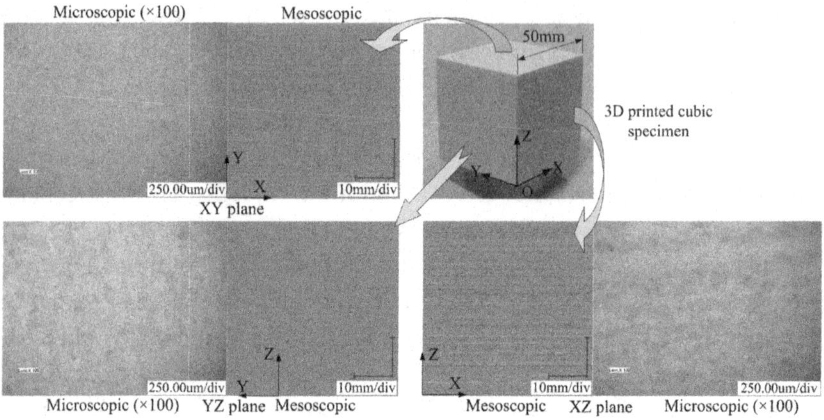

Figure 9.5 Surface structure of a 3D-printed cube.

9.2.2 Microstructure of Printed Objects

The surface structure of a printed 50 mm cubic specimen was inspected by using both a 3D high-depth stereo microscope and "naked-eye" recording using a high-resolution digital camera. Fig. 9.5 shows typical photographs from which the characteristics of the 3D specimen can be summarized:

1. *Layered microstructure*. The 3D-printed sample with cementitious powder material has a clearly layered microstructure. Many parallel lines can be clearly seen in the *XZ* plane. The layered printing structure is also evident in the particle configuration in the *YZ* plane, but it is not as apparent as that in the *XZ* plane.
2. *Striped structure in each layer*. Each layer is composed of many strips, as clearly shown in the *XY* plane which is consistent with the printing process.
3. *Orthotropy*. The *XZ* face is the roughest of the three faces, the *YZ* face is smoother, with the *XY* face being in between. Based on the printing procedure (Fig. 9.3) and observations of the surface structure, it is clear that this material is orthotropic.

9.3 MECHANICAL TESTS

Compression tests of cubes and flexural tests of small beams were conducted to determine the basic mechanical properties of the 3D-printed material, including strength, elastic modulus, and Poisson's ratio. Failure characteristics of the tested specimens are described next.

9.3.1 Compression Test

9.3.1.1 Specimens

Two batches of printed cubes, one with a side length of 70.7 mm and the other with 50 mm were produced and tested for the compression behavior. Each batch was printed in a single printing cycle. Both batches had the same ingredients and were produced using the same printing method with a layer thickness of 0.0875 mm.

The batch of the larger specimens was divided into two groups of three specimens each. The specimens in the first group, designated CZ-B1 to CZ-B3, were loaded in the *Z* direction. Those in the second group, designated CX-B4 to CX-B6, were loaded in the *X* direction. In the specimen designations, the first letter C represents the compression test, the second letter represents the loading direction (either *X* or *Z*), followed by the specimen size (B for the bigger specimens and S for the smaller specimens). Two pairs of foil-strain gauges, each pair consisting of one horizontal and one vertical gauge, were bonded at the center of the two opposite faces of each specimen. The gauges had an active grid length of 3 mm and an electrical resistance of 120 Ω. The specimens were about one month old at the time of testing.

The batch of smaller specimens was divided into three groups, with three specimens in each. They were loaded respectively in *X*, *Y*, and *Z* directions. Based on the above rules, the specimens in the three groups were designated CX1 to CX3, from CY1 to CX3, and from CZ1 to CZ3 respectively. Four pairs (horizontal and vertical) of foil-strain gauges were attached at the center of the four vertical faces for each specimen.

9.3.1.2 Test

The compression test followed the test procedure prescribed in the Chinese "Standard for test method of mechanical properties on ordinary concrete" (GB/T50081-2002) [30]. The loading rate ranged between 0.1 and 0.3 kN/s. The compression strength was obtained from the maximum load divided by the face area of the cubes. The modulus of elasticity and Poisson's ratio were calculated from the initial linear branch of the stress–strain relationship curve. The measurement of modulus and Poisson's ratio using cube specimens is less accurate compared with cylindrical specimens due to frictional constraints at the contact surfaces, but the cylindrical specimen was not used because of limitations of the printing capacity of the 3D printer; it does not print cylindrical specimens well, especially when printing vertically along the longitudinal direction. Each specimen was preloaded twice to a stress of 0.5 MPa, before loaded to failure. In order to minimize loading eccentricity, the position of a specimen was adjusted after each preloading based on the differences between strain readings from gauges on the opposite faces.

9.3.1.3 Failure Mode and Other Observations

When loaded in the *Z* direction, both large and small cubes have similar phenomena. At the beginning of loading, the deformation was small in these specimens. When approaching maximum load, diagonal cracks occurred in the *YZ* plane and they developed quickly. When the ultimate load was reached, the specimens exhibited diagonal failure with two sets of triangular cracks intersecting near the center to form an hourglass shape on the two opposite sides in the *YZ* plane, indicated as *red dashed lines* (black lines in print version) in Figs. 9.6 and 9.7. Some local damage following the printed strips appeared in the contact region between the upper loading platen and the corner of the specimen, as shown in Fig. 9.7

When loaded in the *X* direction, the behavior of both large and small cubes was similar to that under *Z*-direction compression. The only difference is that the hourglass-shaped cracks were now formed on the opposite faces in the *XZ* plane (Figs. 9.6 and 9.8) instead of the *YZ* plane when under *Z*-direction compression. Probably due to printing defects, the large specimen CX-B6 failed in a different mode, namely layer splitting.

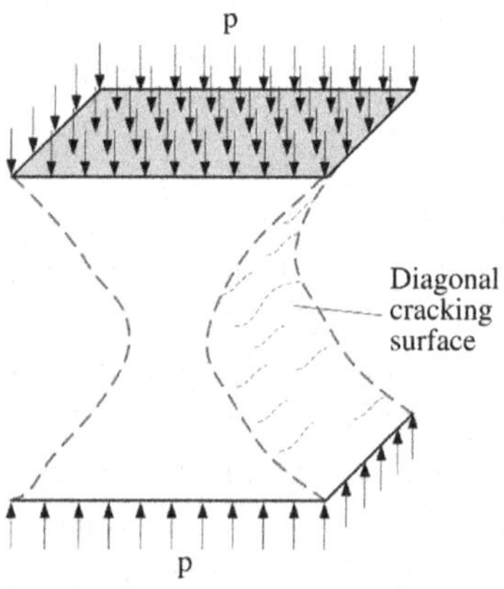

Figure 9.6 Sketch of failure mode of cubic specimens under compression.

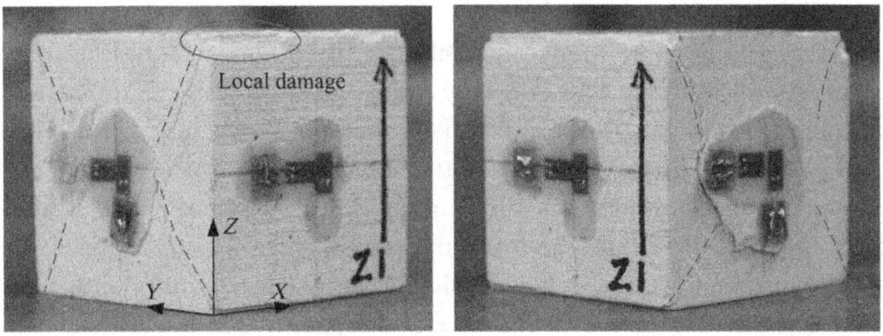

Figure 9.7 Failure of a typical cubic specimen under compression in the *Z* direction.

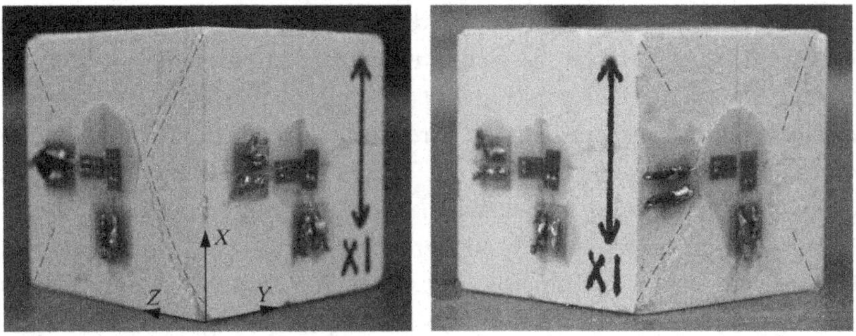

Figure 9.8 Failure of a typical cubic specimen under compression in the *X* direction.

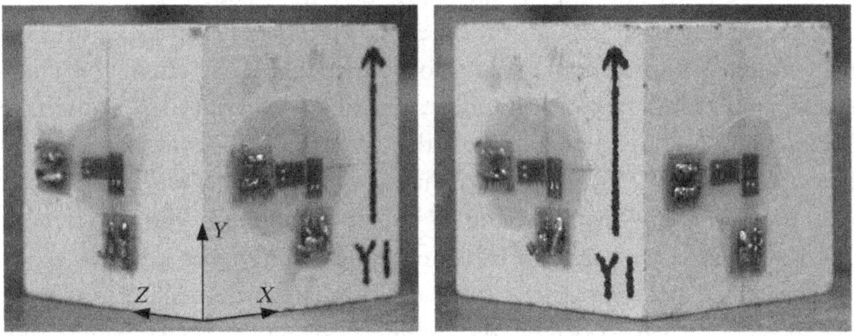

Figure 9.9 Failure of a typical cubic specimen under compression in the *Y* direction.

Under *Y*-direction compression for the small cubes, the behavior was similar initially but the diagonal cracks in the *YZ* plane were finer and developed much more slowly (Fig. 9.9). If the specimens were continuously loaded post peak, the hourglass cracks on the two opposite sides (in the *YZ* plane), as shown in Fig. 9.6, still developed.

9.3.1.4 Results and Analysis

The compressive cube strength and the elastic modulus of the specimens are listed in Table 9.1. The Poisson's ratio is given in Table 9.2. The elastic modulus and the Poisson's ratio of the specimens were determined from the initial linear segment of the experimental stress–strain curve. The elastic modulus was the slope of the linear segment of the vertical stress–strain curve, while the Poisson's ratio was the negative ratio of the horizontal to the vertical strain. Table 9.1 shows that the modulus in the *X* direction is the highest, that in the *Z* direction is the lowest, and with that in the *Y* direction lies in between.

The test results show that the compression strength in the *X* direction is higher than that in both *Y* and *Z* directions. According to the printing process, the specimens can be seen as many thin strips overlaying one after another (Fig. 9.10). The printing speed is high in the *X* direction and the time of printing adjacent strips (*Y* direction) is much shorter than that of printing adjacent layers (*Z* direction). The bond between two parts of the material appears to be higher when they are printed within a shorter period of time, resulting in a higher strength inside a continuous strip than that between strips which is, in turn, higher than that between layers. When loaded in the *X* direction, the cracks pass through interlayer interfaces (vertical component of crack path) and strips (horizontal component of crack path), giving the highest strength in the three directions as the bond strength within strips is the highest. When loaded in the *Y* direction, the cracks pass through the interlayer interfaces (vertical) and interstrip interfaces (horizontal). When loaded in the *Z* direction, the cracks pass through the interstrip interfaces (vertical) and interlayer interfaces (horizontal). As shown in Fig. 9.6, the crack faces have an angle higher than 45 degrees (to horizontal) so that the vertical path is slightly longer than the horizontal path, giving a higher strength in the *Z* direction because the interstrip strength is higher than the interlayer strength.

The tensile behavior in the three directions may be deduced following this analyses. When loaded in the *X* direction, tensile cracks would most likely appear in the *YZ* plane which would be dependent on the bond strength within the strips. Similarly, when loaded in the *Y* direction, cracks would appear in the *XZ* plane which would be dependent on the interstrip bond strength, and when loaded in the *Z* direction, cracks would appear in the *XY* plane which depends on the interlayer bond strength. Therefore, the tensile strength would be the highest in the *X* direction and the lowest in the *Z* direction, with the *Y* direction lies in between.

Table 9.1 Compressive strength and elastic modulus of cubic specimens

	Specimen	Compressive strength (MPa)			Elastic modulus (GPa)		
		Value	Average	Standard deviation	Value	Average	Standard deviation
70.7 mm specimens	CX-B4	11.5	11.2	1.92	3.5	3.6	0.26
	CX-B5	10.9			3.9		
	CX-B6[a]	7.92			3.4		
	CZ-B1	7.89	7.23	0.64	1.8	1.9	0.31
	CZ-B2	7.18			2.2		
	CZ-B3	6.61			1.6		
50 mm specimens	CX1	16.5	16.8	0.38	6.4	7.1	0.61
	CX2	16.6			7.2		
	CX3	17.2			7.6		
	CY1	13.4	11.6	1.63	7.0	5.8	1.15
	CY2	10.2			4.7		
	CY3	11.3			5.7		
	CZ1	12.1	13.2	1.04	4.8	4.9	0.36
	CZ2	13.9			4.6		
	CZ3	13.9	5.3				

[a]Excluded as it had printing defects.

Table 9.2 Poisson's ratio of cubic specimens

	Specimen	Poisson's ratio	Average	Poisson's ratio	Average
70.7 mm specimens		γ_{ZY}		γ_{ZX}	
	CZ-B1	0.138	0.166	Undetermined	
	CZ-B2	0.129			
	CZ-B3	0.232			
		γ_{XY}		γ_{XZ}	
	CX-B4	0.261	0.270	Undetermined	
	CX-B5	0.232			
	CX-B6	0.318			
50 mm specimens		γ_{XY}		γ_{XZ}	
	CX1	0.307	0.285	0.267	0.311
	CX2	0.302		0.303	
	CX3	0.275		0.362	
		γ_{YX}		γ_{YZ}	
	CY1	0.302	0.250	0.232	0.306
	CY2	0.241		0.390	
	CY3	0.201		0.325	
		γ_{ZX}		γ_{ZY}	
	CZ1	0.166	0.147	0.133	0.157
	CZ2	0.147		0.168	
	CZ3	0.127		0.169	

The test results show that the smaller specimens have higher compression strength and elastic modulus than those of the larger ones. There might be two main causes for this phenomenon: curing time and size effect. The curing time was about 1 month for the larger samples but about 6 months for the smaller ones. In terms of size effect, it takes longer to print larger specimens and consequently the stability of 3D printing process may reduce as the nozzle may be slightly blocked which leads to poorer bonding. Additionally, the larger specimens may also have more defects statistically which also leads to lower strength and modulus. However, there was little difference in the failure process and failure modes between the larger and smaller specimens.

There are several "outliers" in the test data, such as the Poisson's ratio of CZ-B3 and the compression strength of CX-B6. This are likely due to

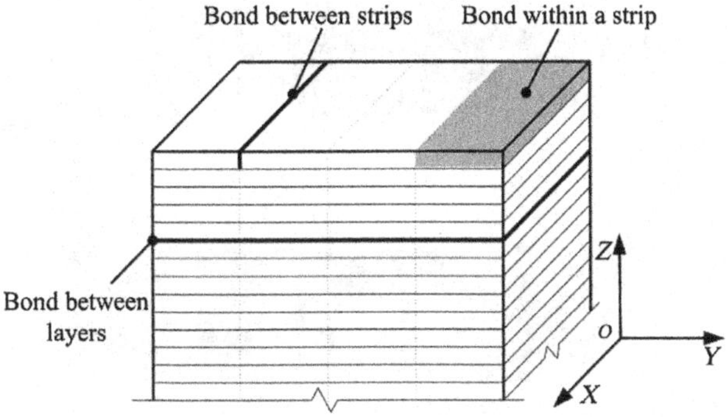

Figure 9.10 Structure of a 3D-printed cubic specimen.

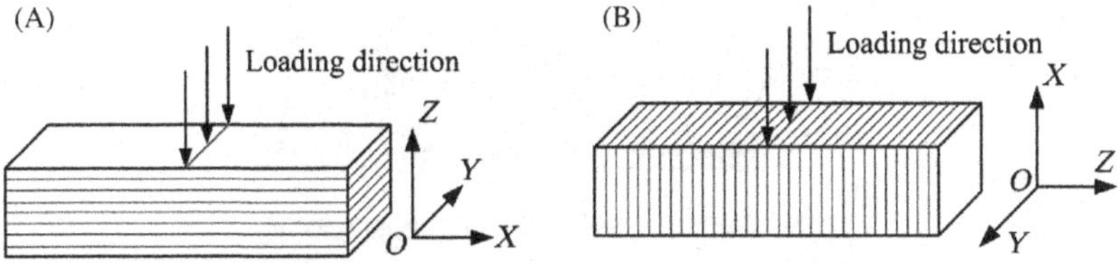

Figure 9.11 Loading direction of flexural specimens: (A) *BZ* loading; and (B) *BX* loading.

defects arising from the printing process, such as inconsistency in the binder content in different parts of the specimen.

From this analysis, it can be concluded that both the size of specimens and loading direction have an effect on the mechanical properties, including strength, elastic modulus, and Poisson's ratio. The 3D-printed materials are apparently anisotropic. The printing procedure ensures that such material has an orthotropic structure.

9.3.2 Flexural Test

9.3.2.1 Specimens

The flexural tests were conducted with 40 mm × 40 mm × 160 mm specimens. The specimens were divided into two groups of three, one loaded in the *Z* direction (Fig. 9.11A) to study the tensile properties within strips and the other loaded in the *X* direction to study the interlayer bond properties (Fig. 9.11B). Each specimen was assigned a unique designation, starting with F (for flexural), followed by either *Z* (for *Z*-direction loading) or *X* (for *X*-direction loading), then followed by "-An" with *n* being the specimen number (1–6).

Figure 9.12 Test setup of the flexural test.

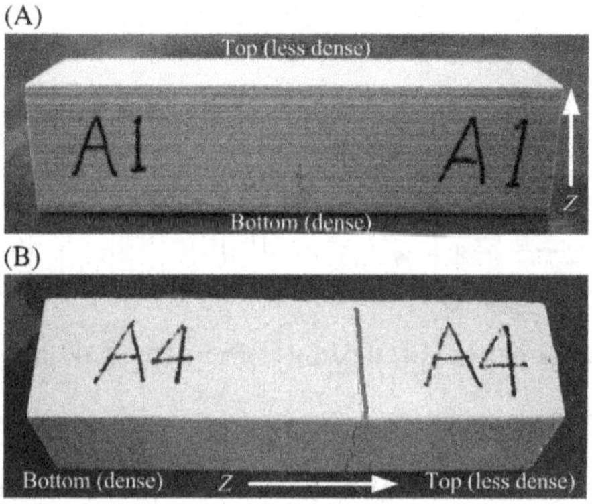

Figure 9.13 Flexural failure of specimens when: (A) loaded in the *Z* direction; and (B) loaded in the *X* direction.

9.3.2.2 Test

The flexural tests were conducted following the Chinese national standard GB/T 17671-1999 for testing the strength of cements [31]. A typical test setup is shown in Fig. 9.12. The applied loading rate was about 50 N/s. The flexural strength was determined based on the linear elastic analysis using the first crack load.

The deformation was not obvious until the maximum load was attained. Once the maximum load was reached, the bottom of the specimen was cracked suddenly leading to a sharp drop of the load and the test was stopped. The failure modes were different for the two groups of specimens loaded in different directions (shown in Fig. 9.13) in two aspects:

1. The development of cracks was different. When loaded in the *X* direction, the cracks propagated suddenly through almost the whole section once it cracked. However, when loaded in the *Z* direction,

the crack developed slower than that in the X direction, with the crack length less than half of the cross-section depth correspondingly.

2. The position of the cracks was also different. When loaded in the Z direction, the cracks appeared at the middle of the specimen. But when loaded in the X direction, the cracks deviated slightly from the middle to the part of the specimen printed late due to nonuniformity caused by the printing process (Fig. 9.13B) which took over ten hours for each specimen.

9.3.2.3 Results and Analysis

The peak loads from the flexural tests were used to obtain the flexural strength, R_f, according to:

$$R_f = \frac{1.5F_fL}{b^3} \tag{9.1}$$

where F_f is the maximum load, L is the distance between two supports, b is the side length of a square cross-section.

The calculated flexural strengths are shown in Table 9.3. It can be seen that when loaded in the Z direction, the flexural strength is one order higher than that loaded in the X direction. The bond strength inside a strip is higher than that between layers. When loaded in the Z direction, the flexural strength depends on the bond strength inside a strip, in contrast, it relies on the interlayer bond strength when loaded in the X direction.

9.3.3 Compression Test After Relevant Flexural Test

9.3.3.1 Specimen

The flexural test specimens were continuously flexed after reaching the peak load until a specimen was separated into two halves completely. The

Table 9.3 Flexural strength

Specimen	Flexural strength (MPa)	Average (MPa)	Standard deviation (MPa)
FZ-A1	4.14	4.12	0.19
FZ-A2	4.30		
FZ-A3	3.93		
FX-A4	0.326	0.365	0.06
FX-A5	0.404		
FX-A6	0.155[a]		

[a]Data excluded as it lies out of ± 10% of the average, reported average is that of the remaining two following the testing standard [31].

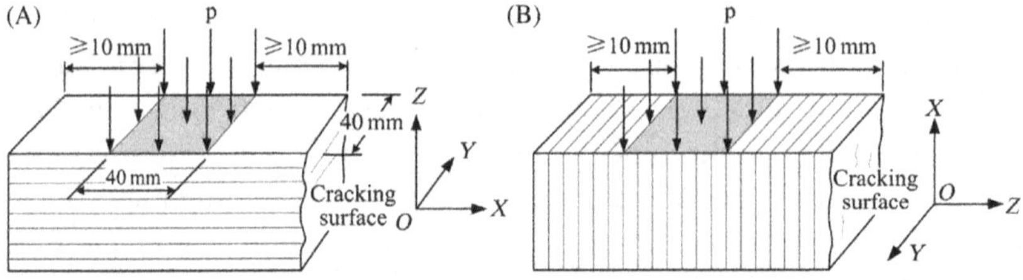

Figure 9.14 Loading direction of half beams: (A) *CZ* loading; and (B) *CX* loading.

two halves of each specimen after each flexural test were used to conduct a compression test. When loaded in the *Z* direction during the flexural test, the crack was located in the middle of the beam so the two halves were almost the same size. When loaded in the *X* direction, the crack was slightly away from the middle, but the tests still satisfied the size requirement that there is at least 10 mm outside the loading area in the longitudinal direction of the specimen (Fig. 9.14).

9.3.3.2 Test

The compression test was conducted following the Chinese standard GB/T 17671-1999 [31]. The loading area was 40 mm × 40 mm with an edge at least 10 mm wide outside the loading area in the longitudinal direction, as shown in Fig. 9.14. The loading rate was about 1.2 kN/s. The compression strength is calculated from the maximum load divided by the compression area.

When loaded in the *Z* direction, there was little deformation at first. As the load increased, gradual local crushing occurred in the loaded area. When approaching the ultimate load, diagonal cracks appeared, beginning from the edges of the loaded zone and propagating diagonally in the *XZ* plane (Fig. 9.15A). However, when loaded in the *X* direction, there was little noticeable compression deformation in the loaded area. The failure mechanism was mostly interlayer shear failure resulting in the loaded portion pushed out of the specimen, with short diagonal cracks occurring only in a few specimens (Fig. 9.15B).

9.3.3.3 Results and Analysis

The compression test results on the half beams are shown in Table 9.4. Note that each flexural specimen was broken into two halves, giving two compression test specimens. Table 9.4 shows that the specimens had higher strength when loaded in the *Z* direction than that in the *X* direction,

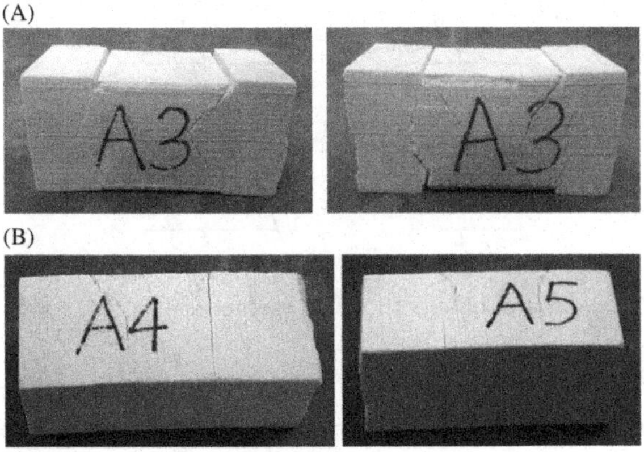

Figure 9.15 Failure mode of half beam tested under compression: (A) loaded in the *Z* direction; and (B) loaded in the *X* direction.

Table 9.4 Compressive strength of the half beams

Specimen	Compressive strength (MPa)		Average (MPa)	Standard deviation (MPa)
CZ-A1	7.84	7.88		
CZ-A2	7.91	7.81	7.79	0.16
CZ-A3	7.83	7.48		
CX-A4	5.38	2.46		
CX-A5	6.38	3.36	N/A[a]	N/A[a]
CX-A6	5.19	1.87		

[a]Scatters too large so not processed.

which is completely opposite to the compression strength test result of cubic specimens. This is mainly because the two tests were conducted following different methods, leading to the different failure modes. In the compression test of half beams, the specimen was partially loaded (Fig. 9.16). This led to interlayer shear failure along the boundaries of the loaded area when loaded in the *X* direction. The failure load mainly depended on the interlayer shear strength which, in turn, relied on the interlayer bond strength, the lowest bond strength of the three directions. Therefore, the failure load in the *X* direction is lower than that in the *Z* direction. However, in the compression test of cubes, the specimens were loaded uniformly, leading to higher compression strength in the *X* direction without shear failure (Fig. 9.16).

Table 9.4 also shows that the test results are stable and consistent when loaded in the *Z* direction, but have very high variability when loaded in

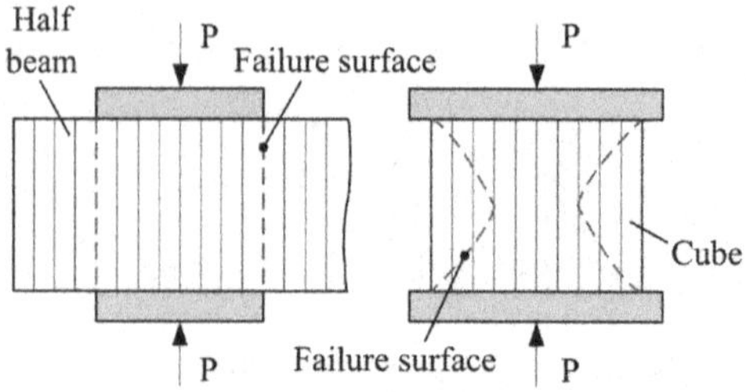

Figure 9.16 Comparison of different loading methods.

the *X* direction. It may be that the bond of the interlayer interface varied a lot more than that of the interstrip bond within a layer and the bond within a strip. Another major contributing factor may be the variation of printing quality. The specimen was 160 mm long in the *Z* direction for the flexural specimens, so it took over 10 hours to print one specimen. At the beginning, the printing quality was very good and the binder content was stable. However, as there are often built-ups in the nozzle as time increases which leads to reduced printing quality as the result of less densely printed material and poorer interlayer bond. This can explain the lower compression strength of the later printed half of the flexural specimens when loaded in the *X* direction.

9.4 MECHANICAL MODELS

9.4.1 Stress–Strain Relationship

The cube test results were analyzed for developing compressive stress–strain relationships in different directions. Fig. 9.17 shows that each curve is almost linear initially, but softens gradually as the stress increases. No apparent yield or horizontal segment is evident. Once it reaches the maximum load, it fails in a brittle manner. No descending branch was obtained because the tests were conducted under load control.

Based on these observations, a parabolic curve may be used to describe the stress–strain relationship:

$$\sigma = \frac{f_0 - E_0\varepsilon_u}{\varepsilon_u^2}\varepsilon^2 + E_0\varepsilon \tag{9.2}$$

where σ is stress under uniaxial compression; f_0 is the maximum stress; E_0 is the initial elastic modulus; ε is the strain under uniaxial compression; and ε_u is the ultimate strain, $\varepsilon_u = 2f_0/E_0$ if ε_u is greater than $2f_0/E_0$.

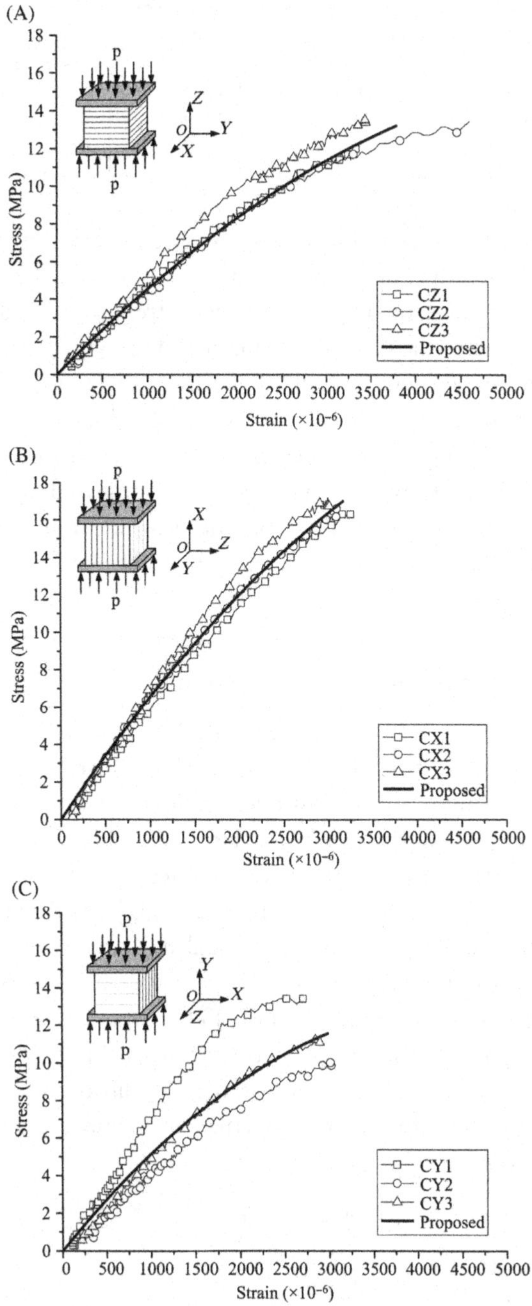

Figure 9.17 The proposed uniaxial stress–strain curve versus test results: (A) loaded in the *Z* direction; (B) loaded in the *X* direction; and (C) loaded in the *Y* direction.

Table 9.5 Values of parameters for the proposed stress–strain curve

	Loading direction	f_0 (MPa)	E_0 (GPa)	ε_u ($\times 10^{-6}$)
50 mm specimens	X	16.8	7.1	3100
	Y	11.6	5.8	3000
	Z	13.2	4.9	3800

Although the frictional restraints of the loading platens cannot be avoided and enhance the compressive strength in the cube test, the cube strength is used as the reference uniaxial compressive strength. The average values of f_0, E_0, and ε_u for the smaller specimen group which was tested in all three directions are shown in Table 9.5. If they are used in Eq. (9.2), the results are in close agreement with the test results (Fig. 9.17).

In tension, the 3D-printed specimens may be assumed linear-elastic brittle, similar to what appeared in the flexural tests. All specimens failed on the tension side in a brittle fashion. The flexural strength is used as the equivalent uniaxial tensile strength in this study before more test data becomes available.

9.4.2 Failure Criterion

The above test results have shown remarkable orthotropy of 3D-printed cementitious products. A failure criterion suitable for orthotropic materials is required for failure analysis of such structures. Several failure criteria are available for orthotropic materials, such as Hoffman's [32] or Tsai-Wu's [33] criterion. Since only uniaxial tests were conducted in this study, the maximum stress criterion may be suitable. Hoffman's [32] criterion can be defined by uniaxial test data, but there are no biaxial test data to verify its validity. Tsai-Wu's [33] criterion requires biaxial tests to determine the parameters. Therefore, the maximum-stress criterion [34] is used here to describe the mechanical behavior of the 3D-printed specimens. In the maximum stress criterion, the stresses in the principal material coordinates must all be smaller than the respective strengths to avoid fracture. It is expressed as:

for tensile stresses,

$$\sigma_1 < X_t,\ \sigma_2 < Y_t,\ \sigma_3 < Z_t \tag{9.3}$$

for compressive stresses,

$$\sigma_1 > X_c,\ \sigma_2 > Y_c,\ \sigma_3 > Z_c \tag{9.4}$$

and for shear stresses,

$$|\sigma_{12}| < S_{12},\ |\sigma_{23}| < S_{23},\ |\sigma_{31}| < S_{31} \tag{9.5}$$

where σ_i ($i = 1, 2, 3$) is the stress in the principal material coordinate (i.e., the coordinate system defined for the 3D printing), where directions 1, 2 and 3 correspond to the *X*-, *Y*- and *Z*- directions respectively in this study, σ_{12}, σ_{23}, σ_{31} are shear stresses in the respective planes of the principal material coordinates, X_i, Y_i, Z_i ($i = t, c$) are the maximum allowable stresses in different directions in tension ($i = t$) and compression ($i = c$), and S_{12}, S_{23}, S_{31} are the maximum allowable shear stresses in the different planes. X_i, Y_i, Z_i ($i = t, c$) and S_{12}, S_{23}, S_{31} can be determined by nine simple uniaxial tests, including tension tests and compression tests in three directions and shear tests in three planes.

The compression strengths of the smaller specimens are used here as an example for the maximum stress criterion because the tests of the smaller cubes were more complete with data from loading in all three directions. To determine the shear strength, a simple direct double shear test was conducted for short rectangular beams as shown in Fig. 9.18 following the standard test for concrete [35]. Test data were validated only when the failure occurred along the preferred failure plane. The resulting shear strengths were 0.628 MPa in the *XY* plane, 4.63 MPa in the *YZ* plane, and 1.264 MPa in the *XZ* plane. The flexural strength in the *X* and *Z* directions may be treated approximately as the direct tensile strengths. The *Y* direction tensile strength was not directly available in this study. However, the tensile strength is greater in the *X* direction than that in the *Y* direction which is, in turn, greater than that in the *Z* direction, so the tensile strength in the *Y* direction was assumed to be 1 MPa

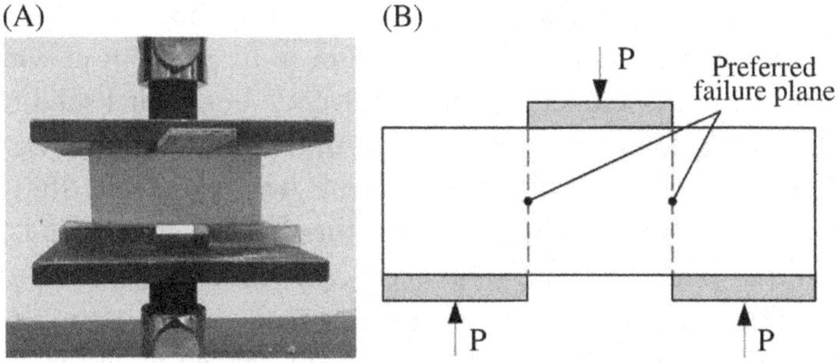

Figure 9.18 Direct double shear test setup and instrumentation: (A) Photograph of specimen; and (B) schematic view.

in the example here. The values of the maximum allowable strengths in the maximum stress criterion are summarized as:

$$X_t = 4.12\text{ MPa}, X_c = -16.8\text{ MPa}, Y_t = 1\text{ MPa}, Y_c$$
$$= -11.6\text{ MPa}, Z_t = 0.365\text{ MPa}, Z_c = -13.2\text{ MPa}$$
$$S_{12} = 0.628\text{ MPa}, S_{23} = 4.632\text{ MPa}, S_{31} = 1.264\text{ MPa}$$

9.4.3 Application of the Constitutive Model

For constructing practical engineering structures, either 3D-printing prefabrication or in situ 3D printing construction may be adopted based on various considerations. For 3D printing prefabrication, it is convenient if the printing process is vertical, that is, the Z direction is perpendicular to the printed layers. However, loading in a given printing direction is not necessarily always ideal, so the loading direction can be controlled by printing elements in different orientations (i.e., printing a column "laying down") and then positioning it vertically in the structure. Because of the orthotropic nature of the 3D printed structures, the relationship between loading direction and printing direction can be important: a careful selection of these angles based on structural analysis may significantly enhance the behavior of the structure. For in situ 3D printing construction, the printing direction is already determined by the printing principle, so it is necessary to optimize the structural design based on the mechanical properties to ensure safety and economy. The relationship between mechanical behavior and printing directions can be evaluated through a simple example.

A finite element analysis (FEA) of an example of a 100 mm thick arch structure was conducted using the maximum stress criterion with the constants listed. The FEA was conducted using MSC Marc 2007 to assess the influence of printing direction on the structural performance. The example structure, as shown in Fig. 9.19A, is 1 m in height, 1 m in width and 3 m in length. It was modeled using the 8-node hexahedral solid element (7 # solid), with the above properties for the orthotropic material constitutive law. The total number of elements was 1560 and the size of elements was about 33 mm. The structure was analyzed under dead weight and a vertical surface (not necessarily uniform) load on the top platform. The boundaries at the supports were fixed (i.e., restrained in all directions). In the model, the dead weight was applied to all the elements as an initial condition and the surface load was applied through

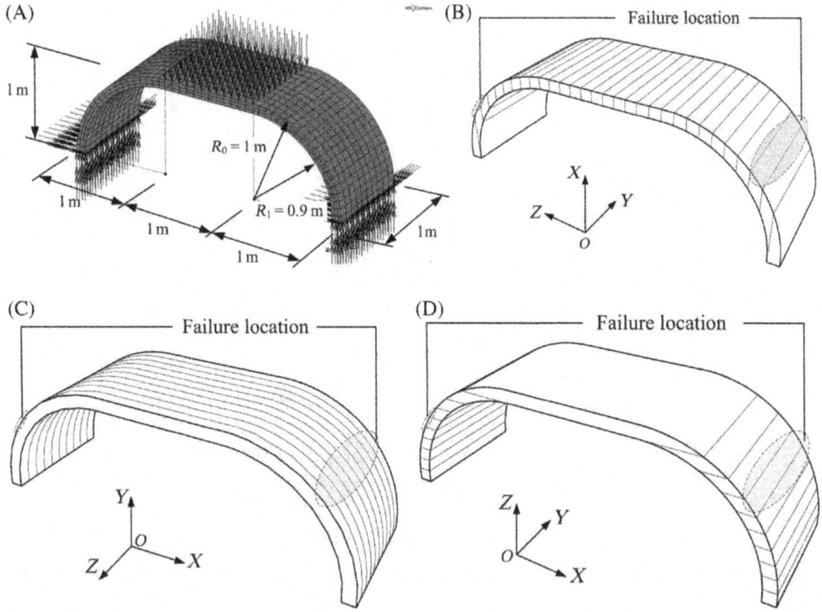

Figure 9.19 Finite element models and different loading directions: (A) boundary conditions and loading; (B) loaded in the *X* direction and failure position; (C) loaded in the *Y* direction and failure position; and (D) loaded in the *Z* direction and failure position.

displacement control to the nodes of the loaded surface. The progressive failure analysis option was turned on (with a residual stiffness coefficient of 0.01) to achieve gradual degradation of the elastic modulus during the failure process. The loading process was conducted using the full Newton-Raphson method. The element coordinate system was changed in each analysis to simulate the different printing directions of the structure, as shown in Fig. 9.19B–D.

Fig. 9.19B–D also illustrate the failure locations, which are similar for all cases. Fig. 9.20 shows that when loaded in the *X* direction, the maximum loading capacity of the structure is the lowest with an average surface load of 24.5 kN/m^2; when loaded in *Z* direction it is 33.3 kN/m^2; and when loaded in *Y* direction, the capacity is the highest at 69.4 kN/m^2. The results show that the printing direction has a very significant effect on the overall load-bearing capacity of the structure. This confirms the importance of choosing a suitable printing direction in the case of 3D prefabrication.

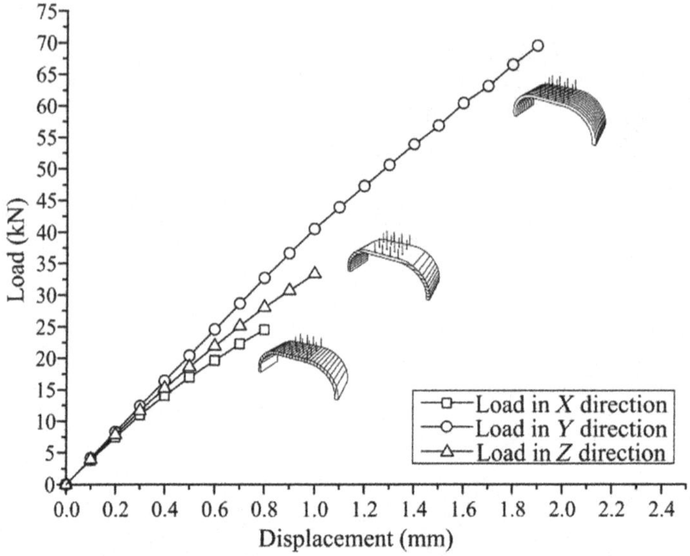

Figure 9.20 Load-displacement curves of the structure when constructed in different directions.

9.4.4 Potential Developments and Applications of 3D Printing

The test data indicate that the structures 3D printed with cementitious plaster powders in this study are not suitable for structural members because of their brittleness and low strength. However, they may be used (and indeed have found applications) for nonstructural components such as printing building decorations and formworks for nonlinear architectural features. Nevertheless, the layered microscopic characteristics and orthotropic mechanical behavior are representative of 3D-printed structures with cementitious powders. Further study should be conducted to use 3D printers using cement or other materials rather than plaster powders to produce high-strength structures. It would also be desirable to either add reinforcements during the printing process or use the 3D printed material in combination with other materials such as steel and Fiber Reinforced Polymer (FRP) composites; the latter has found wide applications in strengthening concrete structures [36]. A preliminary study has shown that the structural capacity and ductility can be significantly improved for 3D-printed columns and beams when strengthened with FRP.

The 3D printer used in this study is an industrial grade machine suitable for printing small objects precisely. Consequently, the cost for printing these specimens was high (about US$0.50 per cubic centimeter). However, with the development of large 3D printers and suitable materials specifically designed for printing structures such as buildings, there is no doubt that the cost will decline dramatically.

In summary, 3D printing technology has the potential to provide an effective method to achieve low-cost, highly efficient, automatic construction with reduced waste, labor cost, and construction risks. It can significantly speed up the construction process.

9.5 CONCLUSIONS

This paper has presented a study on the characteristics of 3D-printed layered cementitious material through microscopic observation, mechanical tests on cubes and small beams, and FE modeling on the effect of the construction process on structural behavior. The basic mechanical parameters and failure characteristics in different directions were obtained through compression tests of cubes and flexural tests of small-scale beams. In addition, a model for the stress–strain relationship is proposed based on the uniaxial test stress–strain curves under compression. On the basis of the test data, parameters for the maximum stress criterion were also defined for the 3D-printed orthotropic cementitious material. These relationships were deployed in a FE analysis of a 3D-printed arch structure to investigate the effect of printing direction on the load-carrying capacity. The main conclusions are:

1. The layer-atop-layer printing process results in the printed material of layered microscopic structures with each layer further consisting of bonded strips, leading to an apparent orthotropic behavior.
2. All 3D printed cubes had a similar failure mode with hourglass-shaped cracking on opposite sides when loaded in the *X*, *Y*, and *Z* directions, but the compression strength and elastic modulus were the highest when loaded in the *X*-direction (printer head travel direction).
3. Based on the uniaxial compression test results, the stress–strain relationship of the 3D-printed cementitious material may be described by a quadratic model.
4. Based on the maximum stress criterion adapted for the 3D-printed orthotropic material, a FEA of a thin shell structure was conducted. The results have shown that the printing direction has a significant effect on the load-bearing capacity of the structure.

ACKNOWLEDGMENTS

The authors acknowledge funding support from the National Natural Science Foundation of China (No. 51278276), the National Basic Research Program of China (973 Program, No. 2012CB026200). Peng Feng also acknowledges funding support from the Beijing Higher Education Young Elite Teacher Project (YETP0078) and Tsinghua University Initiative Scientific Research Program (No.20111081015).

REFERENCES

[1] H. Lipson, M. Kurman, Fabricated: The New World of 3D Printing., John Wiley & Sons, 2013.
[2] D. Dimitrov, K. Schreve, N. De Beer, Advances in three dimensional printing – state of the art and future perspectives, Rapid Prototyping J. 12 (3) (2006) 136–147.
[3] B. Berman, 3-D printing: the new industrial revolution, Business Horizons 55 (2) (2012) 155–162.
[4] C.X.F. Lam, X.M. Mo, S.H. Teoh, D.W. Hutmacher, Scaffold development using 3D printing with a starch-based polymer, Mater. Sci. Eng. 20 (1) (2002) 49–56.
[5] T.F. Wegrzyn, M. Golding, R.H. Archer, Food layered manufacture: a new process for constructing solid foods, Trends Food Sci. Technol. 27 (2) (2012) 66–72.
[6] X. Yan, P. Gu, A review of rapid prototyping technologies and systems, Comput. Aided Des. 28 (4) (1996) 307–318.
[7] M.J. Cima, J.S. Haggerty, E.M. Sachs, P.A. Williams, Three-dimensional printing techniques: U.S. Patent 5,204,055. 1993-4-20.
[8] I. Gibson, D. Shi, Material properties and fabrication parameters in selective laser sintering process, Rapid Prototyping J. 3 (4) (1997) 129–136.
[9] P. Kulkarni, D. Dutta, Deposition strategies and resulting part stiffnesses in fused deposition modeling, J. Manuf. Sci. Eng. 121 (1) (1999) 93–103.
[10] L.J. Hornbeck, Digital light processing update: status and future applications. Electronic Imaging'99, Int. Soc. Opt. Photon. (1999) 158–170.
[11] K. Ikuta, K. Hirowatari, T. Ogata, Three dimensional micro integrated fluid systems (MIFS) fabricated by stereo lithography, in: Micro Electro Mechanical Systems, 1994, MEMS'94, Proceedings, IEEE Workshop on. IEEE, 1994, pp. 1–6.
[12] K.K. Jurrens, Standards for the rapid prototyping industry, Rapid Prototyping J. 5 (4) (1999) 169–178.
[13] B. Utela, D. Storti, R. Anderson, M. Ganter, A review of process development steps for new material systems in three dimensional printing (3DP), J. Manuf. Process. 10 (2) (2008) 96–104.
[14] R. Singh, An overview of three dimensional printing for casting applications, Int. J. Precision Technol. 2 (1) (2011) 93–116.
[15] R. Singh, Three dimensional printing for casting applications: a state of art review and future perspectives, Adv. Mater. Res. 83 (2010) 342–349.
[16] S. Lim, R.A. Buswell, T.T. Le, S.A. Austin, A.G. Gibb, T. Thorpe, Developments in construction-scale additive manufacturing processes, Autom. Constr. 21 (2012) 262–268.
[17] D-Shape. http://www.d-shape.com.
[18] Stone spray project. http://www.stonespray.com/.
[19] Contour crafting. http://www.contourcrafting.org/.
[20] B. Khoshnevis, Automated construction by contour crafting—related robotics and information technologies, Autom. Constr. 13 (1) (2004) 5–19.

[21] B. Khoshnevis, D. Hwang, K. Yao, Z. Yeh, Mega-scale fabrication by contour crafting, Int. J. Ind. Syst. Eng. 1 (3) (2006) 301–320.
[22] B. Khoshnevis, M. Thangavelu, X. Yuan, J. Zhang, Advances in contour crafting technology for extraterrestrial settlement infrastructure buildup, AIAA SPACE 2013 Conference and Exposition, 2013.
[23] 3D concrete printing. http://www.buildfreeform.com/.
[24] R.A. Buswell, R.C. Soar, M. Pendlebury, A.G. Gibb, Investigation of the potential for applying freeform processes to construction, 2005. https://dspace.lboro.ac.uk/2134/10144.
[25] R.A. Buswell, R.C. Soar, A.G. Gibb, A. Thorpe, Freeform construction: mega-scale rapid manufacturing for construction, Autom. Constr. 16 (2) (2007) 224–231.
[26] R.A. Giordano, B.M. Wu, S.W. Borland, L.G. Cima, E.M. Sachs, M.J. Cima, Mechanical properties of dense polylactic acid structures fabricated by three dimensional printing, J. Biomater. Sci., Polym. Ed. 8 (1) (1997) 63–75.
[27] S.H. Ahn, M. Montero, D. Odell, S. Roundy, P.K. Wright, Anisotropic material properties of fused deposition modeling ABS, Rapid Prototyping J. 8 (4) (2002) 248–257.
[28] C.S. Lee, S.G. Kim, H.J. Kim, S.H. Ahn, Measurement of anisotropic compressive strength of rapid prototyping parts, J. Mater. Process. Technol. 187 (2007) 627–630.
[29] Spectrum Z510/DesignmateTM CX 3D Printer, User Manual Rev X, Z Corporation, 2006.
[30] GB/T50081-2002 Standard for test method of mechanical properties on ordinary concrete. Beijing, 2003.
[31] GB/T 17671-1999 Method of testing cements-Determination of strength. Beijing, 1999.
[32] O. Hoffman, The brittle strength of orthotropic materials, J. Compos. Mater. 1 (2) (1967) 200–206.
[33] S.W. Tsai, E.M. Wu, A general theory of strength for anisotropic materials, J. Compos. Mater. 5 (1) (1971) 58–80.
[34] R.M. Jones, Mechanics of Composite Materials., CRC press, 1998.
[35] CECS13: 89 Test Methods Used for Steel Fiber Reinforced Concrete, Beijing, 1996.
[36] J.G. Teng, J.F. Chen, Mechanics of debonding in FRP-plated RC beams, Proc. Inst. Civil Eng. – Struct. Build. 62 (5) (2009) 335–345.

CHAPTER 10

Factors Influencing the Mechanical Properties of Three-Dimensional Printed Products From Magnesium Potassium Phosphate Cement Material

Xiangpeng Cao[1] and Zongjin Li[2]
[1]Shenzhen Mingyuan Building Technology Co., Ltd., Shenzhen, P.R. China
[2]Institute of Applied Physics and Materials Engineering, University of Macau, Macau, P.R. China

10.1 INTRODUCTION

Three-dimensional printing (3D printing), also known as additive manufacturing (AM), are currently widely studied and developed, including the constructional domain, where 3D printing is beginning to move from an architect's modeling tool to delivering full-scale architectural components and elements of buildings such as walls and facades ([1], 262–268). The 3D printing process is an AM method from one dimensional dot/line to a two-dimensional plane, then to a three-dimensional object. All existing printing techniques could be classified into two categories: direct deposition and selective modeling. Direct deposition indicates the material is printed/jetted/extruded directly from the printing nozzle onto the printing bed layer-by-layer. Selective modeling means only part of the material in each layer is selected to react or become solid continuously to form a 3D object by various printing methods such as laser, light, or liquid jetting. The properties of the raw materials are highly relevant to the printing strategy of the selected 3D printing technique. Direct deposition requires the material properties as pumpability, printability, buildability, and open time for bonding of two neighbor layers, indicating the material could be pumped from a container, printed from a printing nozzle, and withstand its self-weight and continuous upper layers;

DOI: https://doi.org/10.1016/B978-0-12-815481-6.00010-5

while selective modeling needs property as open time and shows lower shrinkage ([1], 262–268).

The first attempt at using cement-based materials in an approach toward AM was proposed by Pegna ([2], 427–437) by steaming vapor to the Portland cement powder surface. Currently there are three typical large-scale AM processes aimed at construction and architecture in public domain. They are Contour Crafting ([3], 301–320) from University of South California, by extruding cement paste to build 3D object, D-Shape by Monolite UK Ltd., by jetting liquid to the cement powder, and Concrete Printing [4] from Loughborough University, by extruding cement paste; all three prove the successful manufacture of large-scale components and suitability for construction and/or architecture applications.

Several studies on microscale cementitious material printing have been published. Uwe Klammert printed a small size specimen (less than 10 mm) from powder mix of $Mg_3(PO_4)_2$, K_2HPO_4 and $(NH_4)_2HPO_4$ (with particle size less than 16 μm) by the commercial printer Z-Printer 310 developed by Z-Corporation achieving a maximum compression strength of 10 MPa (after postprocedure) ([5], 2947–2953). A maximum compression strength of 8.0 MPa has been tested from the printed specimens out of tri-calcium phosphate, silica, and zinc oxide by a commercial 3D printer R-1 (R&D Printer) ([6], 113–122). Various research on bone and dental replacement materials have been studied, for example, zinc oxide/acrylate cements([6], 113–122; [7], 99–113), and tri-calcium phosphate powder with bioactive glass ([8], 2563–2567).

Besides the large-scale and microscale 3D printing research, a small-scale homemade 3D printer has been successfully developed, installed, and printed specimens out of the cementitious material magnesium phosphate cement (MPC), which has properties which are very suitable for the 3D printing process, that is, rapid setting, low shrinkage, and high early strength ([9], 447–456; [10], 6281–6288; [11], 5058–5063; [12], 497–502). Major printing parameters in this prototype printer have been tested to find the best parameter combination with which the printing process would be fast and stable as well as (most importantly) the printed product being the same as the design. This chapter discusses the mechanical properties of the successfully printed result. Two parameter sets were used to print two groups of specimens to study the factors of the mechanical properties of the printed products.

10.2 EXPERIMENTAL PROGRAM

The homemade 3D printer, with a system diagram shown in Fig. 10.1, was built mainly from a frame of acrylic plates, five stepper motors, and an embedded controlling motherboard (Printer Control Board). Monitor software was coded and installed in the attached computer to control the printing process. A needle with an inner diameter 0.06 mm was used as the printing nozzle to supply a very stable jetting speed at 0.0150 mL/s under air pressure of 2.0 MPa. In order to print as fast as possible, the printing nozzle moves at the maximum speed of 100 mm/s, which leads to the narrowest printed line and the best printing accuracy for this printer.

The printing mechanism includes one printing nozzle which sprays the printing binder (water was used in this study) by compressed air via a reducing valve and five axes controlled by the control board which execute the printing movements.

1. Binder container: stores the printing binder and delivers the binder to the printing nozzle by the force of compressed air.
2. Printing nozzle: sprays the printing binder into the powder bed with compressed air.
3. X-axis mechanism: moves along x direction in fixed speed (unit: mm/s), carrying the printing nozzle, to print material lines.
4. Y-axis mechanism: moves along y direction, carrying *x*-axis and the printing nozzle to assemble the printed lines into a printed 2D plane

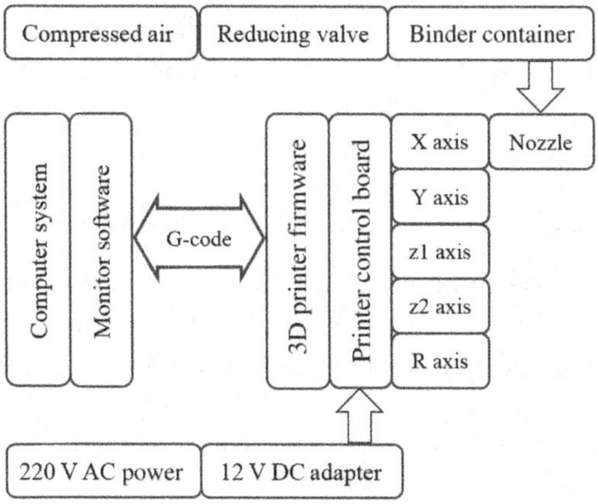

Figure 10.1 System diagram of the homemade 3D powder printer.

and carries a roller to help lay a new layer for the next printing procedure.

5. R-axis mechanism: rotates the roller to create a new powder layer and compresses it with the help of the Y and Z2 axis.
6. Z1-axis mechanism: moves upward to supply powder for R-axis to create a new layer in the printing bed.
7. Z2-axis mechanism: moves downward to receive the new powder layer from Z1-axis.

10.2.1 Material Preparation and Printing Process

Dead burnt magnesia powder, calcined at 1500°C for 5 hours with a purity of 95.1%, powder KDP, and deionized water were the raw materials for printing. The chemical composition of the dead burnt magnesia is shown in Table 10.1. Both magnesia oxygen and KDP were ball milled, oven-dried for 2 hours, and sieved under 300 μm, before dry-mixing for 5 minutes in a vertical-axis planetary mixer with an magnesia-to-phosphate molar ratio (M/P) of 8:1.

The fully mixed powder was then freely poured into the containing box uniformly, but without any compaction or compression (as the side-view of the printer shows in Fig. 10.2A) with a density of $\rho = 0.00155\ g/mm^3$. The deionized water was filled in the printing container under compressed air at 2.0 MPa. The mix powder was then rolled into the printing box to create a smooth powder layer with a thickness of z, from right to left, as shown in Fig. 10.2A. The deionized water was jetted from the printing nozzle to the powder surface with a stable speed v_{water} while the nozzle was moving along x dimension with a stable speed of v_x, as shown in Fig. 10.2B, top view of the printer. The jetted water and the water-selected powder form a line of MPC paste which starts hardening based on the known chemical reaction:

$$MgO + KH_2PO_4 + 5H_2O \rightarrow MgKPO_4 \bullet 6H_2O$$

After one line was printed, the nozzle moved in the Y dimension with a step of y and started to print the next line before the last line hardened, as shown in Fig. 10.2C, eventually forming a 2D layer line-by-line, which

Table 10.1 Chemical composition of the dead burnt magnesia oxygen (mass%)

MgO	SiO_2	CaO	Fe_2O_3	MnO
95.12	3.70	0.78	0.26	0.14

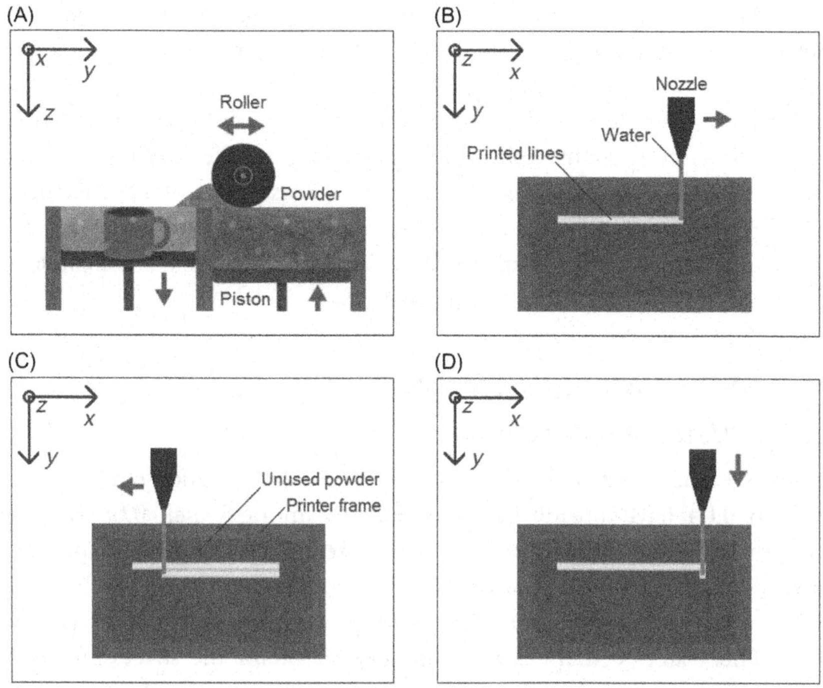

Figure 10.2 Steps of the 3D powder printing process: (A) rolling to create a new powder layer; (B) line printing along the *x* axis; (C) printing nozzle moving to next line along the *y* axis; and (D) the next line printing along the *x* axis.

can be considered as a group of MPC sticks, and then to a 3D object layer by layer. All of the sticks and layers were bonded together during the printing process.

The printed samples were taken out of the printing box half an hour after the printing job finished. Unreacted powder material was then removed thoroughly with a brush and vacuum cleaner and recycled for the next layer of printing.

10.2.2 Postprocedure

3D powder printing technique usually needs a postprocedure (burnt or cured) to improve the mechanical properties of the printed result since the previous printing process mainly focuses on the product as the same as the designed shape. In this study, all the printed samples were then put in the curing room for 1 hour before being cured in air for 3, 7, and 28 days for further tests.

Samples were cut into cubes of 30 mm × 30 mm × 30 mm for compression strength tests and 20 mm × 30 mm × 100 mm with a narrow neck in the middle for direct tensile strength tests. Both of the cutting and placing operations on the universal test machine were carefully carried out in order to guarantee that the compressive/tensile loading direction was parallel or perpendicular to the printed lines/layers of the specimen, when tested along x, y, or z axis. Three samples were tested to get the average data as the final result for each situation: a matrix of *compression/tension*, *layer thickness*, and *loading direction*.

10.3 RESULT AND DISCUSSION

10.3.1 Water–Cement Ratio

A series of designed tests with water–cement (w/c) ratios ranging from 0.15 to 0.04 have already been conducted, among which 0.05 was the best for the being-printed product which retains its designed shape—the printed object collapses when w/c is larger than 0.05 and less bonding ability (hard to take out from the printing box) when w/c is lower than 0.05. The water–cement ratio is the key factor for the successful printability of the material. So, both test parameter sets should guarantee a designed w/c around 0.05. The designed w/c ratio could be calculated according to the printing parameters. Considering in the printing time period t, the x axis moving at speed v_x, the y *axis* step movement y, the designed printing layer z, the binder jetting speed at v_{water}, and the prepared powder density ρ, the designed water cement ratio can be derived from the equation below. The printing parameters for this study and the calculated designed w/c ratio are listed in Table 10.2.

$$\text{designed } w/c = \frac{v_{\text{water}} \cdot t \cdot \rho_{\text{water}}}{(v_x \cdot t) \cdot y \cdot z \cdot \rho} = \frac{v_{\text{water}}}{v_x \cdot y \cdot z \cdot \rho} \quad (\rho_{\text{water}} = 1.0\ \text{g/mL})$$

10.3.2 Structure of the Printed Result

The average printing time for each sample (100 mm × 70 mm × 40 mm) is about 2.5 hours. A sandwich-like layer structure could be observed in Fig. 10.3 (right). A 3D model, shown in Fig. 10.3 (left) is built to help understand the structure and its forming steps: the final printed result is formed from a group of magnesium potassium phosphate cement sticks in one dimension, to a plane of sticks in two dimensions, and eventually to a stack of planes in three dimensions. Each step of the printing is precisely

Table 10.2 Printing parameters and the designed water cement ratio for two test groups

Printing parameters		Test group 1	Test group 2
Water jetting speed	v_{water}	0.0150 mL/s	0.0150 mL/s
Nozzle moving speed	v_x	100.0 mm/s	100.0 mm/s
Powder density	ρ	0.00235 g/mm^3	0.00235 g/mm^3
Layer thickness	Z	2.1 mm	1.5 mm
Y step	Y	0.40 mm	0.65 mm
Designed water cement ratio	$wc_{designed}$	0.050	0.046

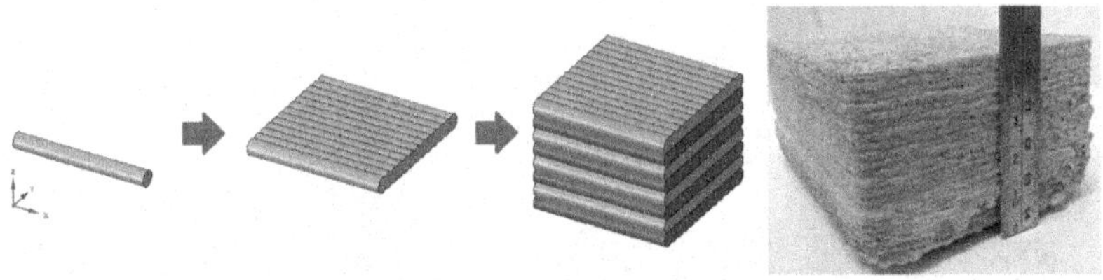

Figure 10.3 A computer aided design simulation of the printing process (left, light gray, and dark gray layers are the same) and the final printed result (right).

controlled by the printer. The dark and light layers in Fig. 10.3 (left) indicate that the different layers reveal same printed result.

The postprocedure helps the printed samples to get well-crystalized by introducing enough water to the specimen in the curing room, as the SEM photo shows in Fig. 10.4 (left). Inside the specimen there was still enough water due to the open and consecutive gaps, as shown in Fig. 10.4 (right), created by the printing process, a stick/layer assembling procedure, as in Fig. 10.3 (left). Meanwhile, gaps are the main reason for the variation of the density of the specimens; with more layers, the more gaps it has. The unit weight of two groups were 1750 kg/m^3 for samples of 2.1 mm layer thickness and 1733 kg/m^3 for samples of 1.5 mm layer thickness. The average layer thickness is 2.09 mm for test group 1, compared to the designed thickness at 2.1 mm, due to the mechanical error with an acceptable error of 0.5%.

10.3.3 Compression Strength

The compression strength of the specimen shows big variety when loading along different axes for both test groups due to the line-layer-body

Figure 10.4 Scanning electron micrograph of the 3D printed product using magnesium potassium phosphate cement. A well-crystalized part (left) and consecutive gaps inside the specimen (right).

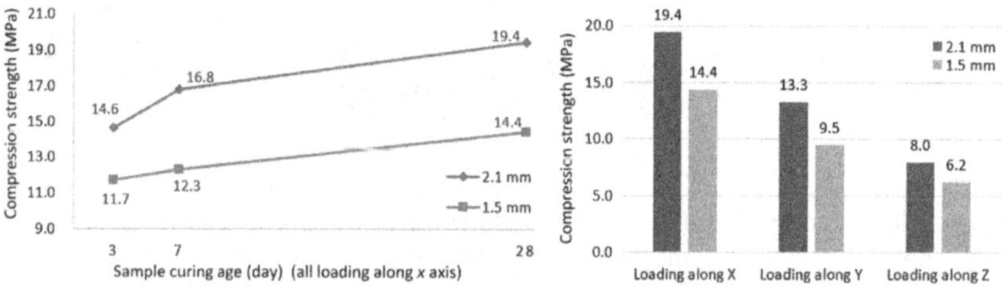

Figure 10.5 Compression strength (loading along *x* axis) at the age of 3-day, 7-day and 28-day (left) and the compression strength when loading along three axes at the age of 28 days (right).

structure, as shown in Fig. 10.5 (right). Compression strength obtains the highest value when the compressing loading is along the x axis. Printed sticks have a lower combination to get lower compression strength when loading along the y axis. The combination of two layers which have rough surfaces lead to the lowest compression strength. The compression strength in all three dimensions increase with curing age as the loading occurs along x axis, as shown in Fig. 10.5 (left).

For each sample group, let the c_{yx} be the compression strength along the y axis divided by that along x axis: $c_{yx} = \sigma_y/\sigma_x \ (0 < c_{yx} < 1)$, as the same as c_{zx}: $c_{zx} = \sigma_z/\sigma_x \ (0 < c_{zx} < 1)$. Both of them seem to be constant at average values of 0.67 and 0.42 with a low standard deviation in spite of different layer thickness, as shown in Table 10.3. This compression strength ratio describes the *combination degree* of the stick/layer joint in this printed structure, which is irrelevant to the layer thickness. Furthermore, under ideal conditions, if the combination degree reached 1, meaning the stick/layer joints were as the same as the stick itself, the whole specimen

Table 10.3 Comparison of compression strength between three dimensions

Layer thickness	$c_{yx} = \sigma_y/\sigma_x$	$c_{zx} = \sigma_z/\sigma_x$
2.1 mm	0.69	0.41
1.5 mm	0.66	0.43
Average	$\overline{c_{yx}} = 0.67$	$\overline{c_{zx}} = 0.42$
Standard deviation	$s_{c_{yx}} = 0.018$	$s_{c_{zx}} = 0.012$

Table 10.4 Comparison of the strength data between two layer thickness groups

Comparison of compression strength					Layer thickness
$\sigma_{x2.1}/\sigma_{x1.5}$	$\sigma_{y2.1}/\sigma_{y1.5}$	$\sigma_{z2.1}/\sigma_{z1.5}$	$\overline{\sigma_{2.1}/\sigma_{1.5}}$	Standard deviation	2.1/1.5
1.35	1.4	1.29	1.35	0.055	1.4

would be uniform in three dimensions as well as having compression strength in three dimensions $(\sigma_z = \sigma_y = \sigma_x)$ and the specimen would be considered as homogeneous as a mold-cast one.

On the other hand, the comparison of the strength of different layer thickness but in the same dimension seem to be equivalent compared to the layer thickness, as shown in Table 10.4, with a standard deviation 0.055, from which the conclusion can be drawn that layer thickness is a key factor to the compression strength of the printed product. When the layer thickness increases, all compressions in three dimensions will increase. Ideally, if the layer thickness increases infinitely, the layer amount would be one in a specimen; in other words, the whole specimen would be considered as a mold-cast one.

10.4 DIRECT TENSILE STRENGTH

The direct tensile strength of the printed specimen at the age of 28 days and tested along 3 dimensions, listed in Table 10.5, are relatively low due to the unconsolidated combination between the sticks and layers.

A similar phenomenon exists in tensile strength when setting tensile strength ratio at $t_{yx}=\sigma_y/\sigma_x(0<t_{yx}<1)$, as well as the $t_{zx}=\sigma_z/\sigma_x$, as the data listed in Table 10.6 shows, meaning that the combination degree of the sticks and layers plays an important role in the 3D tensile strengths. The same analysis of comparison between the two test groups, as shown in Table 10.7, shows less relevancy (by the standard deviation) than that

Table 10.5 Average direct tensile strength of the two specimen groups at 28 days in three dimensions

Loading direction (mm)	σ_x (kPa)	σ_y (kPa)	σ_z (kPa)
2.1	10.6	5.8	3.9
1.5	7.8	4.6	3.2

Table 10.6 Comparison of the tensile strength between three dimensions

Layer thickness	$t_{yx} = \sigma_y/\sigma_x$	$t_{zx} = \sigma_z/\sigma_x$
2.1 mm	0.55	0.37
1.5 mm	0.59	0.41
Average	$\overline{t_{yx}} = 0.57$	$\overline{t_{zx}} = 0.39$
Standard deviation	$s_{t_{yx}} = 0.030$	$s_{t_{zx}} = 0.030$

Table 10.7 Comparison of the tensile strength between two layer thickness groups

Comparison of tensile strength					Layer thickness
$\sigma_{x2.1}/\sigma_{x1.5}$	$\sigma_{y2.1}/\sigma_{y1.5}$	$\sigma_{z2.1}/\sigma_{z1.5}$	$\overline{\sigma_{2.1}/\sigma_{1.5}}$	Standard deviation	2.1/1.5
1.36	1.26	1.22	1.28	0.072	1.4

in compression strength.The reason could be explained as: the pulling operation effects only to the weakest cross section in tensile test, while the compression load would spread inside the whole specimen. The printed specimen would be considered as a mold-cast one when the layer thickness going infinite (combination degree being 1.0) at ideal condition.

10.5 CONCLUSION

3D printing is an integrated system of material science and electromechanical techniques. The printing process manufactures a sandwich-like structure with mechanical differences in three dimensions. This stick-layer structure produces successive gaps to reduce the density of the product and introduces enough water inside the product for further hydration during the postprocedure, indicating that the water—cement ratio is no longer the key factor for the mechanical properties, but to the success of the powder printing technique. The key factor, *layer thickness*, contributes most to the compressive strength and tensile strength when loading along the x axis. The thicker the printed layer is, the higher both the compressive and tensile strengths are. The ideal situation is that the whole product is printed out of one layer, where the strength would be as the same as a mold-casting. The *combination degree* of the bonding area between printed sticks and layers decides the compressive and tensile strengths when loading along the y and z dimensions, with a statistical ratio of c_{yx}, c_{zx}, t_{yx}, and t_{zx}, $(0 < c < 1,\ 0 < t < 1)$. For the ideal situation, when these ratios equaled to 1.0, the printed products would be considered as mold-cast ones. Printing strategy, printing parameters, and the powder formula should be optimized in future to improve the combination degree which is the second key factor of the mechanical properties of the printed product.

ACKNOWLEDGMENT

Special acknowledgment goes to the funding support from the Guangzhou Science Technology and Innovation Commission (project 201508030010) and Shenzhen Mingyuan Building Technology Co., Ltd.

REFERENCES

[1] S. Lim, A.B. Richard, T.T. Le, S.A. Austin, A.G.F. Gibb, T. Thorpe, Developments in construction-scale additive manufacturing processes, Autom. Constr. 21 (2012) 262–268.
[2] J. Pegna, Exploratory investigation of solid freeform construction, Autom. Constr. 5 (5) (1997) 427–437.
[3] B. Khoshnevis, D. Hwang, K.-T. Yao, Z. Yeh, Mega-scale fabrication by contour crafting, Int. J. Ind. Syst. Eng. 1 (3) (2006) 301–320.
[4] S. Lim, A.B. Richard, T.T. Le, R. Wackrow, S.A. Austin, A.G.F. Gibb, et al., Development of a viable concrete printing process, 2011.
[5] U. Klammert, E. Vorndran, T. Reuther, F.A. Müller, K. Zorn, U. Gbureck, Low temperature fabrication of magnesium phosphate cement scaffolds by 3D powder printing, J. Mater. Sci. 21 (11) (2010) 2947–2953.
[6] G.A. Fielding, A. Bandyopadhyay, S. Bose, Effects of silica and zinc oxide doping on mechanical and biological properties of 3D printed tricalcium phosphate tissue engineering scaffolds, Dental Mater. 28 (2) (2012) 113–122.
[7] A. Pfister, U. Walz, A. Laib, R. Mülhaupt, Polymer ionomers for rapid prototyping and rapid manufacturing by means of 3D printing, Macromol. Mater. Eng. 290 (2) (2005) 99–113.
[8] C. Bergmann, M. Lindner, W. Zhang, K. Koczur, A. Kirsten, R. Telle, et al., 3D printing of bone substitute implants using calcium phosphate and bioactive glasses, J. Eur. Ceram. Soc. 30 (12) (2010) 2563–2567.
[9] F. Qiao, C.K. Chau, Z. Li, Calorimetric study of magnesium potassium phosphate cement, Mater. Struct. 45 (3) (2012) 447–456.

[10] Z. Ding, B. Dong, F. Xing, N. Han, Z. Li, Cementing mechanism of potassium phosphate based magnesium phosphate cement, Ceram. Int. 38 (8) (2012) 6281–6288.
[11] A.J. Wang, Zl Yuan, J. Zhang, L.T. Liu, J.M. Li, Z. Liu, Effect of raw material ratios on the compressive strength of magnesium potassium phosphate chemically bonded ceramics, Mater. Sci. Eng. 33 (8) (2013) 5058–5063.
[12] H. Ma, B. Xu, J. Liu, H. Pei, Z. Li, Effects of water content, magnesia-to-phosphate molar ratio and age on pore structure, strength and permeability of magnesium potassium phosphate cement paste, Mater. Des. 64 (2014) 497–502.

CHAPTER 11

Development of Powder-Based 3D Concrete Printing Using Geopolymers

Ming Xia, Behzad Nematollahi and Jay G. Sanjayan
Centre for Sustainable Infrastructure, Faculty of Science, Engineering and Technology, Swinburne University of Technology, Hawthorn, VIC, Australia

11.1 INTRODUCTION

Additive Manufacturing (AM), also known as three-dimensional (3D) printing, is a manufacturing process where a complex 3D object created by computer aided design (CAD) software is built using a layer-by-layer deposition approach through a series of cross-sectional slices [1,2]. Although AM has been extensively adopted in many industries such as automotive, aerospace, medicine, and biological systems [1,3], the construction industry has been rather slow in implementing AM. In recent years, AM has gained a considerable amount of attention from the construction industry due to its remarkable advantages over the conventional approach of casting concrete into a formwork, including enhanced geometrical freedom, increased level of safety and reduction in construction cost, waste, and time [4,5]. In the past few years, a variety of AM technologies have been developed for 3D concrete printing (3DCP) applications. These technologies can be classified into two forms, extrusion-based and powder-based techniques.

11.1.1 Extrusion-Based 3D Concrete Printing

The extrusion-based technique is an AM process where the concrete is extruded from a nozzle mounted on a gantry, crane, or robotic arm and deposited in a layer-by-layer manner. Examples of this technique are Contour Crafting, developed by Khoshnevis [6,7] and concrete printing, designed by Lim et al. [8,9]. The extrusion-based 3DCP technique has been aimed at onsite construction applications such as large-scale building components with complex geometries.

3D Concrete Printing Technology
DOI: https://doi.org/10.1016/B978-0-12-815481-6.00011-7

11.1.2 Powder-Based 3D Concrete Printing

The powder-based 3DCP is another typical AM process that creates accurate structures with complex geometries by depositing binder liquid (or "ink") selectively into to powder bed to bind the powder where it impacts the bed [10]. Two typical examples are the D-shape technique developed by Cesaretti et al. [11] and Emerging Objects by Rael et al. [12] The powder-based 3DCP technique is an offsite process designed to manufacture precast components. The powder-based 3DCP technique is highly suitable for small-scale building components such as panels, permanent formworks, and interior structures that then can be assembled onsite.

The work presented in this chapter focuses on the powder-based 3DCP technique as this method is capable of producing building components with fine details and intricate shapes. There is a demand in the construction industry for such components which can otherwise only be made with expensive formwork systems. The powder-based 3DCP technique needs to be developed so that robust and durable components can be produced at a reasonable speed to satisfy this industrial demand. Hence, there is urgency for research in this area.

A schematic of the powder-based 3DCP process is illustrated in Fig. 11.1. At the start, a roller mounted together with a print head spreads a layer of powder to cover the base of the build plate. The base layer is about 3 mm in thickness. Then, according to the layer thickness setting of the 3D printer, a thin layer of powder (approximately 0.1 mm) is spread and smoothed by the roller over the powder bed surface. Once a layer is completed, the binder solution is delivered from binder feeder to the print

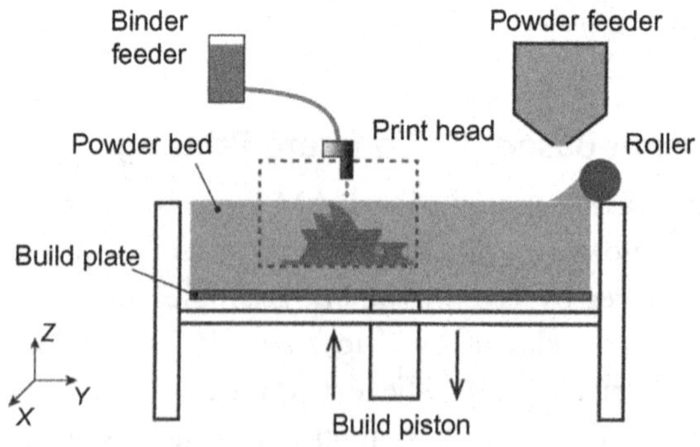

Figure 11.1 Schematic illustrations of the 3D printing process.

head and is jetted by nozzles mounted in the print head. The mechanism of controlling the binder solution is a noncontinuous approach called the drop-on-demand technique which has been widely used in contemporary desktop printing systems. In the print head, the binder solution is pushed out through nozzles either thermally by a thermal bubble or mechanically by a piezoelectric actuator. Then binder droplets are formed and are selectively applied on the powder layer, causing powder particles to bind to each other. By repeating these steps, the built part is completed and removed after a particular drying time and the unbound powder is removed by using an air blower.

The powder-based 3DCP technique has great potential to make a significant and positive contribution to the construction industry. There are, however, a number of challenges to be overcome before the technique can be fully utilized. One of the main challenges is that the proprietary printing materials that are typically used in the commercially available powder-based 3D printers are not suitable for construction applications. As such, it is urgently needed to expand the current severely limited scope of materials that can be used in a powder-based 3D printer.

Geopolymer [13,14], also known as an inorganic polymer, is an alternative material that acts as the binding agent in concrete. Geopolymers are a type of alkali aluminosilicate cement which can have superior mechanical, chemical, and thermal properties compared to Ordinary Portland Cement, and causes significantly lower CO_2 emission. Geopolymers are synthesized by activation of an aluminosilicate precursor (metakaolin, fly ash, and slag) with alkaline activators. In recent years, geopolymers have attracted considerable attention because of its high compressive strength, excellent resistance to sulfate attack, good acid resistance, low creep, and minimal drying shrinkage behavior. The utilization of industrial byproducts such as slag and fly ash in geopolymers is considered particularly beneficial, as the disposition of industrial byproducts is a global concern. However, the applications of geopolymers have been restricted to small areas using conventional manufacturing techniques. A novel method of manufacturing geopolymer component, such as the powder-based 3DCP technique, should be taken into consideration to expand the application of this eco-friendly material.

Fly ash-based geopolymers possess many properties that are better than slag-based geopolymers, including better fire resistance and lower shrinkage and creep properties [15]. Fly ash is abundantly available, and a large part of it still dumped in landfills in many parts of the world.

However, slag has a much better utilization rate and, currently, most of it is consumed in concrete production. Therefore, there are more incentives to use fly ash to produce geopolymer than slag. However, fly ash poses a major limitation in that geopolymers made from fly ash require high-temperature curing (typically 60°C) whereas slag-based geopolymers can set at room temperature. Fly ash alone cannot be used in the powder-based 3DCP technique because it cannot set at room temperature. Therefore, a minimum amount of slag is necessary to promote setting at room temperature.

This chapter presents an innovative methodology for formulating geopolymer-based material for the requirements and demands of commercially available powder-based 3D printers. Different key powder parameters such as particle size distribution, powder bed surface quality, powder true/bulk densities, powder bed porosity, and binder droplet penetration behavior were used to evaluate the printability of prepared geopolymer-based material quantitatively. The formulation is then extended to find the minimum amount of slag required for this purpose and observe the characteristics as slag is reduced in the geopolymer mix. This research is intended for broadening the scope of printable geopolymer materials compatible with the powder-based 3DCP process for construction applications

11.2 EXPERIMENTAL PROCEDURES

11.2.1 Materials

11.2.1.1 Powder

Low calcium (Class F) fly ash and granulated ground blast furnace slag (slag) was supplied by the Gladstone power station in Queensland, Australia, and Building Products Supplies Pty Ltd., Australia, respectively. The chemical composition of the fly ash and slag determined by X-ray Fluorescence analysis are shown in Table 11.1. The total percentages do not add up to 100% due to rounding off.

Anhydrous sodium metasilicate powder (in the form of beads) was supplied by Redox, Australia with a chemical composition of 50.66 wt% Na_2O, 47.00 wt% SiO_2, and 2.34 wt% H_2O was used as the alkaline activator in this study. A high purity silica sand with a median size of 184 μm supplied by TGS Industrial Sand Ltd., Australia was also used.

N Grade sodium silicate and sodium hydroxide (NaOH) solution was used for the postprocessing of green samples. The N Grade sodium silicate solution was supplied by PQ Australia with a modulus (M_s) of 3.22 (where $M_s = SiO_2/Na_2O$, $Na_2O = 8.9$ wt% and $SiO_2 = 28.6$ wt%). The NaOH solution was prepared with a concentration of 8.0 M using NaOH beads of 97% purity supplied by Sigma-Aldrich and tap water.

Five geopolymer powders with different slag to fly ash ratios were investigated in this study, as presented in Table 11.2. To prepare each mix, the alkaline activator beads were firstly ground for 5 minutes using a planetary ball mill with a powder/ball mass ratio of 0.3. Then fly ash, slag, silica sand, and the ground activator powder were thoroughly dry mixed in a Hobart mixer to achieve a homogenous mixture (visually assessed). The process of preparing geopolymer powder is shown in Fig. 11.2.

Table 11.1 The chemical composition of slag and fly ash (wt%)

Chemical	Component	
	Slag	**Fly ash**
Al_2O_3	12.37	25.56
SiO_2	32.76	51.11
CaO	44.64	4.30
Fe_2O_3	0.54	12.48
K_2O	0.33	0.70
MgO	5.15	1.45
Na_2O	0.22	0.77
P_2O_5	0.88	0.01
TiO_2	0.51	1.32
MnO	0.15	0.37
SO_3	4.26	0.24
LOI[a]	0.09	0.57

[a]Loss on ignition.

Table 11.2 Mix proportions of geopolymer precursors

Mix ID	Geopolymer precursors wt%	
	Slag	**Fly ash**
S100FA0	100	0
S75FA25	75	25
S50FA50	50	50
S25FA75	25	75
S0FA100	0	100

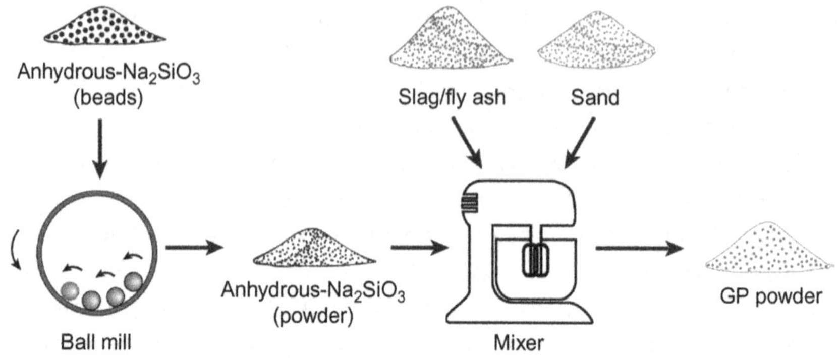

Figure 11.2 The process of preparing geopolymer powder.

11.2.1.2 Binder

An aqueous solvent (Zb 63, Z-Corp, United States) was used as a binder during the 3DCP process. The Zb 63 binder is an aqueous commercial clear solution with a viscosity similar to pure water. The composition of Zb 63 was mainly water with 2-Pyrrolidone by Fourier transform infrared spectroscopy analysis [16].

11.2.2 Printability Characterizations of Geopolymer Powder

The printability and depositability of powder, as defined by Butscher et al. [17], are fundamental requirements for the 3DCP process. Printability is influenced by several parameters of the powder such as particle size distribution, powder true/bulk densities, powder bed porosity, and binder droplet penetration behavior. Before adopting a new powder to a commercially available 3D printer, its printability should be assessed [17].

11.2.2.1 Particle Size Distribution

The particle size and particle size distribution (PSD) of the powders were measured by using a laser diffraction particle analyzer (Cilas 1190, CILAS, France). Each powder was measured three times to find out the D_{10}, D_{50}, and D_{90} values, which indicates 10%, 50%, and 90% of the particles being below those sizes.

11.2.2.2 Powder Bed Surface Quality

After preparing the powder bed in the printer, the build plate with the powder bed was removed from the 3D printer. The surface qualities of the two powder beds were inspected by using a Nikon D810 camera. The inspection direction was 45 degrees to the top of the powder bed.

11.2.2.3 True, Bulk, and In-Process Powder Bed Densities

True ($\rho_{true.powder}$) and bulk ($\rho_{bulk.powder}$) densities of the powders were measured according to Australian Standard AS 1774.6-2001 (R2013) for the determination of true density and AS 1774.2-2001 (R2013) for the determination of bulk density. In-process bed density ($\rho_{bed.powder}$), first proposed by Zhou et al. [18], is the density measured after the powder is spread on the build plate. It is an important characteristic to ensure a high printing quality in the powder-based 3DCP process. Firstly, the powder was spread by a roller from the powder feeder to the build plate. Then the build plate was removed from the 3D printer. The volume of the powder bed and the mass of powder in the build plate were measured to determine the in-process bed density. The powder bed porosity (P_{bed}) can be calculated from Eq. (11.1).

$$P_{bed} = 1 - \frac{\rho_{bed.powder}}{\rho_{true.powder}} \times 100\% \qquad (11.1)$$

11.2.2.4 Drop Penetration Behavior and Wettability

A schematic diagram of the drop penetration behavior measurement apparatus is shown in Fig. 11.3A. After the powder bed was prepared on the build plate, the build plate with the powder bed was removed from the 3D printer. Then a binder (Zb 63, Z-Corp, United States) droplet with a volume of approximately 60 μL was deposited using a 1 mL medical syringe (25-gauge needle) which was positioned above the

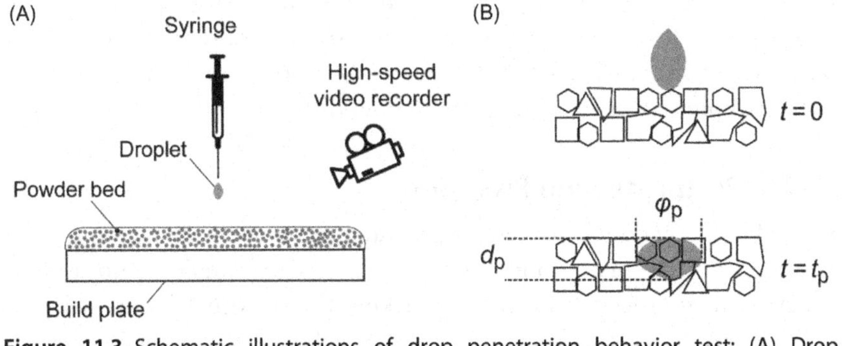

Figure 11.3 Schematic illustrations of drop penetration behavior test: (A) Drop penetration apparatus; and (B) drop penetration parameters.

powder bed surface. The droplet formed at the tip of the needle and grew until its weight exceeded the surface tension force.

In this study, the wettability of the powder was quantified by three parameters (Fig. 11.3B): drop penetration time (t_p), drop penetrating depth (d_p), and drop spreading diameter (φ_p). Drop penetration time (t_p) is defined as the time taken for the droplet to penetrate completely into the powder with no liquid remaining on the powder bed surface. A high-speed video recorder operating at 240 frames/s was applied to track a single binder droplet penetrating into the powder bed. Ten replicates were performed for each powder. The drop penetration time was calculated from the video frame by frame. The starting point was the time when binder droplet firstly contacted the powder bed. The end point was the time when light reflection on the droplet disappeared.

Drop penetrating depth (d_p) and spreading diameter (φ_p) were measured by using a digital caliper with an accuracy of 0.01 mm when the granule was solidified. A sieve and an air blower were used to remove the unbound powder. These two parameters were measured after each drop penetration time measurement. For each powder mix, the t_p, d_p, and φ_p measurements were conducted 10 times.

11.2.3 Powder-Based 3D Concrete Printing Process

A 20 mm cube model designed by using SolidWorks software was used for printing. In this study, the powder-based 3DCP process was carried out on a commercial 3D printer (Zprinter 150, Z-Corp, United States) with an HP11 print head (C4810A). The 3D printer has a specified resolution of 300 × 450 dpi and a 185 × 236 × 132 mm build volume. For simplicity, the default printing parameters setting was used with a layer thickness of 0.1016 mm and a binder/volume ratio of 0.24 for the shell section and 0.12 for the core section of the printed structure. After 6 hours of drying within the powder bed, depowdering was performed using compressed air to remove the unbounded powder.

11.2.4 Postprocessing Procedure

After the depowdering process, the printed cubes were divided into two groups. For the first group, denoted as "green samples," no further postprocessing procedures were undertaken. For the other group, denoted as "postprocessed samples," the printed cubes were immersed in an alkaline solution and placed in the oven at 60 ± 3°C for 7 days. The alkaline solution was composed of N Grade sodium silicate solution with a SiO_2/Na_2O ratio of 3.22 (71.4% w/w) and 8.0 M NaOH solution (28.6% w/w). At the end of the heat-curing period, the postprocessed samples were removed from the oven and kept undisturbed until cool.

11.2.5 Mechanical Properties

The compressive strength of the green and postprocessed samples in both the *X*-direction and *Z*-direction were measured under load control at the rate of 0.33 MPa/s. A population of 10 samples for each testing direction was used.

11.3 RESULTS AND DISCUSSIONS

11.3.1 Powder Characteristics

The particle size and PSD of geopolymer powders are presented in Fig. 11.4. The particle sizes of all geopolymer powders (with different slag to fly ash ratios) range from 0.1 to 100 μm. The raw slag and the raw fly ash used for making the powders have a D_{50} of 12.68 and 3.74 μm, respectively. Thus, the inclusion of fly ash into the geopolymer powder resulted in shifting the cumulative frequency curve to the left-hand side.

In the powder-based 3DCP process, the particle size and PSD are critical for the powder depositability, which enables a smooth and homogeneous powder bed [19,20]. The surface quality of the ZP and GP

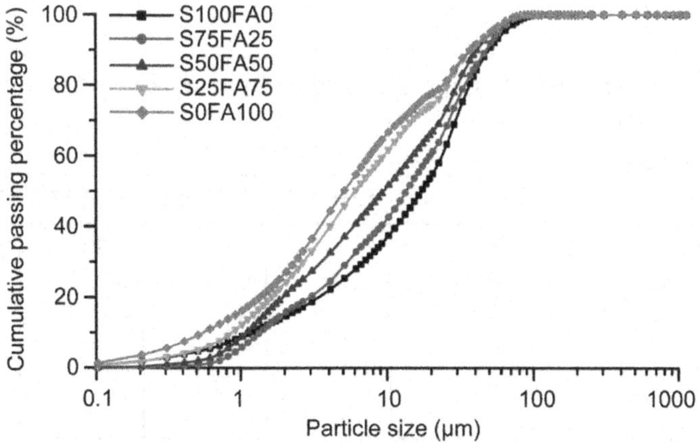

Figure 11.4 Particle size distributions of geopolymer powders.

powder bed after deposition are shown in Fig. 11.5. It was noted that that all the geopolymer powders exhibited sufficient surface qualities for the powder-based 3DCP process.

11.3.2 True/Bulk Densities, In-Process Bed Density, and Powder Bed Porosity

The true/bulk densities, in-process bed density, and powder bed porosity of each geopolymer powder are summarized in Table 11.3.

The increase in the amount of fly ash in the geopolymer powder reduced the true, bulk, and in-process bed densities of the powder, but

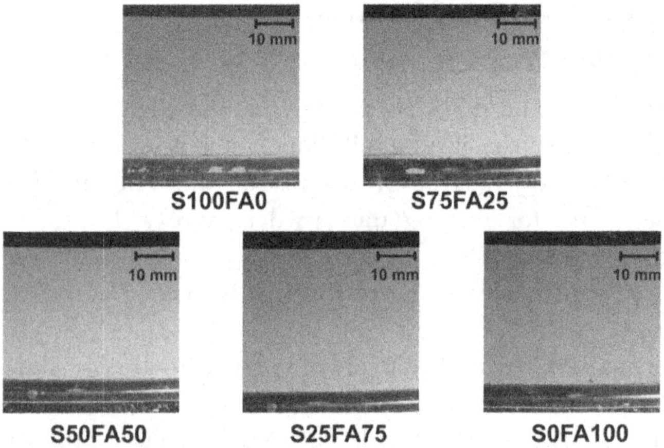

Figure 11.5 Surface qualities of geopolymer powder beds.

Table 11.3 True/bulk densities, in-process bed density, and powder bed porosity of geopolymer powders

Powder properties	Mix ID				
	S100FA0	S75FA25	S50FA50	S25FA75	S0FA100
True density (g/cm^3)	2.81	2.73	2.66	2.59	2.51
Bulk density (g/cm^3)	0.78	0.76	0.75	0.72	0.69
In-process bed density (g/cm^3)	0.83	0.81	0.78	0.75	0.73
Powder bed porosity (%)	70.4	70.4	70.6	71.0	70.9

increased the powder bed porosity. However, the effect was not significant. For each geopolymer powder mix, the marginal difference between bulk density and in-process bed density indicates that the powder has not been significantly compressed during the powder spreading process.

11.3.3 Binder Droplet Penetration Behavior

Fig. 11.6 presents images of the typical binder droplet penetration behavior of the geopolymer powders.

The increase of the fly ash content of the geopolymer powder increased the time needed for the binder droplet to penetrate into the powder bed completely. This implies that the powder properties including powder density as well as the size and shape of the particles influence the binder droplet penetration behavior.

After completion of the penetration of the binder droplet into the powder bed, granules were formed within the powder bed. These solid granules had enough green strength for the measurement of d_p and φ_p, except for the S0FA100 mix (the geopolymer powder with 100% fly ash). The low reactivity of this geopolymer powder containing 100% fly ash caused the granules to be too weak to withstand the pressure during the depowdering process. Table 11.4 summarizes the t_p, d_p, and φ_p of each geopolymer powder. The reported values given in this table for each parameter is the average of 20 measurements, and the reported errors are based on a 95% confidence level.

As shown in Table 11.4, the geopolymer powder containing 100% slag (S100FA0) demonstrated the lowest binder droplet penetration time. The longest t_p was for the S0FA100 powder containing 100% fly ash, which was 73% longer than that of S100FA0 powder. The higher the fly

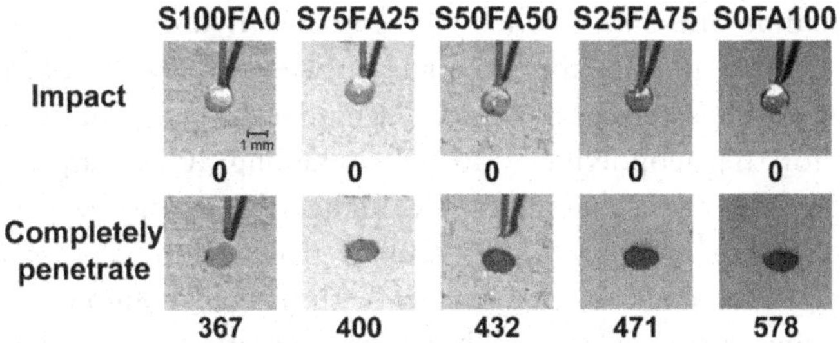

Figure 11.6 Images of binder droplet impacting on surfaces of geopolymer powder beds. Note: The numbers below each image indicate the droplet penetration time (ms).

Table 11.4 Binder droplet penetration behavior results of geopolymer powders

Mix ID	Binder droplet penetration parameters		
	t_p[a] (ms)	d_p[b] (mm)	φ_p[c] (mm)
S100FA0	335 ± 42	2.10 ± 0.21	1.95 ± 0.16
S75FA25	370 ± 64	2.46 ± 0.22	1.82 ± 0.11
S50FA50	440 ± 57	2.78 ± 0.29	1.72 ± 0.15
S25FA75	490 ± 43	2.90 ± 0.18	1.67 ± 0.16
S0FA100	580 ± 76	–[d]	–[d]

[a]Binder droplet penetration time.
[b]Binder droplet penetration depth.
[c]Binder droplet spreading diameter.
[d]Could not be measured due to weak strength of granules made with this powder.

ash content in the geopolymer powder, the finer the average particle size of the powder, which results in a longer binder droplet penetration time. According to Hapgood [21], changing the particle size will alter the pore structure within the powder bed. Fine powder particles tend to agglomerate, thereby creating a large number of macrovoids within the powder bed. The binder liquid tends to flow through microvoids around the macrovoids, which significantly increases the penetration time. Therefore, the binder droplet penetration time will be longer for geopolymer powder containing more fly ash with a finer average particle size.

As shown in Table 11.4, the geopolymer powder containing 100% slag (S100FA0) demonstrated the biggest binder droplet spreading diameter (φ_p). The increase of the fly ash content of the geopolymer powder reduced the φ_p. According to Nefzaoui et al [22], φ_p depends on the physicochemical characteristics of the binder droplet and the powder bed surface hydrophilicity. Therefore, the lower φ_p of the geopolymer powder with higher fly ash content might be attributed to the reduction in the powder bed surface hydrophilicity, resulting from the reduction in the average particle size of the geopolymer powder. On the other hand, the lowest binder droplet penetration depth (d_p) was for the geopolymer powder containing 100% slag (S100FA0). The higher the fly ash content, the higher the d_p of the geopolymer powder. In general, the binder droplet penetration takes place after spreading. For a single binder droplet with a fixed-volume, the smaller the spreading diameter, the higher the penetration depth (i.e., binder droplet has to penetrate deeper into the powder bed to complete the penetration process). Thus, it is reasonable that the S100FA0 powder had the lowest d_p, but the biggest φ_p.

11.3.4 Mechanical Properties

Fig. 11.7 presents the 3D printed cubes using the geopolymer powders with different slag/fly ash ratios. No sample could be printed from the geopolymer powder containing 100% fly ash (S0FA100) due to the low reactivity of this powder with the binder liquid at ambient temperature. Thus, no data is available for the S0FA100 powder with regards to the linear dimensional accuracy and compressive strength results.

The 7-day uniaxial compressive strengths of the green and postprocessed 3D printed cubes are presented in Fig. 11.8. It should be pointed that the S25F75 samples were dissolved in the curing solution during the postprocessing process, therefore, no data is available for the compressive strength of the S25FA75 postprocessed samples.

11.3.4.1 Green Samples

As shown in Fig. 11.8, the green samples exhibited relatively low compressive strength ranging from 0.24 to 0.91 MPa, depending on the testing direction and fly ash content. However, it is necessary to note that this strength level is sufficient for the depowdering process.

In both directions, the increase of fly ash content significantly decreased the compressive strength of the green samples. This is attributed to the low reactivity of fly ash at ambient temperature.

An anisotropic phenomenon was also observed regarding the compressive strength of the green samples depending on the loading directions. Regardless of the content of fly ash, the compressive strength was always higher in the *X*-direction than in the *Z*-direction.

Figure 11.7 Powder-based 3D printed green cubes using geopolymer powders with different slag/fly ash ratios.

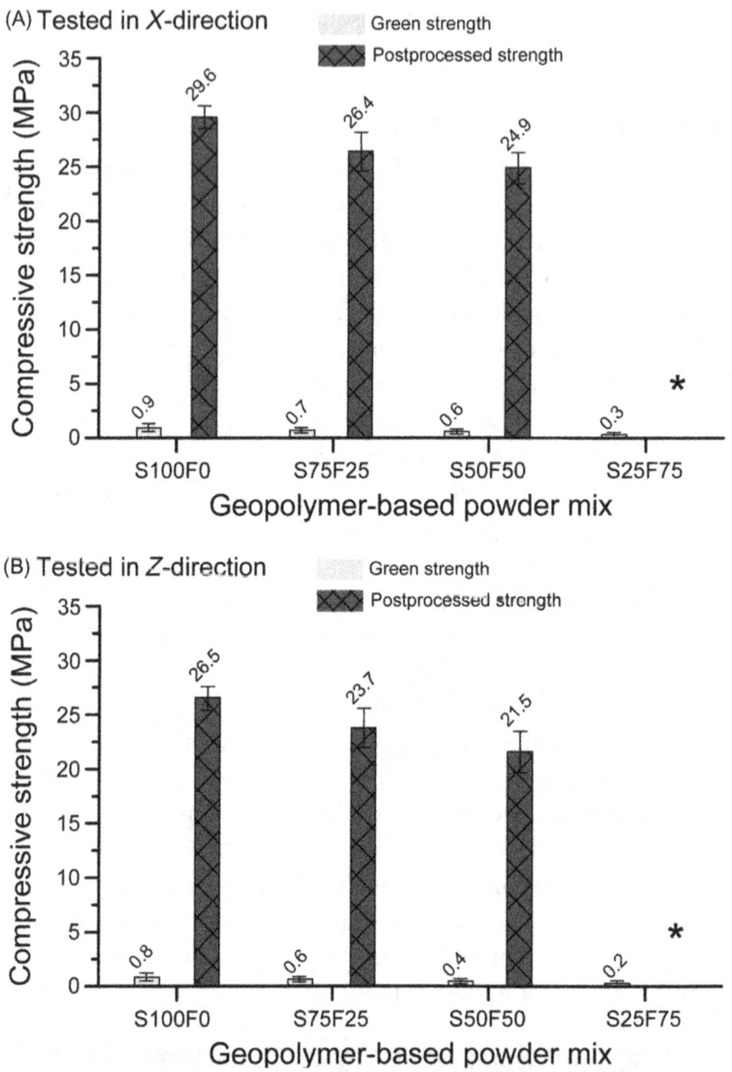

Figure 11.8 The 7-day uniaxial compressive strength of both green and postprocessed 3D printed cubes measured in: (A) the *X*-direction; and (B) the *Z*-direction. Note: No data is available for S25FA75 postprocessed samples as they were dissolved in the curing solution.

The green compressive strength of the S100FA0, S75FA25, S50FA50, and S25FA75 samples in the *X*-direction were 20%, 25%, 41%, and 42%, respectively, higher than that of the corresponding samples in the *Z*-direction. This anisotropic phenomenon might be related to the preferential orientation of the powder particles during the powder spreading process [23]. Within the powder layers, larger facets of the particles oriented parallel to the $X-Y$ plane, while smaller facets with associated sharper edges were connected in the Z-direction [23]. The gel formed between the larger facets of powder particles was stronger than that formed between the smaller facets. More energy is required for the fracture between the larger facets of powder particles.

11.3.4.2 Postprocessed Samples

According to Fig. 11.8, in both directions the compressive strength of postprocessed samples was significantly higher than that of the green samples. This is true regardless of fly ash content. The 7-day postprocessed compressive strength was in the range of 21.5–29.6 MPa, depending on the fly ash content and testing direction, meeting the compressive strength requirements for many applications in the construction industry.

The postprocessed compressive strengths of the S100FA0, S75FA25, S50FA50, and S25FA75 samples in the X-direction were 32, 37, and 41 times that of the corresponding green samples' compressive strengths. In the Z-direction, the postprocessed samples' compressive strengths were 33, 39, and 54 times of that of the corresponding green samples' compressive strengths. This was due to the continued geopolymerization process resulting from the immersion of the samples for seven days in alkaline solutions. More geopolymeric products were formed and developed, resulting in the densification of the porous structures of the green samples.

Similar to the compressive strength of the green samples, there was an anisotropic phenomenon regarding the compressive strength of the postprocessed samples depending on the loading directions. Irrespective of the fly ash content, the compressive strength of the postprocessed samples was always higher in the X-direction than in the Z-direction.

Regardless of the testing direction and type of curing solution, S100FA0 samples consistently exhibited the highest postprocessed compressive strength. The increase of fly ash content decreased the compressive strength of the postprocessed samples in both directions.

11.4 CONCLUSIONS

An innovative methodology is developed in this study for formulating geopolymer-based materials to be used in commercially available powder-based 3D printers for construction applications. The influences of several key powder parameters and incorporation of fly ash on the printability and properties of powder-based 3D printable geopolymers were quantitatively evaluated. The following conclusions can be drawn:

1. The prepared geopolymer-based materials gained sufficient depositability and wettability and they are applicable to replace commercially available printable material in a powder-based 3D printer.
2. The amount of fly ash did not have any significant effect on the true, bulk, and in-process bed densities and powder bed porosity of the 3D printable geopolymer powders. Visual observations revealed that all geopolymer powders exhibited sufficient surface qualities for the powder-based 3DCP process.
3. The new postprocessing method introduced in this study increased the strengths many times over the original strength. The specimens printed with 100% slag-based geopolymer powder exhibited the highest postprocessed compressive strength of up to 29.6 MPa at 7 days. The minimum amount of slag required to prepare fly ash/slag-blended geopolymer powder for powder-based 3DCP process was 50 wt%. The 7-day compressive strength of up to 24.9 MPa was achieved for the postprocessed specimens printed with 50 wt% slag/50 wt% fly ash powder.
4. The compressive strength of both green and postprocessed samples exhibited anisotropic properties depending on the loading directions, where the strengths were always higher in the *X*-direction than in the *Z*-direction. This was true regardless of the fly ash content.

The method developed in this study is readily scalable to produce large structural components.

ACKNOWLEDGMENT

The authors acknowledge the support by the Australian Research Council Discovery Grant DP170103521 and Linkage Infrastructure Grant LE170100168, and Discovery Early Career Researcher Grant DE180101587.

REFERENCES

[1] W. Gao, Y. Zhang, D. Ramanujan, K. Ramani, Y. Chen, C.B. Williams, et al., The status, challenges, and future of additive manufacturing in engineering, Comput. -Aided Des. 69 (2015) 65–89.
[2] B. Berman, 3-D printing: the new industrial revolution, Business Horizons 55 (2012) 155–162.

[3] I. Hager, A. Golonka, R. Putanowicz, 3D printing of buildings and building components as the future of sustainable construction? Procedia Eng. 151 (2016) 292–299.
[4] S. Lim, R.A. Buswell, T.T. Le, S.A. Austin, A.G. Gibb, T. Thorpe, Developments in construction-scale additive manufacturing processes, Autom. Constr. 21 (2012) 262–268.
[5] T. Wangler, E. Lloret, L. Reiter, N. Hack, F. Gramazio, M. Kohler, et al., Digital concrete: opportunities and challenges, RILEM Tech. Lett. 1 (2016) 67–75.
[6] B. Khoshnevis, S. Bukkapatnam, H. Kwon, J. Saito, Experimental investigation of contour crafting using ceramics materials, Rapid Prototyping J. 7 (2001) 32–42.
[7] B. Khoshnevis, D. Hwang, K.-T. Yao, Z. Yeh, Mega-scale fabrication by contour crafting, Int. J. Ind. Syst. Eng. 1 (2006) 301–320.
[8] S. Lim, T. Le, J. Webster, R. Buswell, S. Austin, A. Gibb, et al., Fabricating construction components using layer manufacturing technology, 2009.
[9] S. Lim, R.A. Buswell, T.T. Le, R. Wackrow, S.A. Austin, A.G. Gibb, et al., Development of a viable concrete printing process, 2011.
[10] E.M. Sachs, J.S. Haggerty, M.J. Cima, P.A. Williams, Three-dimensional printing techniques, U.S. Patent No. 5,204,055. Washington, DC: U.S. Patent and Trademark Office., 1993.
[11] G. Cesaretti, E. Dini, X. De Kestelier, V. Colla, L. Pambaguian, Building components for an outpost on the Lunar soil by means of a novel 3D printing technology, Acta Astronaut. 93 (2014) 430–450.
[12] R. Rael, V. San Fratello, Developing concrete polymer building components for 3D printing, in: ACADIA. 31st Annual Conference of the Association for Computer Aided Design in ArchitectureBanff, 2011.
[13] J. Davidovits, Geopolymers, J. Thermal Anal. Calorim. 37 (1991) 1633–1656.
[14] P. Duxson, A. Fernández-Jiménez, J.L. Provis, G.C. Lukey, A. Palomo, J.S.J. Deventer, Geopolymer technology: the current state of the art, J. Mater. Sci. 42 (2006) 2917–2933.
[15] B. Nematollahi, J. Sanjayan, F.U.A. Shaikh, Synthesis of heat and ambient cured one-part geopolymer mixes with different grades of sodium silicate, Ceram. Int. 41 (2015) 5696–5704.
[16] M. Asadi-Eydivand, M. Solati-Hashjin, S.S. Shafiei, S. Mohammadi, M. Hafezi, N. A.A. Osman, Structure, properties, and in vitro behavior of heat-treated calcium sulfate scaffolds fabricated by 3D printing, PLoS One 11 (2016) e0151216.
[17] A. Butscher, M. Bohner, C. Roth, A. Ernstberger, R. Heuberger, N. Doebelin, et al., Printability of calcium phosphate powders for three-dimensional printing of tissue engineering scaffolds, Acta Biomater., 8 (2012) 373-385.
[18] Z. Zhou, F. Buchanan, C. Mitchell, N. Dunne, Printability of calcium phosphate: Calcium sulfate powders for the application of tissue engineered bone scaffolds using the 3D printing technique, Mater. Sci. Eng. 38 (2014) 1–10.
[19] B. Utela, D. Storti, R. Anderson, M. Ganter, A review of process development steps for new material systems in three dimensional printing (3DP), J. Manuf. Process. 10 (2008) 96–104.
[20] B.R. Utela, D. Storti, R.L. Anderson, M. Ganter, Development process for custom three-dimensional printing (3DP) material systems, J. Manuf. Sci. Eng., Trans. ASME 132 (2010) 0110081–0110089.
[21] K.P. Hapgood, J.D. Litster, S.R. Biggs, T. Howes, Drop penetration into porous powder beds, J. Colloid Interface Sci. 253 (2002) 353–366.

[22] E. Nefzaoui, O. Skurtys, Impact of a liquid drop on a granular medium: inertia, viscosity and surface tension effects on the drop deformation, Exp. Thermal Fluid Sci. 41 (2012) 43–50.

[23] Y. Shanjani, Y. Hu, R.M. Pilliar, E. Toyserkani, Mechanical characteristics of solid-freeform-fabricated porous calcium polyphosphate structures with oriented stacked layers, Acta Biomater. 7 (2011) 1788–1796.

CHAPTER 12

Interlayer Strength of 3D Printed Concrete: Influencing Factors and Method of Enhancing

Taylor Marchment, Jay G. Sanjayan, Behzad Nematollahi and Ming Xia
Centre for Sustainable Infrastructure, Faculty of Science, Engineering and Technology, Swinburne University of Technology, Hawthorn, VIC, Australia

12.1 INTRODUCTION

Additive manufacturing (AM), commonly known as three-dimensional (3D) printing is a group of emerging techniques for fabricating 3D structures directly from a digital model in successive layers. The American Society for Testing and Materials (ASTM) defines AM as "the process of joining materials to make objects from 3D model data, usually layer upon layer" [1]. AM technologies were initially developed in the 1980s and have already been successfully applied in a wide range of industries including aerospace and automotive manufacturing, biomedical, consumer, and food [2]. Currently, AM techniques have become an integral part of modern product development.

3D printing technology is gaining popularity in construction industry. Unlike the conventional approach of casting concrete into a formwork, 3D concrete printing (3DCP) combines digital technology and innovative insight from materials technology to allow freeform construction without the use of expensive formwork. Freeform construction would enhance architectural expression, where the cost of producing a structural component will be independent of the shape, providing much-needed freedom from rectilinear designs [3]. When compared to conventional construction processes, the application of 3D printing techniques in the construction industry may offer excellent advantages, including: (1) reduction of construction costs by eliminating expensive formwork [3,4]; (2) reduction of injury rates by eliminating dangerous jobs (e.g., working at heights), which would result in an increased level of safety in construction [3]; (3)

3D Concrete Printing Technology
DOI: https://doi.org/10.1016/B978-0-12-815481-6.00012-9

creation of high-end, technology-based jobs [3]; (4) reduction of onsite construction time by operating at a constant rate [3,4]; (5) minimizing the chance of errors by highly precise material deposition [3]; (6) increasing sustainability in construction by reducing wastages of formwork [3,4]; (7) increasing architectural freedom, which would enable more sophisticated designs for structural and esthetic purposes [3,4]; and (8) enabling the potential of multifunctionality for structural/architectural elements by taking advantage of the complex geometry [5].

Over the past few years, different 3DCP technologies have been developed to adopt AM in concrete construction. These technologies are mainly based on two techniques, namely extrusion-based and powder-based techniques. The powder-based technique is a typical AM process that can make structures with fine details and intricate shapes by jetting binder liquid selectively through a nozzle(s) on the powder layer, causing powder particles to bind to each other. Examples of 3DCP technologies developed based on the powder-based technique include the D-shape technique [6] and Emerging Objects [7]. The powder-based technique is an offsite process, which is suitable for manufacturing building components with complex geometries such as panels, permanent formworks, and interior structures which can be later assembled onsite. There is a demand in the construction industry for such components, which currently can only be manufactured with the use of expensive formworks using the available construction systems. The powder-based technique has the potential to make robust and durable building components at a reasonable speed to satisfy this industrial demand. However, the very limited scope of cement-based printing materials that can be used in commercially available powder-based 3D printers prevent this technique from performing at its maximum potential for application in the construction industry. To tackle this limitation, the authors of this study developed an innovative methodology to adopt geopolymer-based materials to meet the requirements of commercially available powder-based 3D printers [8,9].

The extrusion-based technique is like the fused deposition modeling method, which extrudes cementitious material from a nozzle mounted on a gantry, crane, or a 6-axes robotic arm to print a structure layer by layer. This technique has been aimed at onsite construction applications such as large-scale building components with complex geometries. Examples of 3DCP technologies developed based on the extrusion-based technique include contour crafting [10–12], concrete printing [13,14], and CONPrint3D [15]. Although extrusion-based 3DCP has great potential to make a significant and positive contribution to the construction industry, there are several technological challenges that need to be overcome before this technology can be used.

One of the main challenges is the weak interlayer strength of printed concrete. The interlayer strength is seen as a major weakness of printed concrete in the extrusion-based 3D printing process as potential flaws can be created between extruded layers, which induce stress concentration [14]. In conventional mold-cast concrete, the bond strength between existing and new concretes typically depends on the surface and moisture conditions of the existing concrete surface [16]. According to Gillette [17], free water on the existing concrete surface decreases the bond strength. However, Pigeon and Saucier [18] concluded that the moisture condition of the existing concrete surface does not influence the bond strength. Austin et al. [19] reported that higher bond strength could be obtained when the existing concrete surface is in a saturated surface dry condition. These controversies in the reported results may be due to different materials being used for existing and new concretes, along with different environmental conditions and test methods adopted in each research study. In 3DCP, unlike conventional mold-cast concrete, the extruded layers are still in the fresh state. Therefore, there is a need to investigate the parameters influencing the bond strength between two subsequent layers in the fresh state. The interlayer strength depends on the adhesion between extruded layers, which is a function of the print-time interval between layers, referred to as delay time. There is a need to optimize the delay time which, on the one hand, should be long enough so the printed layers can gain adequate green strength to support the upper layers without significant deformation or collapsing. On the other hand, the delay time should be short enough to guarantee adequate bond strength between printed layers and to provide an economically feasible rate of construction. Therefore, the delay time becomes a critical parameter in the extrusion-based 3DCP process. Le et al. [14] investigated the effect of delay time (in increments of 15, 30 minutes, 1, 2, 4, 8, 18 hours, and 1, 3, 7 days) on the interlayer strength of a printed, high-performance, fiber-reinforced mortar using direct tension tests on cylindrical cored specimens. It was concluded that the interlayer strength reduced as the delay time increased. The researchers stated that "this reduction with increasing gap in printing time was expected as the adhesion reduced," [14] but they did not investigate factors such as surface moisture affecting the adhesion between the layers. Therefore, our study aims to fill this knowledge gap. The first part of this chapter (Part I) focuses on investigating the effect of delay time on the strength properties of extrusion-based 3D printed concrete including compressive, flexural, and interlayer strengths.

The second part of this chapter (Part II) investigates a method of increasing the bond strength between layers. Currently, there are two assumptions as to why the interface bond remains the weakest point of 3D printed elements: (1) mechanical bond: the interlayer contains voids due to the stiffness of the fresh concrete layers. This is mechanical bond problem of not being malleable enough to penetrate and anchor into the surface pores; (2) chemical bond: the chemical hydration process across layers is hindered and compromised. These assumptions would therefore be exacerbated, succumbed to the stiffening effects of cement subject to time gap intervals in layer depositions [12,20]. In Part II, we hypothesize that the mechanical bond is the primary issue and aim to increase bond strength by investigating the mechanical effects of increased contact area at the layer interface. Taking inspiration from the construction techniques of bricks and mortar, an experimental process was performed through the application of malleable cementitious pastes on top of the 3D printed specimens prior to subsequent overlay layer deposition.

12.2 EXPERIMENTAL PROCEDURES

12.2.1 Materials and Mix Proportions

Ordinary Portland cement (OPC) conforming to the Australian Standard, AS 3972 [21] general purpose cement (Type GP) was used in this study. Table 12.1 presents the chemical composition and loss on ignition (LOI) of the OPC determined by X-ray Fluorescence. The total does not add up to 100% because of rounding-off of the percentages. The percentages of C3S, C2S, C3A, and C4AF as the main constituents of OPC were 57.59%, 14.87%, 4.10%, and 13.94%, respectively, calculated from the chemical composition shown in Table 12.1 using the Bogue formula [22]. In Part I of the study, sieve-graded high silica purity sands with two different particle sizes were used. The finer silica sand denoted as "FS" with a maximum particle size of 300 mm was supplied by TGS Industrial Sand Ltd., Australia. The coarser silica sand denoted as "CS" with a maximum particle size of 500 mm was supplied by Building Products Supplies Pty Ltd., Australia. Fig. 12.1 presents the particle size distribution of the OPC and silica sands used in this study determined by using a CILAS

Table 12.1 Chemical composition of the ordinary Portland cement

Chemical	Component (wt%)
Al_2O_3	4.47
SiO_2	20.34
CaO	62.91
Fe_2O_3	4.58
K_2O	0.29
MgO	1.24
Na_2O	0.31
P_2O_5	–
TiO_2	–
MnO	–
SO_3	2.58
LOI[a]	3.27

[a]Loss on ignition.

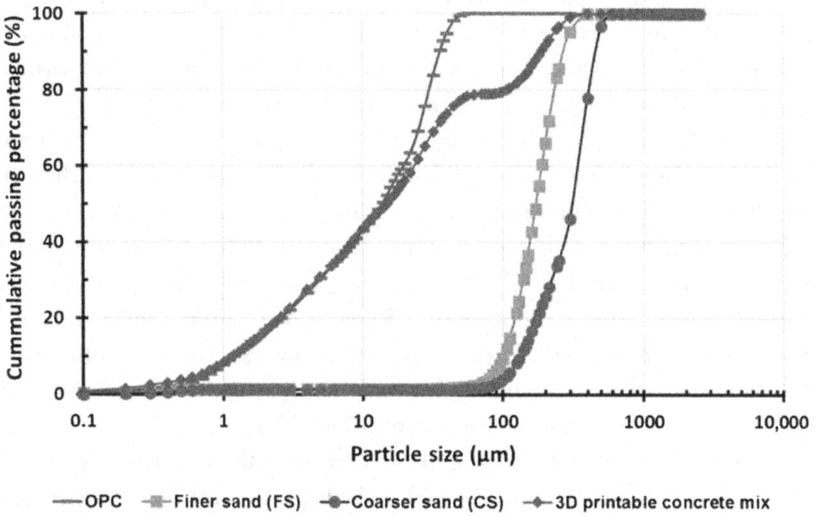

Figure 12.1 Particle size distribution of ordinary Portland cement, silica sands, and 3D printable concrete mix used in Part I.

particle size analyzer model 1190. In Part II, a mixture of two types of premium graded sand (supplied by Sibelco Pty Ltd., namely the 16/30 and the 30/60) was used. Sand Type-1 (16/30) is a course sand with a particle size of 700 μm and sand Type-2 (30/60) is a finer sand with a particle size of 300 μm.

Table 12.2 Mix proportions of the 3D printable concrete used in Part I

OPC	FS	CS	Water
1.0	0.375	1.125	0.38

Note: All numbers are mass ratios of the OPC weight.
CS, coarser sand; *FS*, finer sand; *OPC*, ordinary Portland cement.

Table 12.2 presents the mix proportions of the 3D printable concrete used in Part I of this study. According to Le et al. [13], the most critical fresh properties of a 3D printable concrete mixture are extrudability and buildability, which have mutual relationships with workability and open time. In other words, the fresh concrete needs to be able to be extruded through a nozzle to form small filaments (i.e., extrudability requirement). In addition, the fresh concrete needs to be stiff enough to retain its shape after being extruded and support further layers without substantial deformation and collapsing due to self-weight and yet provide a suitable bond between layers (i.e., buildability requirement) [13]. The methods used to measure the fresh properties are described in detail in Le et al. [13]. Several trials were performed before arriving at the final mix proportions (presented in Table 12.2) which satisfied the extrudability and buildability requirements. Initially, a mix with water to cement ratio (w/c) of 0.5 and CS sand to cement ratio (s/c) of 3.0 was trialed. It was observed that the mix was too flowable and did not hold its shape when extruded. Therefore, the w/c of the mix was reduced to 0.38. It was found that some of the extruded filaments were fractured during extrusion. Therefore, the s/c was reduced to 1.5. However, some cracks were still observed on the surface of the extruded filaments. To tackle this issue, 25% of CS sand was replaced by FS sand. The particle size distribution of the 3D printable concrete mix is given in Fig. 12.1.

In Part II of this study, the mix ratio in reference to weight was adopted as 1 (OPC): 1 (Type-1 sand): 0.5 (Type-2 sand): 0.36 (water) and four various OPC paste mixtures with a w/c of 0.36 were used as the bonding interface for the 3D printed layers. Of these four paste mixtures, three contained concrete additives (supplied by BASF Pty Ltd.). Paste-1 employed the use of a retarder (MasterSet RT 122), Paste-2 contained a viscosity modifying agent (MasterMatrix 362), and Paste-3 adopted a slump retainer (MasterSure 1008). Paste-1 and Paste-3 consisted of 0.3 mL of additive per 100 g of cement, while Paste-2 contained 0.9 mL of additive per 100 g cement.

12.2.2 Mixing, Printing, Curing, and Testing of Specimens

The mixture in Part I was prepared in a 3 L Hobart mixer. The OPC and silica sands were dry-mixed for about 1 minute at low speed. Then, tap water was gradually added and the mixing was continued for about 4 minutes. After the ingredients were thoroughly mixed to achieve a consistent fresh state (visually assessed), the workability and setting time of the fresh mixture were measured.

A custom-made 3D printer was designed and manufactured to simulate the extrusion-based 3DCP process (Fig. 12.2A). A piston-type extruder was used in this printer, where the material is extruded by means of a piston through a cylinder measuring 50 mm in diameter and 600 mm in length. The fresh mixture was gradually loaded into the extruder. Moderate external vibration was applied to the extruder to ensure adequate compaction of the mixture inside the extruder. A nozzle with a 25 mm × 15 mm opening and having an angle of 45 degrees to the build platform was designed and attached to the end of the extruder (Fig. 12.2B). The extruder could move in horizontal and vertical directions.

The specimens were extruded by moving the extruder in the horizontal direction at a constant speed. In Part I, the printed specimens consisted of two extruded layers. Three different delay times, namely 10, 20, and 30 minutes were chosen as the print-time intervals between layers. The first layer was extruded with the dimensions of 250 mm (L) × 25 mm × (W) 15 mm (H) (Fig. 12.2C). Then, after the given delay times, the second layer was extruded on top of the first layer.

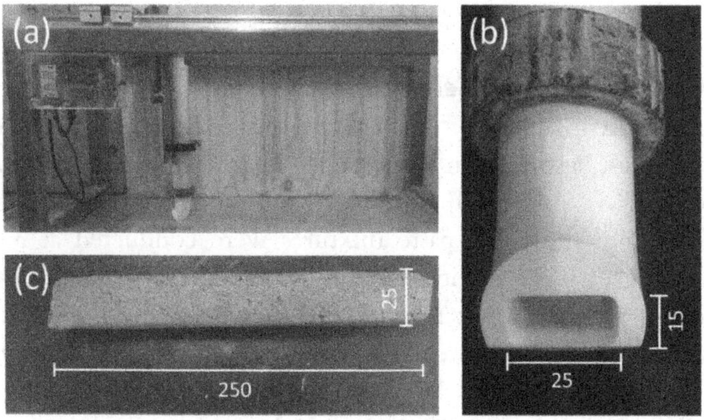

Figure 12.2 (A) Custom-made extrusion-based 3D concrete printer, (B) 45 degrees nozzle with a 25 mm (W) × 15 mm (H) opening, (C) One layer of extruded concrete with dimensions of 250 mm (L) × 25 mm (W) × 15 mm (H).

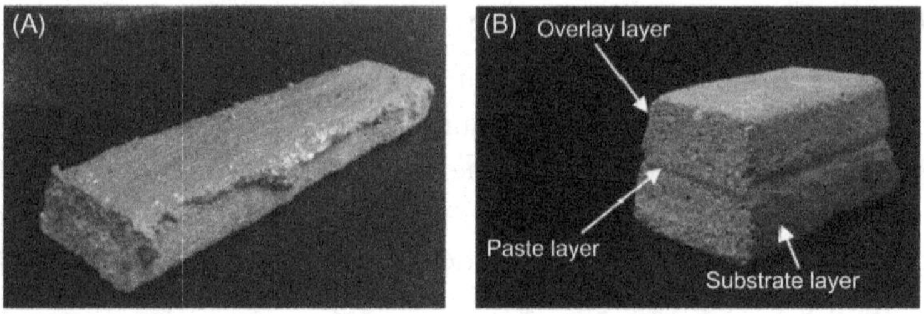

Figure 12.3 (A) Application of the paste layer on top of substrate layer. (B) A cured and cut 50 mm long specimen.

In Part II, samples were fabricated of two printed layers with an OPC paste applied in-between and with a constant deposition rate. The process began by loading the 3D printing apparatus with the mortar mix and then depositing the first layer (substrate layer) for all samples. Immediately after deposition, the application of the relevant OPC paste mixture was applied by means of a 25 mm wide brush (ideally replicating the application of a second nozzle attached and simultaneous with the extruder) (see Fig. 12.3A). In creating the second layer (overlay layer) another batch of the mortar mix was loaded into the printing apparatus cylinder and deposited, maintaining a 15-minute time gap interval from the substrate layer deposition.

All specimens were covered with damp hessian and left in the laboratory at ambient temperature (23°C ± 3°C) for 7 days. All specimens were tested 7 days after printing. The tests conducted in this study are described next.

12.2.2.1 Workability and Setting Time

To determine workability of the fresh 3D printed mixture used in Part I, a minislump test, also known as spread-flow test, was conducted in accordance to ASTM C1437 [23]. In Part II, minislump tests of the fresh 3D printed mix and the OPC paste mixtures were conducted at 3 minutes from the time of water addition to the dry materials. Materials from the same batches were then left to sit for 15 minutes before being subjected to the same testing procedure. This was to measure the change in flow/stiffness after a 15-minute delay time, replicating the 3D printing procedure outlined in Section 12.2.2. In Part I, the initial and final setting times of the fresh mixture were also measured in accordance to ASTM C191 [22] using Vicat needle apparatus.

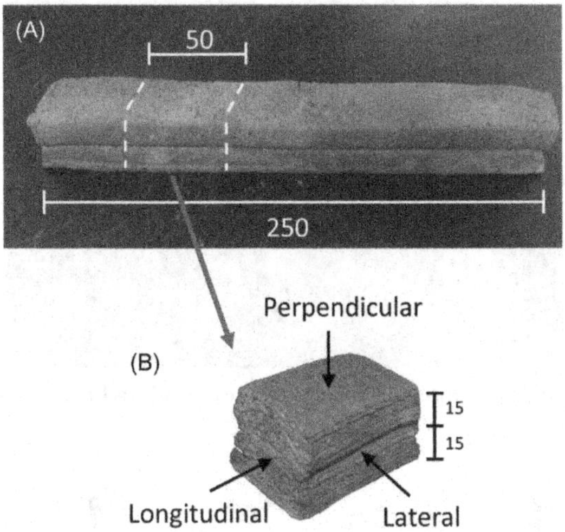

Figure 12.4 Cutting diagram and testing directions for compression testing of printed concrete used in Part I.

12.2.2.2 Compressive Strength

In Part I, to measure the compressive strength of printed concrete, 50 × 25 × 30 mm specimens were sawn from the 250 × 25 × 30 mm printed samples (Fig. 12.4A) and tested in one of the three directions, namely perpendicular, lateral, and longitudinal (Fig. 12.4B). At least six specimens were tested for each testing direction and delay time. In Part II, 25 mm cube samples were cast to measure the compressive strength of the 3D printable mix and the OPC paste mixtures. Samples were cured for 7 days under the same conditions as the 3D printed samples. All specimens were tested in uniaxial compression under load control at the rate of 20 MPa/min.

12.2.2.3 Flexural Strength

To measure the flexural strength of printed concrete in Part I, at least three 250 × 25 × 30 mm specimens were loaded in one of two directions, namely perpendicular and lateral (Fig. 12.5). All specimens were tested in a three-point bending test setup with a span of 150 mm under displacement control at the rate of 1.0 mm/min.

12.2.2.4 Interlayer Strength

To measure the interlayer strength of printed concretes in both Part I and Part II, 50 × 25 × 30 mm specimens were sawn from the

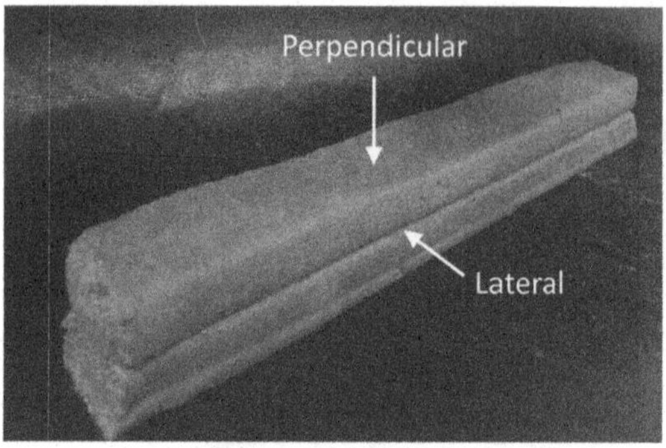

Figure 12.5 Testing directions for flexural testing of printed concrete used in Part I.

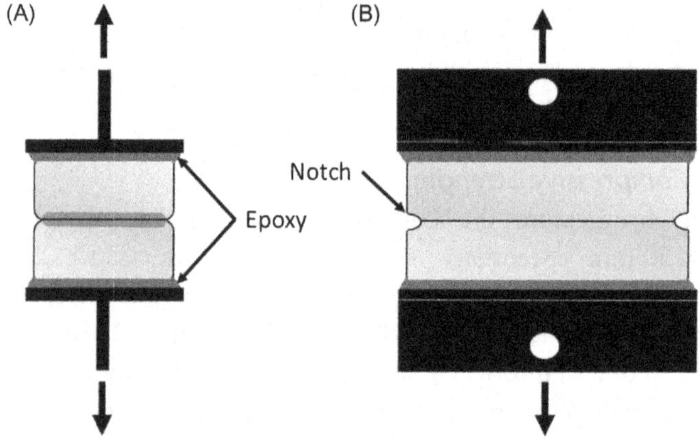

Figure 12.6 (A) Longitudinal view of test specimen for interlayer strength measurement. (B) Perpendicular view of test setup.

250 × 25 × 30 mm printed samples and loaded in uniaxial tension. The test setup and specimen for the interlayer strength measurement are shown in Fig. 12.6. Two metallic brackets were epoxy glued on the top and bottom of the printed specimen. At the interface of the layers, a small notch with an approximate depth of 5 mm was made on both cross-sections of the specimen to ensure failure of the specimen at the interface. The interlayer strength test was conducted using MTS testing machine under displacement control at the rate of 1 mm/min. Care was taken to align the specimen in the machine to avoid any eccentricity. At least six specimens were tested for each delay time in Part I and each of the four OPC paste mixtures in Part II.

12.3 RESULTS AND DISCUSSIONS

12.3.1 Workability and Setting Time

The spread diameter of the fresh mixture used in Part I was 182 mm "after the standard 25 drops of the flow table" in accordance to ASTM C1437. It should be noted that before the initiation of the 25 drops of the flow table the spread diameter was 100 mm only, equal to the bottom diameter of the flow cone. This means that the fresh mixture was almost a zero-slump mix. The initial and final setting times were measured to be 142 and 284 minutes, respectively.

Fig. 12.7 presents the results from the flow table tests of the mortar and paste mixtures used in Part II of this study. The extruded mortar mix was the overall stiffest mix in the group. A spread diameter of 166 mm was obtained when tested at 3 minutes, and a spread diameter of 151 mm was obtained when tested at 15 minutes. Among the paste mixtures and considering the varied accuracy of the testing regime, spread diameter results were relatively similar. However, it was observed that the OPC paste had a greater rate of stiffening after 15 minutes compared to the other paste mixtures (214–184 mm). Although Pastes-1, -2, and -3 contained admixtures, these pastes expressed minimal change in flow after the 15-minute period, with Paste-3 exhibiting slightly higher flow. The

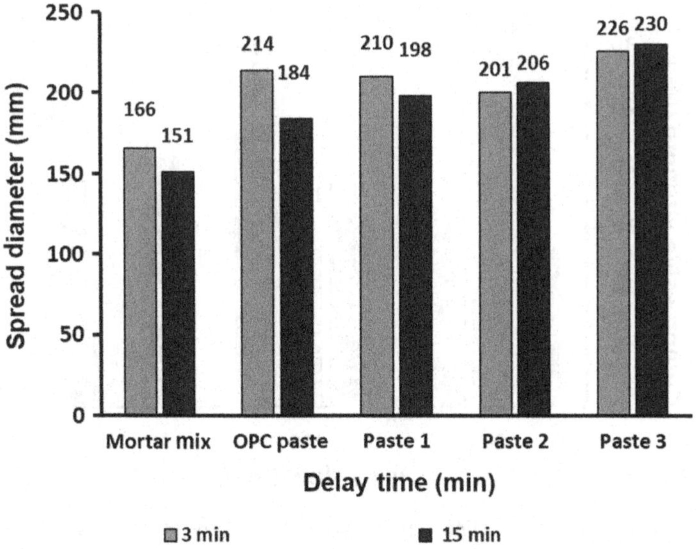

Figure 12.7 Flow table test results for the paste mixtures.

main observation that can be drawn from these results confirm the aims of the experiment; that paste mixtures with additives slow the stiffening effects of cement to maintain a malleable bond interface.

12.3.2 Compressive Strength

Fig. 12.8 presents the average compressive strength of the printed concrete used in Part I for different delay times and testing directions. As can be seen, the anisotropic phenomenon was observed in the compressive strength of the printed concrete depending on the orientation of the loading relative to the layers. It should be noted that geometry, dimensions, and surface quality may have contributed to this difference. This may well be the reason for the high variability observed in the results. However, Le et al. [14] and Panda et al. [24] also reported anisotropic phenomenon for the compressive strength of the printed concrete. In this study for both compression and flexural tests, surfaces of the printed specimens loaded in different directions were ground to have a smooth and flat surface while testing. Therefore, the influence of surface quality was mitigated. As can be seen in Fig. 12.8, the average compressive strength of the specimens tested in the lateral direction was lower than that of the specimens tested in longitudinal and perpendicular directions. This pattern was true irrespective of the delay time. A similar pattern was reported by Panda et al. [24]. The highest mean compressive strength was for the longitudinal

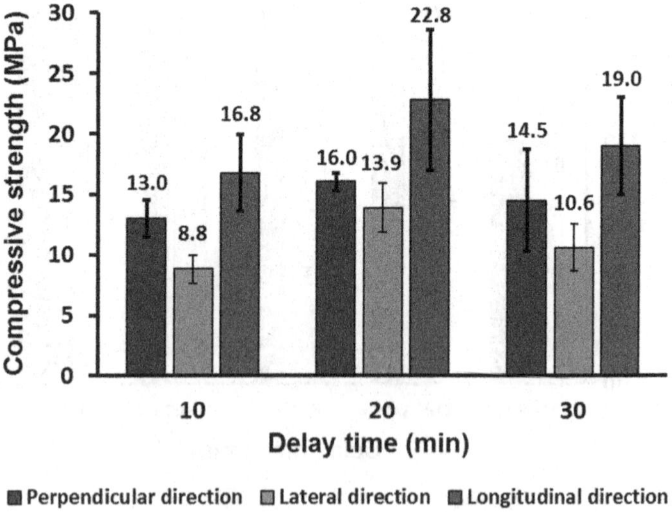

Figure 12.8 Compressive strength of the printed concrete used in Part I.

direction. This is likely due to high pressure exerted on the material in this direction during the extrusion process [25]. The lowest mean compressive strength was consistently found for the lateral direction. The lateral direction has the least amount of pressure during the setting process. Without any mold or formwork preventing the lateral flow and settling of the material, the fresh concrete is free to settle and expand causing weakness in this direction. The perpendicular direction is found to have mean strengths in between the longitudinal (extrusion direction) and lateral direction (free to expand and settle), since there is some level of pressure in the perpendicular direction during the setting process due to the self-weight of the concrete.

Regarding the effect of delay time on the compressive strength of the printed concrete in Part I, as can be seen in Fig. 12.8, the mean compressive strength of the specimens printed with a 20 minutes delay time was higher than that of the specimens printed with 10 and 30 minutes delay times. This pattern is true irrespective of the testing directions. The reason for this phenomenon is not apparent at this stage. Further research is necessary to understand this consistently expressed behavior of printed concrete.

The compressive strengths of the mixtures used in Part II are presented in Fig. 12.9. The results show that the extruded mortar mix had an average compressive strength of 34 MPa which was the lowest among the samples tested. This result can be drawn to being the only mix containing sand, thus, is assumed to have slightly different properties. The OPC paste and Pastes-1 and -2 had a similar compressive strength of 42–45 MPa. Paste-3 was observed to have the highest compressive strength. From

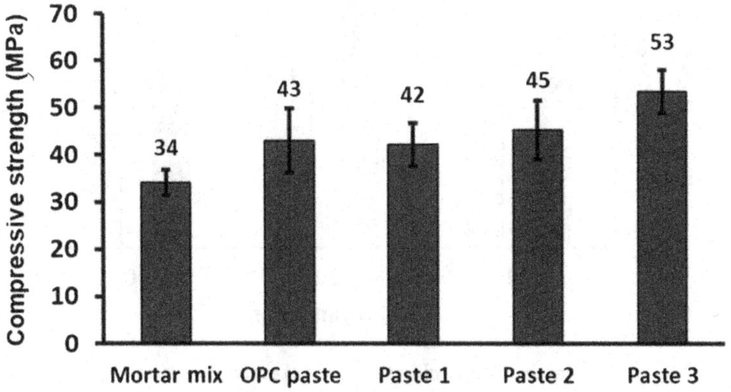

Figure 12.9 Compressive strength results of the 25 mm cube samples used in Part II.

these results we can draw that there are some major strength differences between mixtures. Since the mortar mix is the weakest of the samples, it can partly explain the weakness at the interface based on its strength. In the case of the no-paste mixture, the interface relies on the strength of the mortar mix. We must also assume strength as a factor of the lack of contact area due to the stiffness of the fresh mix. The pastes exhibit similar compressive strengths, except Paste-3 which is comparatively higher. When assessing the range of standard deviation, variations between these pastes were minimal.

12.3.3 Flexural Strength

Fig. 12.10 presents the average flexural strength of the printed concrete used in Part I for different delay times and testing directions. Like the compressive strength results, the orthotropic phenomenon was also observed in the flexural strength of the printed concrete depending on the testing direction. The directional dependency of the compressive and flexural strengths of the printed concrete is believed to be an inherent feature of the layer-by-layer 3D printing process [25]. As can be seen in Fig. 12.10, for each delay time the mean flexural strength of the specimens tested in the perpendicular direction was higher than that of the specimens tested in the lateral direction. This result is consistent with the

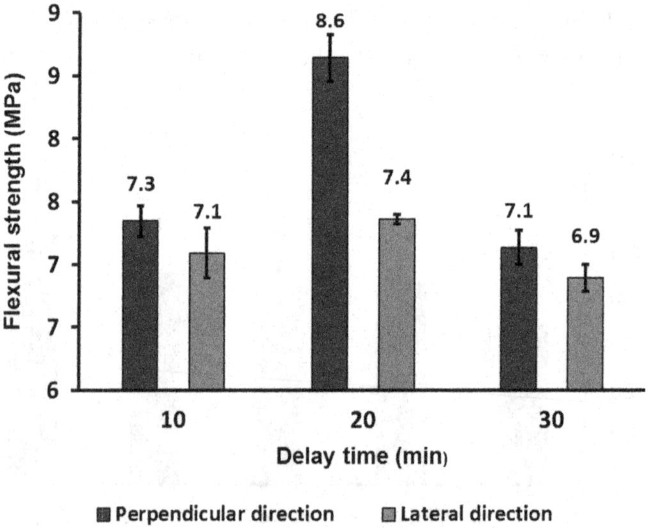

Figure 12.10 Flexural strength of the printed concrete used in Part I.

compressive strength results. It can be argued that at 10 and 30 minutes delay times, the difference in flexural strengths between the perpendicular and lateral directions are within the scatter plot of the results. However, Le et al. [14] and Panda et al. [24] also reported a orthotropic phenomenon for the flexural strength of the printed concrete. In the case of testing flexural specimens in the perpendicular direction, the mid-span of the bottom layer governs the flexural strength where the maximum tensile stress takes place. The bottom layer was most likely well-compacted due to the self-weight of the top layer. In addition, the water to cement ratio of the bottom layer may have been reduced if some water bled out when printing the top layer [14]. The combined effects result in a higher load capacity of the bottom layer, thereby the higher flexural strength in the perpendicular direction compared to the lateral direction.

Comparisons between Figs. 12.8 and 12.10 reveal that the mean flexural strength in the perpendicular direction was 3%–16% higher than that in the lateral direction depending on the delay time. However, the mean compressive strength in the perpendicular direction was 15%–48% higher than that in lateral direction depending on the delay time. Therefore, it can be said that the extrusion process introduced relatively little orthotropic behavior in terms of the flexural strength as compared to the compressive strength of the printed concrete.

Regarding the effect of delay time on the flexural strength of the printed concrete, the flexural strength of the specimens printed with 20 minutes delay time was higher than that of the specimens printed with 10 and 30 minutes delay times, which is like the pattern observed for the compressive strength results. This pattern is true irrespective of the testing directions. This consistently expressed phenomenon by 3D printed concrete needs further research to identify the reasons for these results.

12.3.4 Interlayer Strength

12.3.4.1 Results of Part I

Fig. 12.11 presents the average interlayer strength of the printed concrete used in Part I for different delay times. All specimens printed with different delay times failed at the interface between the first and second layers (Fig. 12.11(B)). As can be seen in Fig. 12.11(A), the inter-layer strength of the specimens printed with 10 and 30 min delay times were surprisingly identical, but higher than that of the specimens printed with 20 min delay time. This pattern is opposite to the pattern observed for the compressive and flexural strengths of the printed concrete with respect to

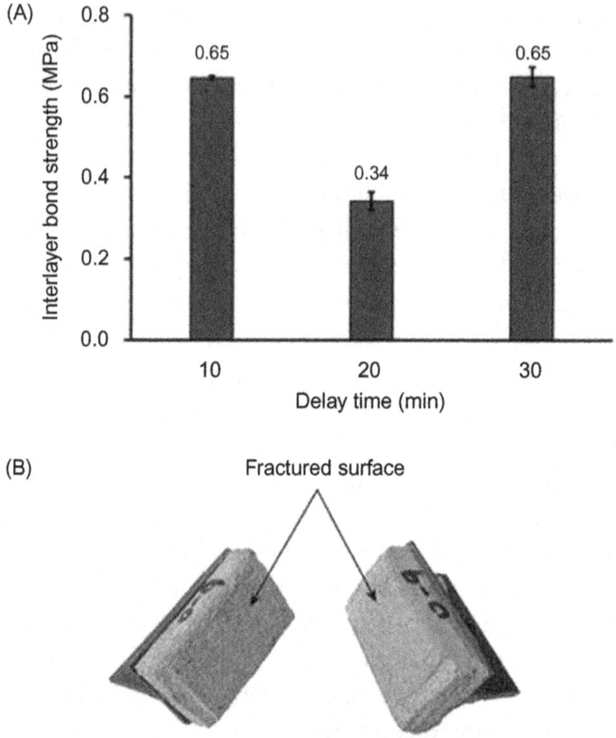

Figure 12.11 (A) Interlayer strength of the printed concrete. (B) Failure mode (fractured surfaces) of interlayer strength specimens.

different delay times. The reason for this behavior was further investigated as inter-layer strength is an important parameter in printing concrete, and an in depth understanding of the factors affecting the inter-layer strength is critical for the success of the 3D concrete printing process. As mentioned in Introduction, Le et al. [14] reported that theinter-layer strength is expected to decrease with increasing delay time. This is intuitive because as the adhesion between old and new layers reduces when the setting times of the layers are moving further apart. However, the pattern observed in this study is in contrast to the widely believed trend as reported by Le et al. [14]. It was hypothesized that some moisture on the surface of the concrete is required to promote the interlayer bond. The surface moisture is necessary to maintain adequate workability (or malleability) of the concrete surface to develop the interlayer bond. As soon as the concrete is extruded, significant moisture can be observed on the surface of concrete, which is a result of the extrusion process that brings

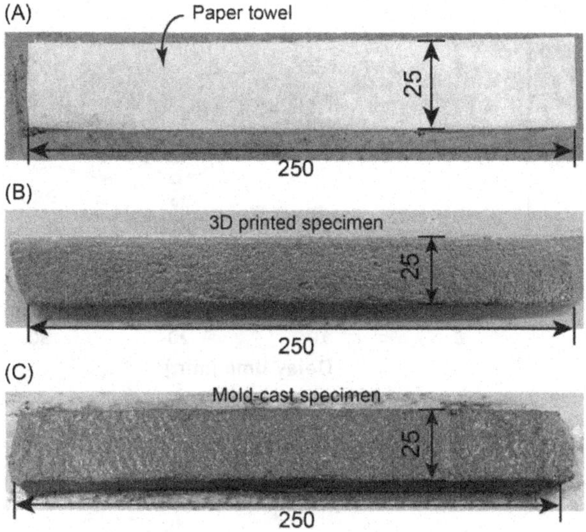

Figure 12.12 Experimental setup for surface moisture contents measurement. (A) A cut-to-size paper towel to be placed on the surface of the single layers and cast specimens for 20 seconds. (B) A 3D printed specimen. (C) A mold cast specimen.

additional moisture to the surface. After the extrusion process, the concrete is exposed to the air during the delay time when the surface moisture evaporates leaving the surface dry and rigid. However, further moisture is brought to the surface by bleeding of water from the concrete when the delay time is sufficiently large (30 minutes). Further investigation was carried out to test the effect of bleed rate on the interlayer strength.

To investigate this hypothesis, the surface moisture contents of the printed and conventional mold-cast concrete were measured for different delay times. The experimental setup is shown in Fig. 12.12. For the printed concrete, three single layers measuring 250 × 25 × 15 mm were extruded. For conventional mold-cast concrete, three specimens measuring 250 × 25 × 15 mm were cast and compacted using a vibrating table. After each given delay time, a cut-to-size paper towel measuring 250 × 25 mm was placed on the surface of the single layers and cast specimens for 20 seconds. The surface moisture was absorbed by the paper towel during this time. The weight change of the paper towel was measured. The temperature and relative humidity in the laboratory were maintained at 24°C and 65%, respectively, during the measurements. For each delay time, at least three measurements were recorded. The results of

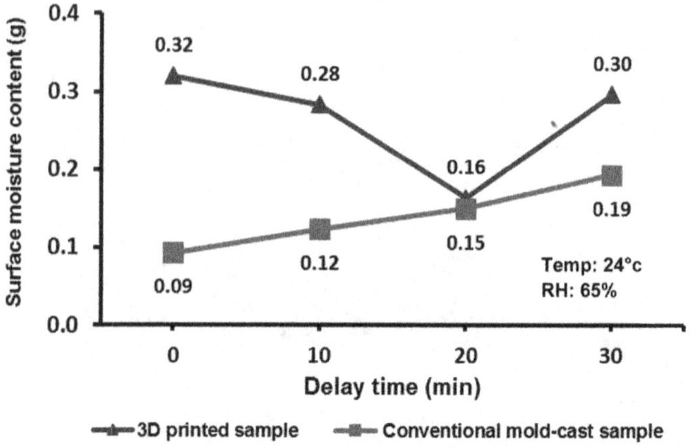

Figure 12.13 Surface moisture contents of the 3D printed and conventional mold-cast samples.

surface moisture contents are presented in Fig. 12.13. As can be seen, in the conventional mold-cast concrete the surface moisture contents continuously increased with increasing delay time. This is believed to be due to bleeding of cast concrete which caused a continual rise of the moisture on the concrete surface. However, in the case of the printed concrete, from 0 to 20 minutes the surface moisture contents reduced with increasing delay time; but from 20 to 30 minutes the surface moisture contents increased with increasing delay time. The pattern observed for the surface moisture content of the printed concrete from 10 to 30 minutes delay times were similar to that observed for the interlayer strength of the printed concrete (Fig. 12.11A). The surface moisture content for 20 minutes delay time was significantly lower than that for 10 and 30 minutes delay times, providing a lower level of moisture at the interface of the old and new layers, resulting in lower interlayer strength compared to 10 and 30 minutes delay times.

The surface moisture content is affected by several factors including printing process (e.g., type and size of the extruder, pressure) and material condition (such as mix composition, bleeding rate, and evaporation rate of the mix). Thus, as discussed in the following, further research was conducted to understand the underlying reasons for the pattern observed for the surface moisture content of the printed concrete with different delay times.

The bleeding rate of the concrete mix used for printing was measured in accordance with AS1012.6 [26]. The temperature and relative humidity in the laboratory were maintained at 24°C and 65%, respectively, during

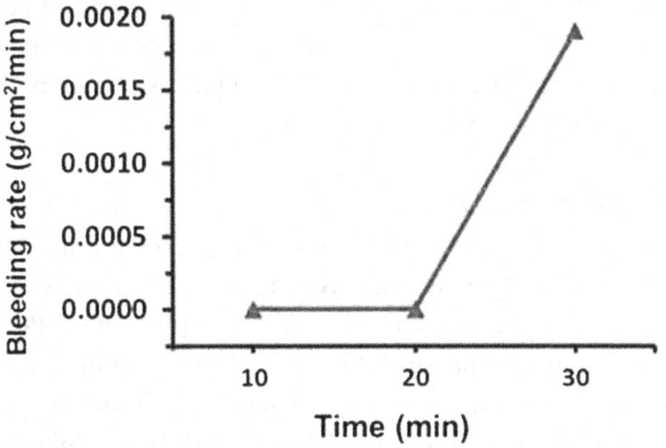

Figure 12.14 Bleeding rate from the surface of the printed concrete.

the measurements. As can be seen in Fig. 12.14, no bleeding was observed from 0 to 20 minutes after extruding the single layer. A sudden increase in the bleeding rate was observed from 20 to 30 minutes delay times, which clearly explains the reason for the increase observed in Fig. 12.13 for the surface moisture contents from 20 to 30 minutes delay times.

The reduction in the surface moisture content from 0 to 20 minutes delay times (Fig. 12.13) is because there was no bleeding within this time, as shown in Fig. 12.14, but there was constant evaporation of moisture from the surface of the printed concrete. The rate of evaporation of moisture from the surface of concrete is widely studied. For concrete at 24°C and 65% relative humidity, the rate of evaporation can be calculated to be 0.08 $kg/m^2/h$ [27]. This is assuming no wind is present, which was the case inside the laboratory. The loss of moisture from the 250 × 25 mm surface over a period of 20 minutes at this moisture loss rate calculates to be 0.16 g, which is precisely the amount of moisture loss observed from 0 to 20 minutes (Fig. 12.13).

The reason for the higher surface moisture content at zero-minute delay time (i.e., at the time when the first layer was extruded) is due to pressurized bleeding, which is caused by the high pressure exerted on the fresh mixture during the extrusion process. When the fresh mixture inside the extruder is subjected to a high pressure exerted by the piston, a lubricating layer is formed near the inner surface of the extruder which is caused by the shear-induced particle migration [28]. In other words, when the fresh mixture flows in the extruder under high pressure, the shear strain is minimum at the center of the cross-section and increases as it approaches the inner surface of the extruder. Thereby, aggregates tend to concentrate toward the center and a proportion of the cement paste phase forms the lubricating layer [29]. Therefore, when the mixture exists the extruder, the moisture content on the surface of the printed layer is the highest, which is caused by the lubricating layer forming on the surface.

12.3.4.2 Results of Part II

Fig. 12.15 presents the average interlayer strength of the mixtures used in Part II. The samples with no interlayer paste were observed to produce the lowest bond strength at an average of 0.44 MPa. The OPC paste samples attained a higher bond strength of 0.72 MPa, while the pastes containing admixtures saw an even greater increase in bond strength. Paste-1 generated the highest strength results of 1.26 MPa, while Paste-2 and Paste-3 mixtures presented similar results of 0.98 and 1.00 MPa. All samples conformed to a uniform fracture through the interlayer, with paste samples fracturing from the overlay interface.

Out of the sample size it can be noted that a large scatter of results is present within the specimen range, which is common in tensile testing concrete [30]. Although there is a wide variation within the data range, a trend can be drawn between samples with a paste layer at the interface and the control samples utilizing typical 3D printed construction (no paste). Applying an OPC paste without admixtures provided a notable bond strength increase of approximately 60%. While pastes mixed with an additive were observed to produce a notably higher bond strength of approximately 120%–180% higher than the no paste samples. Although the addition of a paste layer proved to provide an interlayer

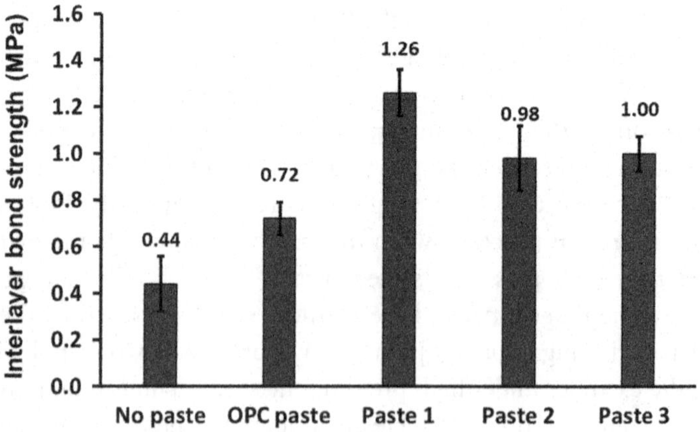

Figure 12.15 Bond strength test results of the mixtures used in Part II.

bond strength increase, the variations are investigated further through the testing of material properties such as paste strength and flow.

When making a comparison of the bond strength to the material strength we can see a trend within the data of Fig. 12.9 that the paste mixtures are stronger than the mortar mixture. Although Paste-3 had the strongest compressive strength, the bonding strength was not exhibited as the strongest. The strength of the material is only one component in the equation of bond strength. Beushausen [30] found through his study on concrete to concrete bonding, that the workability in the overlaying substance is a greater influencing factor in bonding than the material strength. Observing the trend in the mixture flow results shown in Fig. 12.7, the correlation of workability/flowability to bond strength has similarities. The paste samples exhibit higher bond strength to no paste samples due to overall workability. While comparing the OPC paste to the pastes containing admixtures, the paste with admixtures display a maintained flow-ability after a 15-minute time gap. This confirms Beushausen's [30] findings as the paste mixtures containing additives attain a higher bond strength attributed to the higher degree of workability. The conclusions that cannot be drawn through these test results are to how Paste-1 samples achieved the highest interlayer bond strength, while Paste-3 displayed superior material properties regarding workability and strength.

12.4 CONCLUSIONS

The effect of delay time (i.e., the printing time interval between layers) on the mechanical properties of extrusion-based 3D printed concrete was investigated in Part I of this study. Specimens with different delay times were printed and the compressive, flexural, and interlayer strengths were measured. The effect of the testing direction on the compressive and flexural strengths of the printed specimens was also investigated. The following conclusions were drawn from Part I of this study:

1. One of the major factors affecting interlayer strength is the moisture level at the surface between the layers. If the surface is dry, it does not have the workability (or malleability) for the bond to develop.
2. The moisture level between the layers is a function of many parameters including the printing process, evaporation rate, and bleeding rate of the mixes, as well as the level of moisture expelled to the surface during the extrusion process. However, the relationship between moisture level and interlayer strength is universal.

3. For the mix tested in this study, the moisture level on the surface first decreased and then increased with increasing delay times. Exploring this phenomenon has helpful to identify the major influence of surface moisture level on the interlayer strength.
4. The compressive strengths of printed concrete layers are different in all three directions, exhibiting an orthotropic strength behavior. The highest mean compressive strength was observed in the longitudinal direction where the extrusion pressure is likely to increase compaction and strength. The lowest mean compressive strength was observed in the lateral direction. The mean compressive strength in the perpendicular (vertical) direction was in between the strength of the other directions. These trends are true irrespective of the delay time.
5. Flexural strengths measured in the longitudinal and lateral directions also showed orthotropic behavior. Irrespective of the delay time, the mean flexural strength in the perpendicular (vertical) direction was higher than in lateral direction.
6. Both compressive and flexural strengths first increased and then decreased with increasing delay times between the layers. This behavior is found to be consistent although the reason behind this behavior is not clear at this stage and further research is needed.

In Part II of this study, increasing interlayer bond strength by the application of cementitious paste mixtures applied at the bonding interface between extrusion-based concrete 3D printed layers were investigated. A conventional printed sample with no paste layer was compared with samples employing the use of cementitious pastes, mixed with and without concrete additives. These methods where undertaken to increase the interlayer contact area and mechanical strength through surface pore anchorage. The effectiveness of these pastes was measured through uniaxial tensile tests, providing the ultimate interlayer bond strength. The material properties such as flow characteristics and compressive strengths were also measured to investigate any correlations with bond strength to further understand the bond mechanism. The following conclusions were drawn from Part II of this study:

1. Compressive strength of the mixtures was used to outline any strength differences in the layered mixtures. With a maintained w/c ratio of 0.36 used for all mixtures, it was observed that the paste mixtures were relatively similar in strength, but were comparatively stronger than that of the mortar mixture.
2. The flow characteristics of the mixtures show a similar trend to that of the strength characteristics. Although pastes containing additives

maintained more workability over a 15-minute time interval compared to the OPC paste.

3. Considering the strength of the pastes and the correlating workability of each, a relationship can be drawn concerning the interlayer bond strength. The use of a workable paste layer displayed a higher bond strength compared to conventional layer-on-layer deposition. With the addition of additives to offset the stiffening effects of cement over time, the bond strength is further enhanced. Thus, it is concluded that our assumption of increasing the surface area and mechanical anchorage are prime factors in generating strong interlayer bonds. Further research needs be undertaken into the hydration and chemical effects across layer interfaces.

ACKNOWLEDGMENTS

The authors acknowledge the support by the Australian Research Council Discovery Grant DP170103521, Linkage Infrastructure Grant LE170100168 and Discovery Early Career Researcher Award DE180101587. Taylor Marchment also acknowledges the Australian Government Research Training Program Scholarship for his PhD study.

REFERENCES

[1] ASTM F42, Standard Terminology for Additive Manufacturing Technologies, American Society for Testing and Materials, PA, 2015.
[2] T. Wohlers, 3D Printing and Additive Manufacturing State of the Industry, Wohlers Associates Inc., Colorado, 2014.
[3] B. Nematollahi, M. Xia, J. Sanjayan, Current progress of 3D concrete printing technologies, in: Proceedings of 34th International Symposium on Automation and Robotics in Construction, Taiwan, 2017, pp. 260–267.
[4] R.A. Buswell, R. Soar, A.G. Gibb, A. Thorpe, Freeform construction: mega-scale rapid manufacturing for construction, Autom. Constr. 16 (2) (2007) 224–231.
[5] C. Gosselin, R. Duballet, P. Roux, N. Gaudillière, J. Dirrenberger, P. Morel, Largescale 3D printing of ultra-high performance concrete – a new processing route for architects and builders, Mater. Des. 100 (2016) 102–109.
[6] G. Cesaretti, E. Dini, X. De Kestelier, V. Colla, L. Pambaguian, Building components for an outpost on the Lunar soil by means of a novel 3D printing technology, Acta Astronaut. 93 (2014) 430–450.
[7] R. Rael, V. San Fratello, Developing concrete polymer building components for 3D printing, in: Proceedings of the 31st Annual Conference of the Association for Computer Aided Design in Architecture, (ACADIA 11), Banff, Canada, 2011, pp. 152–157.
[8] M. Xia, J. Sanjayan, Method of formulating geopolymer for 3D printing for construction applications, Mater. Des. (2016) 382–390.
[9] M. Xia, J. Sanjayan, Post-processing methods for improving strength of geopolymer produced using 3D printing technique, in: Proceedings of International Conference on Advances in Construction Materials and Systems (ICACMS-2017), India, 2017, pp. 350–358.

[10] B. Khoshnevis, S. Bukkapatnam, H. Kwon, J. Saito, Experimental investigation of contour crafting using ceramics materials, Rapid Prototyping J. 7 (1) (2001) 32–42.
[11] A. Kazemian, X. Yuan, E. Cochran, B. Khoshnevis, Cementitious materials for construction-scale 3D printing: laboratory testing of fresh printing mixture, Constr. Build. Mater. 145 (2017) 639–647.
[12] B. Zareiyan, B. Khoshnevis, Interlayer adhesion and strength of the structures in contour crafting – effects of aggregate size, extrusion rate, and layer thickness, Autom. Constr. 81 (2017) 112–121.
[13] T.T. Le, S.A. Austin, S. Lim, R.A. Buswell, A.G.F. Gibb, T. Thorpe, Mix design and fresh properties for high-performance printing concrete, Mater. Struct. 45 (8) (2012) 1221–1232.
[14] T.T. Le, S.A. Austin, S. Lim, R.A. Buswell, R. Law, A.G.F. Gibb, et al., Hardened properties of high-performance printing concrete, Cem. Concr. Res. 42 (3) (2012) 558–566.
[15] V.N. Nerella, M. Krause, M. Näther, V. Mechtcherine, CONPrint3D – 3D printing technology for onsite construction, Concr. Australia 42 (3) (2016) 36–39.
[16] H. Beushausen, M.G. Alexander, Bond strength development between concretes of different ages, Magaz. Concr. Res. 60 (2008) 65–74.
[17] R.W. Gillette, Performance of bonded concrete overlay, ACI J. 60 (3) (1963) 39–49.
[18] M. Pigeon, F. Saucier, Durability of repaired concrete structures, in: Malotra (Ed.), Advances in Concrete Technology, CANMET, Ottawa, 1992, pp. 741–773.
[19] S. Austin, P. Robins, Y. Pan, Tensile bond testing of concrete repairs, Mater. Struct. 28 (1995) 249–259.
[20] B. Zareiyan, B. Khoshnevis, Effects of interlocking on interlayer adhesion and strength of structures in 3D printing of concrete, Autom. Constr. 83 (2017) 212–221.
[21] AS 3972, General Purpose and Blended Cements, Australia Standards, Australia, 2010.
[22] ASTM C150/C150M, Standard Specification for Portland Cement, American Society for Testing and Materials, PA, 2012.
[23] ASTM C1437, Standard Test Method for Flow of Hydraulic Cement Mortar, American Society for Testing and Materials, PA, 2007.
[24] B. Panda, S. Chandra Paul, M. Jen Tan, Anisotropic mechanical performance of 3D printed fiber reinforced sustainable construction material, Mater. Lett. 209 (2017) 146–149.
[25] N. Oxman, E. Tsai, M. Firstenberg, Digital anisotropy: a variable elasticity rapid prototyping platform, Virt. Phys. Prototyping 7 (4) (2012) 261–274.
[26] AS1012.6, Methods of Testing Concrete – Method for the Determination of Bleeding of Concrete, Australia Standards, Australia, 2016.
[27] P.J. Uno, Plastic shrinkage cracking and evaporation formulae, ACI Mater. J. 95 (4) (1998) 365–375.
[28] R.D. Browne, P.B. Bamforth, Tests to establish concrete pumpability, J. Proc. 74 (5) (1977) 193–203.
[29] J.H. Kim, S.H. Kwon, S. Kawashima, H.J. Yim, Rheology of cement paste under high pressure, Cem. Concr. Compos. 77 (2017) 60–67.
[30] H. Beushausen, The influence of concrete substrate preparation on overlay bond strength, Magaz. Concr. Res. 62 (11) (2010) 845–852.

CHAPTER 13

Properties of Powder-Based 3D Printed Geopolymers

Ming Xia, Behzad Nematollahi and Jay G. Sanjayan
Centre for Sustainable Infrastructure, Faculty of Science, Engineering and Technology, Swinburne University of Technology, Hawthorn, VIC, Australia

13.1 INTRODUCTION

Additive manufacturing, commonly known as three-dimensional (3D) printing is a group of emerging techniques for fabricating 3D structures directly from a digital model in a layer-wise form with less waste material. The current construction processes are labor intensive combined with a high rate of injuries [1], thus, making automation of construction processes a necessity. Development of new construction techniques such as 3D concrete printing (3DCP) opens many thresholds into the future of construction industry. Unlike the conventional approach of casting concrete into a formwork, 3DCP techniques will combine digital technology and new insights from materials technology to allow freeform construction without the use of expensive formwork. Freeform construction would enhance architectural expression, where the cost of producing a structural component will be independent of the shape, providing the much-needed freedom from the rectilinear designs [1].

In the past few years, different 3DCP techniques have been explored and adopted in the construction industry [2–5]. One of the techniques is the extrusion-based technique, similar to the fused deposition modeling method, which extrudes cementitious material from a nozzle mounted on a gantry, crane or a 6-axes robotic arm to print a structure layer by layer. This technique has been aimed at onsite construction applications such as large-scale building components with complex geometries. The weak interlayer strength between printed layers is one of the limitations of this technology when compared with cast-in-the-mold concrete.

Another technique is powder-based 3DCP, which is an offsite process designed to manufacture precast building components with fine details and intricate shapes. Powder-based 3DCP involves a sequential layering

3D Concrete Printing Technology
DOI: https://doi.org/10.1016/B978-0-12-815481-6.00013-0

process. Once a thin powder layer is spread, binder droplets are selectively jetted on the layer through print-head nozzles, causing the powder particles to bind each other. Repeating the described steps, the built part is completed and removed after a certain drying time and the unbound powder is removed by using an air blower.

There is a demand in the construction industry for building components which can only be made by using expensive formwork systems with the currently available construction processes. The powder-based 3DCP technique is highly suitable for producing small-scale building components such as panels, permanent formworks, and interior structures which can later be assembled onsite. One of the main limitations of powder-based 3DCP is the limited range of printable construction materials. The authors of this chapter have developed an innovative methodology for formulating geopolymer-based material for the requirements and demands of commercially available powder-based 3D printers. Geopolymers are aluminosilicate binders which can have superior properties, yet significantly lower CO_2 emissions to ordinary Portland cement. The utilization of industrial byproducts such as slag and fly ash in geopolymers is considered particularly beneficial. The 3D printed geopolymer models have shown good dimensional accuracy and mechanical properties which are comparable to the conventional mold-casting process (Fig. 13.1).

In the powder-based 3DCP process, it is necessary to control various parameters including powder properties, binder properties, and the

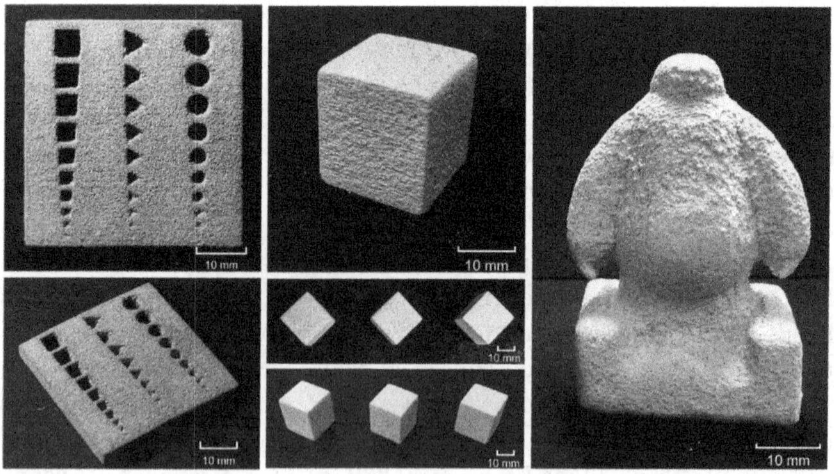

Figure 13.1 3D printed structures using the developed geopolymer powder.

printing parameters, all of which could substantially affect the accuracy and strength of the printed specimens. Among the controllable printing parameters, binder saturation is a decisive parameter that affects the binder/powder interaction directly in the powder-based 3DCP process [6], which is somewhat similar to water-to-cement (w/c) ratio in the conventional concrete casting process.

Binder saturation is defined as the ratio between the volume of deposited liquid binder (V_{Binder}) and the volume of pores in the powder bed (V_{Pores}). Two subvariables of binder saturation are employed by the 3D printer, namely shell and core. As can be seen in Fig. 13.2, the shell refers to the region that comprises the edges of the sample and parts of the interior area within the edges of the sample. The core refers to the remaining interior areas within the edges of the sample. The aim of using these two subvariables is to obtain printed samples in a short time, providing enough stability and preventing oversaturation which can lead to distortion [7,8]. If the binder saturation level is too high, the binder will bleed into the surrounding powder, while too low binder saturation will result in extremely weak green strength due to the poor bonding between powder particles [9]. Thus, having a good understanding and scientific insight of the quantitative effect of binder saturation in the powder-based 3DCP is of great importance.

The result of a successful powder-based 3DCP process after depowdering is a bound part ("green part") in which the powder particles are bound together by a weak binding force between powder particles. The green part is usually porous and has low strength and, thus, need to be further postprocessed to form a densified part with geometric precision [10]. In the previous study, the printed cubic structure gained a green

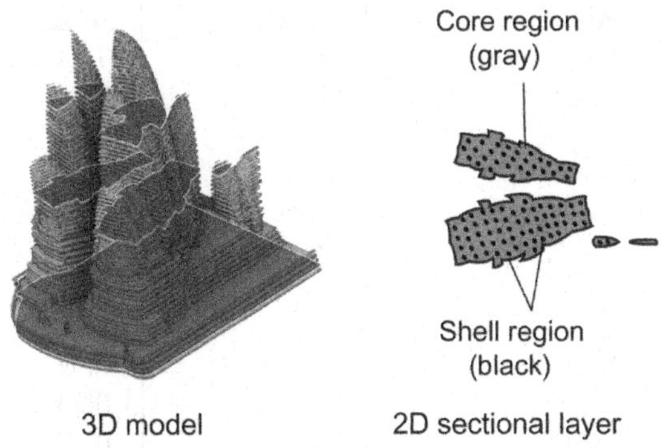

Figure 13.2 Schematic illustration of shell and core regions.

compressive strength of up to 0.9 MPa, which was sufficiently strong enough to handle depowdering and postprocessing. However, this strength is too low for construction applications and further improvement is necessary.

This study therefore aims to investigate the effect of binder saturation level on dimensional accuracy and compressive strength of 3D printed geopolymers. The knowledge gained can be employed for tailoring the printing parameters and improving the properties of the 3D printed geopolymer components depending on their specific applications. In the meantime, a number of postprocessing methods are explored to evaluate the effect of curing temperatures and curing mediums on the compressive strength of geopolymers produced via the powder-based 3DCP process.

13.2 EXPERIMENTAL PROCEDURES

13.2.1 Materials

Low calcium (Class F) fly ash and granulated ground blast furnace slag (slag) was supplied by Gladstone power station located in Queensland, Australia and Building Products Supplies Pty Ltd., Australia, respectively. The chemical composition of the fly ash and slag determined by X-ray Fluorescence analysis are shown in Table 13.1. The total percentages do not add up to 100% due to rounding off.

Table 13.1 The chemical composition of slag and fly ash (wt%)

Chemical	Component	
	Slag	Fly ash
Al_2O_3	12.37	25.56
SiO_2	32.76	51.11
CaO	44.64	4.30
Fe_2O_3	0.54	12.48
K_2O	0.33	0.70
MgO	5.15	1.45
Na_2O	0.22	0.77
P_2O_5	0.88	0.01
TiO_2	0.51	1.32
MnO	0.15	0.37
SO_3	4.26	0.24
LOI[a]	0.09	0.57

[a]Loss on ignition.

Saturated anhydrous sodium metasilicate solution was prepared using anhydrous sodium metasilicate powder and tap water. The N Grade sodium silicate solution was supplied by PQ Australia with a modulus ratio (M_s) equal to 3.22 (where $M_s = SiO_2/Na_2O$, $Na_2O = 8.9$ wt%, and $SiO_2 = 28.6$ wt%). The NaOH solution was prepared with a concentration of 8.0 M using NaOH beads of 97% purity supplied by Sigma-Aldrich and tap water.

13.2.2 Powder-Based 3DCP Process

Zprinter 150 which is a commercial powder-based 3D printer manufactured by Z-Corp, United States, was used in this study. The Zprinter 150 has a specific resolution of 300 × 450 dpi, a build volume of 182 × 236 × 132 mm, and a build speed of 2–4 layers/min. The Zprinter 150 uses an HP11 print head (C4810A) which has an array of 304 nozzles with a single droplet size of 18 pL. The mechanism of controlling the binder solution is a noncontinuous approach called the drop-on-demand technique which has been widely used in contemporary desktop printing systems.

A 20 mm cube model designed by using SolidWorks software was used for printing. The powder layer thickness was set to 0.1016 mm with different binder saturation levels, as detailed in Table 13.2. It should be pointed out that the Zprinter 150 allows for binder saturation levels in the range of 21%–170%. The initial trials revealed that geopolymer samples printed using binder saturation level below 75% exhibited extremely low strength and the samples could not hold their shapes during the depowdering process. Therefore, binder saturation levels ranging from 75% to 170% were selected in this study.

After printing, the cubes were left inside the powder bed at room temperature for 6 hours before the depowdering process to remove any

Table 13.2 Binder saturation levels used in powder-based 3D printed geopolymer

Sample ID	Saturation level (%)		V_{Binder}/V_{Pores}	
	Shell	Core	Shell	Core
S75C75	75	75	0.183	0.092
S100C100	100	100	0.244	0.122
S125C125	125	125	0.305	0.153
S150C150	150	150	0.366	0.183
S170C170	170	170	0.415	0.208

unbound powder particles. The green linear dimensional accuracy and green compressive strength were then tested for all specimens.

13.2.3 Postprocessing Procedures

For the S100C100 samples, different postprocessing procedures were conducted. The specimens were divided into two groups. In the first group, the effect of curing temperatures on compressive strength was investigated. In this group, the green samples were immersed in saturated anhydrous sodium metasilicate solution for 3 and 7 days at different curing temperatures. The curing temperatures were 25°C, 40°C, 60°C, and 80°C with a variation of ±1°C. The optimum temperature obtained from the result of the first group was used as the curing temperature for the second group to study of the effect of curing mediums on strength. In the second group, the green samples were immersed in seven curing mediums, namely, one tap water, three alkaline solutions, and three fly ash based geopolymer slurries.

The following alkaline solutions were used:

1. Solution I: composed of saturated anhydrous sodium metasilicate solution with SiO_2/Na_2O ratio of 1.0 (100% w/w).
2. Solution II: composed of saturated anhydrous sodium metasilicate solution with SiO_2/Na_2O ratio of 1.0 (71.4% w/w) and 8.0 M NaOH solution (28.6% w/w).
3. Solution III: composed of N Grade sodium silicate solution with SiO_2/Na_2O ratio of 3.22 (71.4% w/w) and 8.0 M NaOH solution (28.6% w/w).

The fly ash-based geopolymer slurries were prepared by mixing fly ash and saturated anhydrous sodium metasilicate solution (Solution I) with mass ratios of 0.1:1(Slurry I), 0.5:1 (Slurry II), and 1:1 (Slurry III).

13.2.4 Characterizations of 3D Printed Specimens

13.2.4.1 Linear Dimensional Accuracy

A digital caliper with an accuracy of 0.01 mm was used to measure dimensions of green cube samples in three directions, namely *X*-direction (the direction of binder jetting), *Y*-direction (the direction of powder layer spreading), and *Z*-direction (the direction of layer stacking). The green linear dimensional error was calculated by Eq. (13.1).

$$\text{Error} = L_{\text{actual}} - L_{\text{nominal}} \tag{13.1}$$

where L_{actual} is the measured length, whereas $L_{nominal}$ is the length of the digital model.

A population of 10 samples for each testing direction was used. For each sample, three measurements were taken for each testing direction and the mean errors were calculated. The mean errors were used for the assessment of green linear dimensional accuracy.

13.2.4.2 Compressive Strength

The compressive strength of both green and postprocessed 3D printed geopolymer samples in both the *X*-direction and *Z*-direction were measured under load control at the rate of 0.33 MPa/s. A population of 10 samples for each testing direction was tested.

13.3 RESULT AND DISCUSSIONS

13.3.1 Effect of Binder Saturation Level

13.3.1.1 Green Linear Dimensional Accuracy

Fig. 13.3 presents the results of the linear dimensional accuracy of green samples.

For the printed geopolymer specimen, the mean error values in all directions were always greater than zero. This indicates that the measured

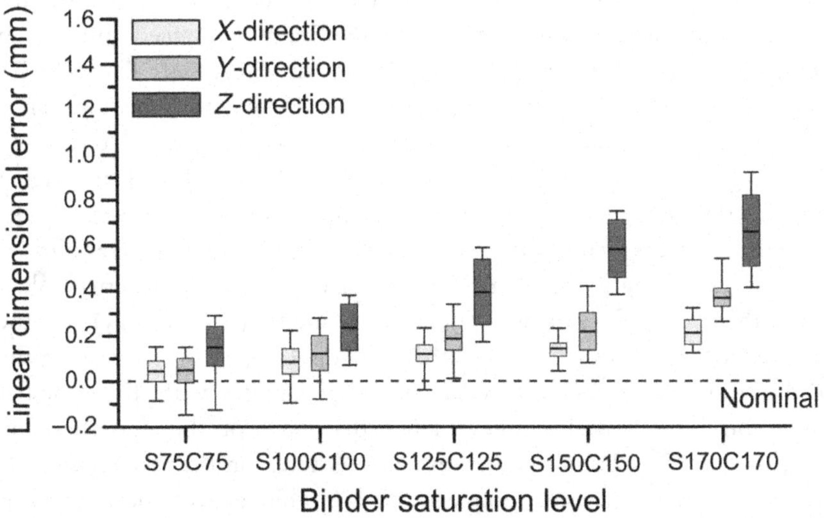

Figure 13.3 Linear dimensional accuracy results of green samples. (Box: mean ± standard deviation; whisker: min to max).

dimensions of green samples in all directions were more than those of the digital model. This pattern is true regardless of the binder saturation level. However, it is necessary to note that an anisotropic phenomenon was observed regarding the linear dimensional accuracy of the green samples depending on the directions. For all binder saturation levels, the *Z*-direction had the highest mean error and standard deviation values. In other words, regardless of the binder saturation level, the *Z*-direction had always the lowest linear dimensional accuracy. This can be attributed to the different rates of binder spreading in vertical (*Z*) direction and lateral (*X* and *Y*) directions. The results of the previous study by the authors demonstrated that the penetration of liquid binder into the geopolymer powder bed was higher in the vertical direction than lateral directions. On the other hand, for all binder saturation levels, the *X*-direction had the lowest mean error and standard deviation values and, thereby, the highest linear dimensional accuracy. This might be because *X*-direction (i.e., the binder jetting direction) is not affected by the powder spreading which takes place in *Y*-direction.

According to Fig. 13.3, the increase in binder saturation level considerably increased the mean error and standard deviation values in all directions. For instance, the mean error value in the *Z*-direction increased significantly from 0.15 ± 0.08 mm in S75C75 to 0.66 ± 0.08 mm in S170C170, compared with 0.04 ± 0.05 mm (S75C75) and 0.21 ± 0.06 mm (S170C170) in the *X*-direction. In other words, the increase in binder saturation level significantly reduced the linear dimensional accuracy of green samples in all directions. The S75C75 samples exhibited the lowest mean error and standard deviation values and, thereby, the highest linear dimensional accuracy in all directions. This might be explained by the bleeding mechanism [11] since at a higher binder saturation level the excess binder spreads outside the edges of the printed sample, which results in reduction of linear dimensional accuracy.

It is interesting to note that the rate of increase in mean error values (i.e., the rate of reduction in linear dimensional accuracy) in the *Z*-direction was lower than the other two directions. For instance, in the *Z*-direction the mean error value of green samples with 170% binder saturation level was 4.4 times higher than that of the corresponding samples with 75% binder saturation level. Whereas in the *X*-direction the mean error value of green samples with 170% binder saturation level was 5.25 times higher than that of the corresponding samples with 75% binder saturation level. In general, it can be concluded that the binder saturation level has a significant effect on the linear dimensional accuracy of the 3D printed geopolymer.

13.3.1.2 Green Compressive strength

Fig. 13.4 shows the green compressive strengths of printed geopolymer samples in both the *X*-direction and *Z*-direction.

The increase in binder saturation level resulted in a significant increase in green compressive strength in both directions. For instance, in the *X*-direction the compressive strength of green samples with 170% binder saturation level was 3.25 times higher than that of the corresponding samples with 75% binder saturation level.

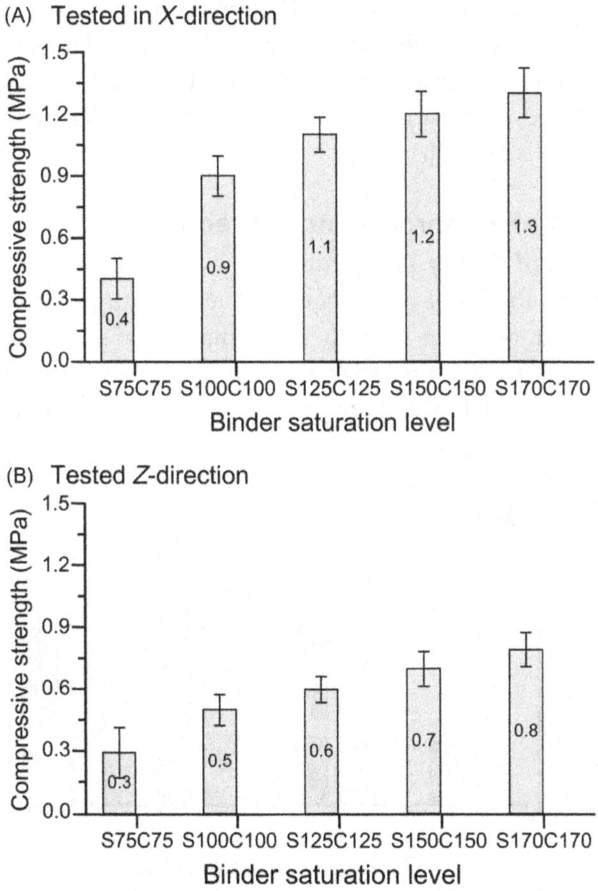

Figure 13.4 Green compressive strength of 3D printed geopolymer tested in (A) the *X*-direction; and (B) the *Z*-direction.

The inferior compressive strength of samples with lower binder saturation level is probably due to incomplete geopolymerization process which results in a weak bond between powder particles. Higher binder saturation levels resulted in a higher volume of binder during the printing process which subsequently led to superior bonding between powder particles. It should be noted that this trend is opposite to the trend observed in the conventional concrete making process, that is, increase in liquid binder content decreases the compressive strength. In the powder-based 3DCP process, the powder particles are loosely packed and are partially connected by powder/binder reaction product. The highly porous property might be the reason for the trend.

As is also shown in Fig. 13.4, at a constant binder saturation level the green compressive strengths 3D printed geopolymer samples in *X*-direction were higher than those in *Z*-direction. This anisotropy in compressive strength might be related to the week interlayer bond strength and the preferential orientation of the powder particles in the deposition process [12,13].

13.3.2 Effect of Postprocessing Procedures

13.3.2.1 Effect of Curing Temperatures

Fig. 13.5 shows the result of the compressive strength of S100C100 samples immersed in Solution I which was composed of saturated anhydrous sodium metasilicate solution with SiO_2/Na_2O ratio of 1.0 at four different temperatures. Uncertainty in the standard deviation of the test results was

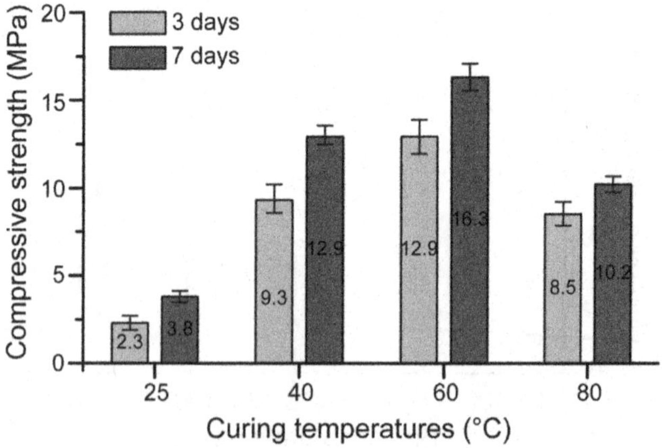

Figure 13.5 Effect of curing temperatures on compressive strength of S100C100 samples.

represented using error bars in the figure. The error in the present results was based on a 95% confidence level.

As shown in Fig. 13.5, the compressive strength of the printed samples cured at elevated temperatures increased with respect to samples cured at room temperature. Increasing curing temperature from 25°C to 60°C increases compressive strength for both curing times. For 3-day curing samples, the rates of increase of compressive strength from 25°C to 60°C were 304% and 461%, respectively, for 40°C and 60°C compared to the room temperature result. A similar trend was observed for the 7-day curing results. The higher reaction rate at elevated temperature condition might explain this ascending trend. Heat is a reaction accelerator for the geopolymerization process. An increase in temperature increases the nucleation rates and polycondensation of geopolymer and reduces the setting time [14]. Meanwhile, before being immersed in the alkaline solution, the geopolymer powder in the printed cubic structure is partially reacted. Elevated temperature will accelerate dissolution of silica and alumina species from the unreacted geopolymer powder and a larger amount of Si and Al will be available for the geopolymerization process.

However, a reverse trend was shown when increase the curing temperature from 60°C to 80°C. The compressive strengths cured at 80°C were 34.1% and 37.4% less than the values cured at 60°C for 3-day and 7-day curing, respectively. According to previous studies [14,15], there is a threshold temperature in the geopolymerization process at which temperature-controlled kinetics is inhibited. When the curing temperature is higher than this point, more silica and alumina species are dissolved and reacted. Geopolymer gel forms rapidly and deposits on the surface of unreacted geopolymer powder, which will inhibit further dissolution. As a result, the compressive strength decreases substantially. Therefore, 60°C was selected as the optimum curing temperature for the second group of samples.

The result in Fig. 13.5 also clearly indicates that the compressive strength increases as the time of curing increases irrespective of the curing temperature.

13.3.2.2 Effect of Curing Mediums

Fig. 13.6 shows the results of compressive strength of S100C100 samples after immersion in seven curing mediums at 60°C. There is no data for Slurry III at 60°C because this curing medium was setting and hardened. Uncertainty in the standard deviation of the test results was represented

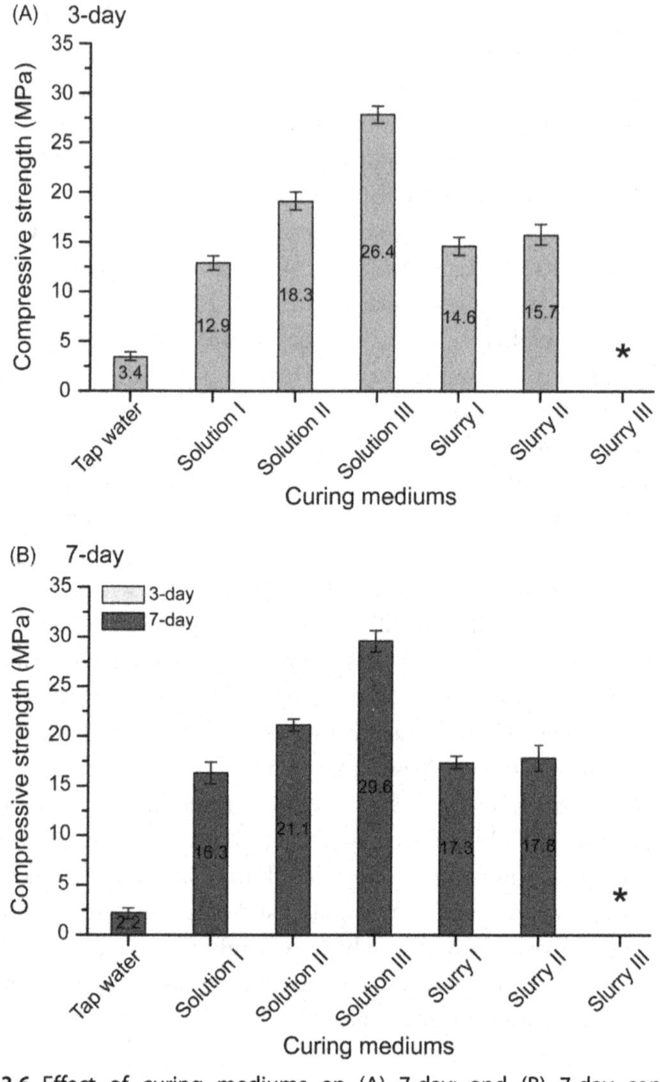

Figure 13.6 Effect of curing mediums on (A) 7-day; and (B) 7-day compressive strength of S100C100 samples at 60°C. (Note: *no data for slurry III at 60°C due to curing medium's self-setting and hardening).

using error bars in the figure. The error in the present results was based on a 95% confidence level.

As can be seen in Fig. 13.6, the compressive strengths of the S100C100 samples cured in alkaline solutions and fly ash-based geopolymer slurries were higher than that of tap water curing.

The compressive strength of tap water curing was quite low indicating that very little reaction occurred. This is due to the low concentration of OH^- ions in the curing medium, which acts a reaction catalyst in the geopolymerization process. The low presence of OH^- ions will cause inefficient dissolution and formation of hydroxyl species and oligomers [16]. Therefore, the polymerization or condensation reaction could not be established properly, leading to low compressive strength. Meanwhile, tap water curing at elevated temperature had an adverse effect on the compressive strength development and this is due to heat accelerating the leaching of silica and alumina species from the aluminosilicate gel existing in the 3D printed geopolymer samples.

In the case of alkaline solutions curing, the 3- and 7-day compressive strengths of printed samples cured in Solution II were 41.8% and 29.4% higher than those cured in Solution I. Printed samples cured in Solution III gained the highest compressive strength of 26.4 and 29.6 MPa after 3 and 7 days curing, respectively. This strength is already sufficient for a wide range of construction applications. The 8.0 M NaOH solution in Solution II and Solution III provided a strong alkaline environment which promoted the dissolution of silica and alumina species and surface hydrolysis of the partially unreacted slag particles.

Meanwhile, the compressive strength increment rate of slurry curing was higher than that of Solution I and, as the content of fly ash increased, the increment rate increased. The increment rate of 3-day curing was higher than that of 7-day curing. In the fly ash-based geopolymer slurry, Al-O bonds present in fly ash will break under high OH^- concentration, releasing aluminum ions into the solution and forming a high number of alumina species. According to previous studies [16,17], alumina species will favor the formation of Al-rich phase (*Gel* I) in the early stage of the geopolymerization process, resulting in higher early strength increment.

Therefore, it can be concluded that tap water was not an effective curing medium for 3D printed geopolymers. Alkaline solution and fly ash-based geopolymer slurry curing improved the early mechanical properties of 3D printed geopolymers.

13.4 CONCLUSIONS

The quantitative influence of binder saturation level on linear dimensional accuracy and compressive strength of green 3D printed geopolymers before any postprocessing was investigated. A set of possible postprocessing methods was developed and adopted to improve the final mechanical properties of powder-based 3D printed geopolymers. The effects of curing temperature and curing media on compressive strength gain were investigated. The following conclusions can be drawn from this study:

1. The linear dimensional accuracy of green geopolymer specimens significantly decreased in all directions as the binder saturation level increased. However, the rate of reduction in linear dimensional accuracy in the *Z*-direction was lower than the other two directions. This can be attributed to the bleeding mechanism, since at higher binder saturation level the excess binder spreads outside the edges of the printed sample, which results in reduction of linear dimensional accuracy.
2. An anisotropic phenomenon was observed in terms of the linear dimensional accuracy of the green samples depending on the directions. Regardless of the binder saturation level, the *Z*-direction always showed the lowest linear dimensional accuracy. This can be attributed to the higher rate of penetration of liquid binder into the geopolymer powder bed in the *Z*-direction compared to the other two directions. On the other hand, regardless of the binder saturation level, the *X*-direction had the highest linear dimensional accuracy. This might be due to the fact that the *X*-direction (i.e., the binder jetting direction) is not affected by the powder spreading which takes place in *Y*-direction.
3. At a constant binder saturation level, the compressive strengths of green 3D printed geopolymer samples in the *X*-direction were higher than those in the *Z*-direction. This anisotropy in compressive strength might be related to the week interlayer bond strength and the preferential orientation of the powder particles in the deposition process.
4. The postprocessing procedures investigated greatly improved the final mechanical properties of 3D printed geopolymers. Appropriate elevation in curing temperature (60°C) speeded up the hardening process and improved the mechanical properties of the 3D printed geopolymer samples.
5. Tap water was an ineffective curing medium for 3D printed geopolymers and caused an adverse effect on the compressive strength development at elevated temperature. This is likely due to leaching out of the silica and alumina species.

6. Alkaline solution and fly ash-based geopolymer slurry were both effective curing mediums for 3D printed geopolymers. Increment rate in compressive strength at earlier ages (3 days) was faster when cured in fly ash-based geopolymer slurry.
7. The use of fly ash-based geopolymer slurry as the curing medium provided an indirect way to expand the application of this industrial byproduct.

ACKNOWLEDGMENT

The authors acknowledge the support by the Australian Research Council Discovery Grant DP170103521 and Linkage Infrastructure Grant LE170100168, and Discovery Early Career Researcher Grant DE180101587.

REFERENCES

[1] B. Nematollahi, M. Xia, J. Sanjayan, Current progress of 3D concrete printing technologies, in: ISARC. Proceedings of the International Symposium on Automation and Robotics in Construction, Vilnius Gediminas Technical University, Department of Construction Economics & Property, Taipei, 2017.
[2] B. Khoshnevis, X. Yuan, B. Zahiri, J. Zhang, B. Xia, Construction by contour crafting using sulfur concrete with planetary applications, Rapid Prototyping J. 22 (2016) 848–856.
[3] S. Lim, R.A. Buswell, T.T. Le, R. Wackrow, S.A. Austin, A.G. Gibb, et al., Development of a Viable Concrete Printing Process, 2011.
[4] G. Cesaretti, E. Dini, X. De Kestelier, V. Colla, L. Pambaguian, Building components for an outpost on the Lunar soil by means of a novel 3D printing technology, Acta Astronaut. 93 (2014) 430–450.
[5] R. Rael, V. San Fratello, Developing concrete polymer building components for 3D printing, in: ACADIA. 31st Annual Conference of the Association for Computer Aided Design in ArchitectureBanff, 2011.
[6] M. Vaezi, C.K. Chua, Effects of layer thickness and binder saturation level parameters on 3D printing process, Int. J. Adv. Manuf. Technol. 53 (2011) 275–284.
[7] J. Suwanprateeb, R. Sanngam, T. Panyathanmaporn, Influence of raw powder preparation routes on properties of hydroxyapatite fabricated by 3D printing technique, Mater. Sci. Eng. 30 (2010) 610–617.
[8] M. Castilho, B. Gouveia, I. Pires, J. Rodrigues, M. Pereira, The role of shell/core saturation level on the accuracy and mechanical characteristics of porous calcium phosphate models produced by 3Dprinting, Rapid Prototyping J. 21 (2015) 43–55.
[9] G.A. Fielding, A. Bandyopadhyay, S. Bose, Effects of silica and zinc oxide doping on mechanical and biological properties of 3D printed tricalcium phosphate tissue engineering scaffolds, Dent. Mater. 28 (2012) 113–122.
[10] J.J. Beaman, J.W. Barlow, D.L. Bourell, R.H. Crawford, H.L. Marcus, K.P. McAlea, Solid Freeform Fabrication: A New Direction in Manufacturing, vol. 2061, Kluwer Academic Publishers, Norwell, MA, 1997, pp. 25–49.
[11] S. Stopp, T. Wolff, F. Irlinger, T. Lueth, A new method for printer calibration and contour accuracy manufacturing with 3D-print technology, Rapid Prototyping J. 14 (2008) 167–172.

[12] Y. Shanjani, Y. Hu, R.M. Pilliar, E. Toyserkani, Mechanical characteristics of solid-freeform-fabricated porous calcium polyphosphate structures with oriented stacked layers, Acta Biomater. 7 (2011) 1788–1796.
[13] W. Zhang, R. Melcher, N. Travitzky, R.K. Bordia, P. Greil, Three-dimensional printing of complex-shaped alumina/glass composites, Adv. Eng. Mater. 11 (2009) 1039–1043.
[14] Sindhunata, J.S.J. van Deventer, G.C. Lukey, H. Xu, Effect of curing temperature and silicate concentration on fly-ash-based geopolymerization, Ind. Eng. Chem. Res. 45 (2006) 3559–3568.
[15] E. Altan, S.T. Erdoğan, Alkali activation of a slag at ambient and elevated temperatures, Cem. Concr. Compos. 34 (2012) 131–139.
[16] P. Duxson, A. Fernández-Jiménez, J.L. Provis, G.C. Lukey, A. Palomo, J.S.J. Deventer, Geopolymer technology: the current state of the art, J. Mater. Sci. 42 (2006) 2917–2933.
[17] L. Weng, K. Sagoe-Crentsil, Dissolution processes, hydrolysis and condensation reactions during geopolymer synthesis: Part I-Low Si/Al ratio systems, J. Mater. Sci. 42 (2007) 2997–3006.

CHAPTER 14

Design 3D Printing Cementitious Materials Via Fuller Thompson Theory and Marson-Percy Model

Yiwei Weng[1,2], Mingyang Li[1], Ming Jen Tan[1] and Shunzhi Qian[1,2]
[1]Singapore Centre for 3D Printing, School of Mechanical and Aerospace Engineering, Nanyang Technological University, Singapore, Singapore
[2]School of Civil and Environment Engineering, Nanyang Technological University, Singapore, Singapore

14.1 INTRODUCTION

3D printing, also referred to as Additive Manufacturing (AM), is a technology which builds a solid object via a layer-by-layer process [1]. Due to its advantages such as customized production, reduced waste, and diminished lead-time of rapid prototype [2], 3D printing has attracted much attention from various fields including the building and construction industry [3]. In the past decade, much research has been conducted in the field of 3D printing for building and construction, especially in the development of printing systems such as mortar printing [3–5], FreeFab [6], contour crafting [7–9], and robotic printing systems [10,11]. While most of the 3D printing for building and construction processes can be classified as 3D cementitious materials printing (3DCMP), little research has been done on how to design materials for 3DCMP [12–14], especially from the aspect of material design methods.

Materials used in 3DCMP need to meet certain specific rheological requirements [3,15]. The most essential steps in 3DCMP are conveying mixed materials to the nozzle via a delivery system and depositing materials to build the solid object in a layer-by-layer manner. In the conveying step, the materials are required to have good pumpability which indicates how easily material can be conveyed; and in the deposition step, the materials are required to have good buildability which indicates how well the materials can be stacked stably. Both pumpability and buildability are closely related to the rheology performance of materials, namely static/dynamic yield stress and plastic

3D Concrete Printing Technology
DOI: https://doi.org/10.1016/B978-0-12-815481-6.00014-2

viscosity. Static yield stress is the minimum shear stress required to initiate the flow and dynamic yield stress is the critical shear stress below which the shear stress is insufficient to maintain the flow. Plastic viscosity is the resistance of a fluid to flow when the fluid is flowing. All these rheological properties are attributed to the inter-particle force [16]. Typically, higher static/dynamic yield stress and plastic viscosity would enhance buildability and hinder pumpability [17] and, thus, seeking a balance between buildability and pumpability is critical in material rheology design for 3DCMP.

It is well-known that aggregates take up 60%–80% of the total volume of cementitious materials, the most commonly used building and construction material worldwide [18]. Rheological properties of cementitious material are highly affected by the gradation of aggregate [18–20]. Good aggregate gradation contributes to high density and proper rheological properties of materials. Fuller Thompson theory is a classic theory for gradation design. In 1907, W.B. Fuller and S.E. Thompson proposed a theory for gradation design based on experimental results [21]. Later, the Federal Highway Administration (FHWA) proposed a modified Fuller Thompson equation [22]. Fuller Thompson theory has been widely used in producing high-performance concrete [20,23], designing sustainable concrete with minimum content of cement [24], and optimizing rheology [25].

Rheological performance is significantly affected by the packing fraction of a system as well, which is defined as the ratio of solids to the total volume. Theoretically, materials designed by continuous gradation can achieve maximum packing fraction which describes a condition where the void volume reaches minimum for a given system. According to the Marson-Percy model [26], the highest packing fraction of materials results in the lowest plastic viscosity.

To fulfill the material rheological property requirements in 3DCMP, Fuller Thompson theory and Marson-Percy model were applied in materials design of 3DCMP. Six different mixtures were prepared with different gradation approaches using five different sands (0.6–1.2 mm, 0.25–0.6 mm, 0.15–0.25 mm, less than 0.15 mm, and natural river sand), that is, mixture A designed by Fuller Thompson theory, mixture B and C designed by uniform-gradations, mixture D and E designed by gap-gradations, and mixture F using natural river sand without special gradation design. Rheological tests were carried out to investigate the fresh performance of all mixtures and printing tests for buildability were conducted among different mixtures via a gantry printer. Density and mechanical performance (compressive and flexural properties) were characterized as well.

It should be noted that the time and temperature effect of rheological/material properties is not considered due to the limited scope of this study. Furthermore, all tests including printing were conducted within 30 minutes after water addition (21 minutes after finishing mixing) and under a constant lab temperature of 26°C.

14.2 THEORY

14.2.1 Built-Up Theory and the Bingham Plastic Model

Perrot et al. established the relation between static yield stress and buildability of 3DCMP [17]. The simplified relation can be expressed as:

$$H = \frac{\alpha}{\rho g}\tau_s \tag{14.1}$$

where H (m) and α are the printed height (buildability) and the geometric factor of printed structure, respectively; ρ (g/cm^3) and g (m/s^2) are the density of materials and gravitational constant, respectively. τ_s (Pa) is the static yield stress which corresponds to static torque in Fig. 14.1.

Fig. 14.1 presents the typical rheological test result via a Schleibinger Viskomat XL rheometer with a vane probe. The static yield stress can be

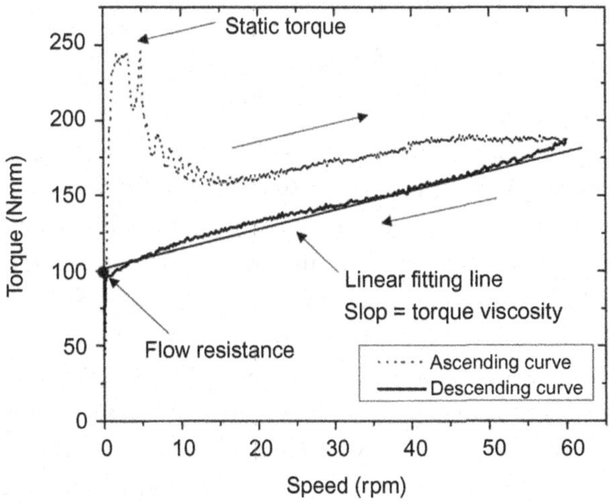

Figure 14.1 Typical rheological test result.

converted from the maximum torque (also referred to as static torque) in the ascending curve. Based on linear fitting of descending curve, its slop and intersection with Y axis are defined as torque viscosity and flow resistance, respectively, which can then be converted into plastic viscosity and dynamic yield stress using the Bingham plastic model.

The Bingham plastic model is widely accepted to describe the rheological property of cementitious materials [27–33] which has the shear stress τ (Pa) expressed as

$$\tau = \tau_d + k\dot{\gamma} \tag{14.2}$$

where τ_d (Pa) and k (Pa · s) are dynamic yield stress and plastic viscosity, respectively. $\dot{\gamma}$ (1/s) is the shear rate.

From Eq. (14.1), the buildability is proportional to the static yield stress. It should be noted that in this study, the rheological properties are assumed to be constant during the whole printing process as the time consumed for the printing test (around 2 minutes) is much shorter than the rheological test (10 minutes). The effect of time on the rheological performance, printability, and buildability will be investigated in the future. It also worth noting that in this study we focused on the effect of yield stress and plastic viscosity on the printability and buildability and, thus, the thixotropic behavior and the rate of reflocculation are out of the scope of this chapter.

Generally, materials' pumpability is reflected by the pumping pressure during the pumping process, which is related to Bingham plastic parameters as well. The relationship can be described by [34]:

$$P = \left(\frac{8\tau_d}{3R} + \frac{8k}{\pi R^4}Q\right)L \tag{14.3}$$

where P (Pa) is the pumping pressure. L (m) and R (m) are the length and radius of the pipe, respectively. Q (m^3/s) is the average flow rate. In this case, R is much less than 1 m, which means that plastic viscosity (k) has a more significant effect on the pumping pressure compared to the dynamic yield stress (τ_d). Eq. (14.3) shows that the increases of dynamic yield stress and plastic viscosity augment the pumping pressure of cementitious material for specific pipe and flow conditions, which reduces the pumpability of cementitious material. Therefore, a low plastic viscosity is required for cementitious materials to decrease the pumping pressure, which will facilitate a safe and economic operation of the pumping/printing process.

14.2.2 Fuller Thompson Theory and the Marson-Percy Model

Gradation is an important attribute to produce economical cementitious materials with maximum density and minimum void content. The sand gradation formula proposed by W.B. Fuller and S.E. Thompson [21] is expressed as:

$$p_i = \left(\frac{d_i}{D}\right)^{0.5} \tag{14.4}$$

where p_i is the percent pass ith sieve, d_i (mm) is the opening size of ith sieve, and D (mm) is the maximum particle size. Later, in 1962, FHWA published a modified version for Fuller's equation:

$$p_i = \left(\frac{d_i}{D}\right)^{0.45} \tag{14.5}$$

Three types of gradation are schematically shown in Fig. 14.2. Applying the Fuller Thompson equation in sand gradation design forms a continuous gradation in which interaction between particles exists at many points of contact and realizes a maximum densified system. Gap gradation is designed via missing a certain range of sand particle size in the paste, which possesses more voids content and less contact points compared with the continuous grading system. Using single sized sand in the mixture can achieve a uniform gradation mixture which has minimum contact points between particles. Since contact points contribute to the friction in the mixture, increasing the contact points between particles can raise the friction and, thus, results in increased static yield stress to initiate flow and dynamic yield stress to maintain the flow.

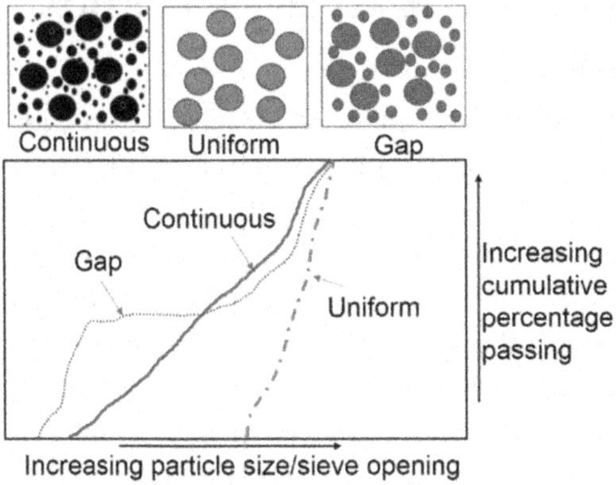

Figure 14.2 Three types of sand gradation.

Shear viscosity μ is affected by packing fraction ϕ, maximum packing fraction ϕ_m, and shear rate $\dot{\gamma}$. The relationship between packing fraction, shear rate, and shear viscosity is commonly expressed by the Marson-Percy model [26]

$$\mu(\phi, \dot{\gamma}) = \left(1 - \frac{\phi}{\phi_m}\right)^{-2} \mu_b(\dot{\gamma}) \tag{14.6}$$

where $\mu_b(\dot{\gamma})$ (Pa · s) is the shear viscosity of the binder, which is the same among all samples. In our study, ϕ and $\dot{\gamma}$ are kept as constants and ϕ_m is variable for mixtures with different gradation. Theoretically, a mixture designed by the Fuller Thompson theory can achieve a maximum density and minimum void system, which means ϕ_m is at maximum among all mixtures and it should possess the lowest shear viscosity μ.

The relation between shear viscosity and plastic viscosity of Eq. (14.2) can also be expressed as:

$$\mu = k + \frac{\tau_d}{\dot{\gamma}} \tag{14.7}$$

The Fuller Thompson theory indicates that when the mixture is designed by Eq. (14.4), it possesses the highest dynamic yield stress (τ_d) and lowest shear viscosity (μ from Eq. (14.6)) among all mixtures. Therefore, Eq. (14.7) indicates that the plastic viscosity (k) of the mixture designed by Eq. (14.4) is the smallest as well when the shear rate is the same among different mixtures. A low plastic viscosity is required for cementitious materials to decrease the pumping pressure which will facilitate a safe and economic operation of the pumping/printing process.

14.3 MATERIALS AND MIXTURE PROPORTION

14.3.1 Materials

Mixtures in this study consisted of ordinary Portland cement (OPC, ASTM type I, Grade 42.5), silica fume (SF, undensified, Grade 940, Elkem company), silica sand, fly ash (FA, Class F), natural river sand, water, and superplasticizer. Fig. 14.3 illustrates the particle size distribution of cement, SF, and FA which were analyzed by Mastersizer 2000. Table 14.1 illustrates the chemical composition of OPC and FA, respectively. Silica sand with four different sizes, that is, 0.6–1.2 mm, 0.25–0.6 mm, 0.15–0.25 mm, and less than 0.15 mm, were used in this study either as mono-sized ingredients or a mixture of all four.

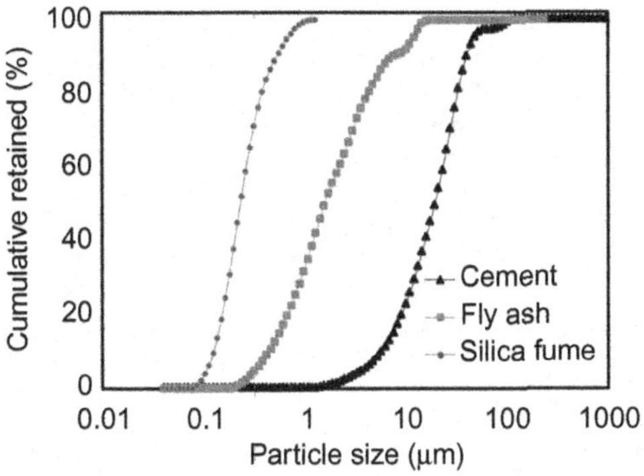

Figure 14.3 Particle size distribution of ordinary Portland cement, fly ash, and silica fume.

Table 14.1 Chemical composition of fly ash and ordinary Portland cement

Formula	**Concentration (%)**	
	Fly Ash	**Cement**
SiO_2	58.59	24.27
Al_2O_3	30.44	4.56
Fe_2O_3	4.66	3.95
TiO_2	2.02	0.55
K_2O	1.51	0.61
CaO	1.21	62.2
MgO	0.776	3.34
P_2O_5	0.531	0.15
Na_2O	–	0.21
SO_3	0.0914	–
ZrO_2	0.04	–
MnO	0.0351	–
Cr_2O_3	0.027	–
CuO	0.0254	–
ZnO	0.0229	–

14.3.2 Mixture Proportion

Six different mixtures, including five mixtures with silica sands following different gradations and one mixture with natural rive sand were used in this study. The sand gradation curves are shown in Fig. 14.4. Sand for mixture A followed the Fuller Thompson theory to achieve a continuous gradation system. The continuous gradation mixture used in this study

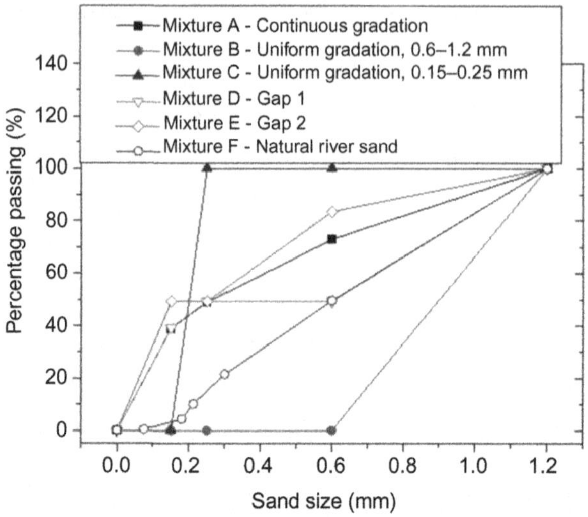

Figure 14.4 Sand gradation curves for different mixtures.

was 39.23% of sand size less than 0.15 mm, 10.14% of sand with 0.15−0.25 mm, 23.84% of sand with 0.25−0.6 mm, and 26.80% of sand with 0.6−1.2 mm. Sand for mixture B was a uniform gradation mixture with sand size ranging from 0.6 to 1.2 mm. Sand for mixture C was another uniform gradation mixture with sand size ranging 0.15−0.25 mm.

As can been seen from Fig. 14.4, the gap gradation 1 (mixture D) used in this study removed the sand particle size between 0.25−0.6 mm and was composed of 39.23% of sand particle size less than 0.15 mm, 10.14% of sand particle size with 0.15−0.25 mm, and 50.63% of sand particle size with 0.6−1.2 mm. Gap gradation 2 (mixture E) removed sand particles sized between 0.15−0.25 mm and is composed of 49.37% of sand particles sized less than 0.15 mm, 23.84% of sand particles sized between 0.25 and 0.6 mm, and 26.69% of sand particles sized between 0.6 and 1.2 mm.

Mixture F used natural river sand as the raw material and sand particles sized bigger than 1.2 mm were sieved out. Sieved natural river sand was then analyzed by a sieving machine. The sieved natural rived sand consists of 0.86% of sand particles sized less than 0.075 mm, 6.99% of sand ranging from 0.075 to 0.18 mm, 3.68% of sand ranging from 0.18 to 0.212 mm, 11.24% of sand ranging from 0.212 to 0.3 mm, 32.87% of sand ranging from 0.3 to 0.6 mm, and 44.36% of sand ranging from 0.6 to1.2 mm, as shown in Fig. 14.4.

Table 14.2 Mixture proportion

OPC	Sand	W	FA	SF	Superplasticizer (g\L)
1	0.5	0.3	1	0.1	1.3

Note: all ingredients content are expressed as weight proportion of cement content. *OPC*, ordinary Portland cement; *FA*, fly ash; *SF*, silica fume; *W*, water.

All the mixtures, except silica sand itself and natural river sand, followed the same mixture proportion, as shown in Table 14.2.

14.4 MIXING AND TESTING PROCEDURES

14.4.1 Mixing Procedures

A hobart mixer X200L was used for mixing. Since many factors can affect the rheological properties of cement slurries, such as mixing time, mixing speed and temperature [35], the mixing procedures in this study were fixed to minimize the differences among batches. Firstly, the powder of all solid ingredients was dry mixed for 3 minutes in stir speed, then water was added and the mixing process continued for 3 minutes in stir speed followed by 2 minutes in speed I. Finally, the superplasticizer was added, the mixing process continued for 1 minute in stir speed followed by 3 minutes in speed I.

14.4.2 Rheological Characterization

Rheological properties of mixed materials were evaluated via Viskomat XL and mini-slump. The six-blade vane probe and cage were used for the rheological test to avoid slippage of cement paste. Both the diameter and the height of the vane probe are 69 mm and the gap between probe and cage, and the bottoms of probe and barrel are both 40 mm. During the rheological test, the speed of the rheometer is increased linearly from 0 to 60 rpm in 5 minutes. Afterwards, it is decreased linearly to 0 rpm in another 5 minutes, as shown in Fig. 14.5. Then the static/dynamic yield stress and plastic viscosity can be computed by Chhabra et al.'s formula [34]:

$$\Gamma = \frac{4\pi R_1^2 R_2^2 l \eta}{R_2^2 - R_1^2}\omega_2 - \frac{4\pi R_1^2 R_2^2 l \tau_0}{R_2^2 - R_1^2}\ln\frac{R_1}{R_2} \tag{14.8}$$

where Γ (N · m) is the torque, ω_2 (rad/s) is the rotational speed of outer barrel, l (m) and R_1 (m) are the length and radius of the probe,

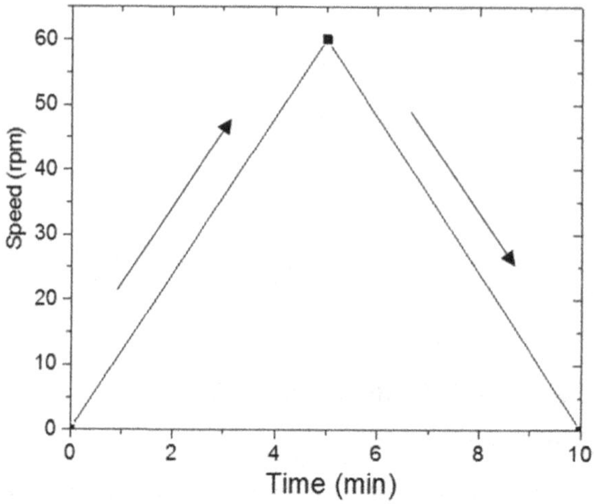

Figure 14.5 Rheological testing programs.

respectively, and R_2 (m) is the radius of the outer barrel. Static yields stress can be directly obtained by applying static torque in Eq. (14.8).

Mini-slump test is a classic method to measure materials' workability which is based on the spread diameter of slurry placed into a cone-shape mold [36]. The mini-slump cone typically has a height of 70 mm, an internal diameter of 50 and 100 mm at the top and bottom, respectively. The cone is placed on a flat, level, and nonabsorbent hard surface. During the test, it is filled with fresh material and then raised after a certain amount of time to allow the material to spread. After spreading ceases, two diameters of material mass are measured in the orthorginal direction and the workability of cementitiou material can be characterized by the average of the two diameters.

14.4.3 Printing Investigation

Finally, printing tests were conducted to investigate the buildability of different mixtures. As shown in Fig. 14.6, a gantry printer with a 1.2 m × 1.2 m × 1.0 m ($L \times W \times H$) printing volume was used to control the nozzle position for printing. As shown in Fig. 14.7, the materials were pumped by a MAI Pictor pump from the black funnel to the nozzle head through a hose 3 m in length and 2.54 cm in diameter. Fig. 14.8 illustrates the 3D model used in the printing test. The circular column with 10 cm inner diameter is composed of 50 layers. Each layer is theoretically 20 mm in width and 10 mm in height.

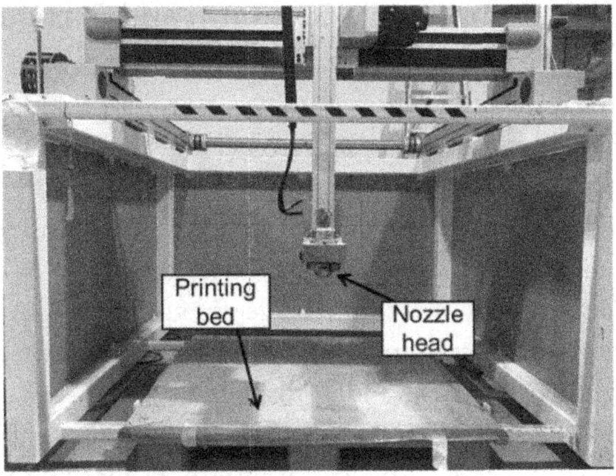

Figure 14.6 Gantry concrete printer used for printing test.

Figure 14.7 MAI Pictor pump to convey materials.

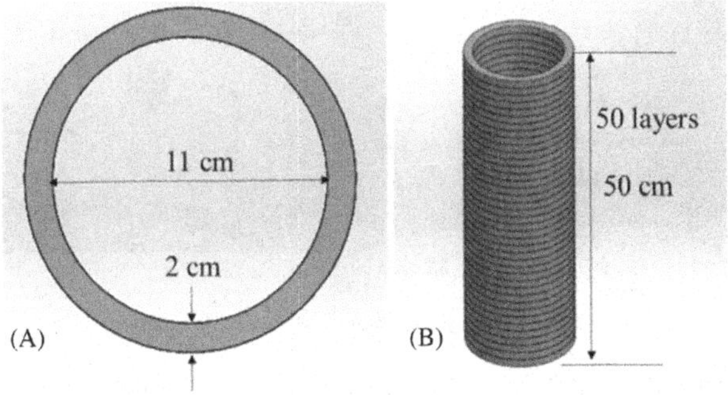

Figure 14.8 3 D model for printing test.

14.4.4 Density, Compressive Strength, and Flexural Performance

All six mixtures were casted into 50 mm × 50 mm × 50 mm cubic molds, consolidated via vibration table and trowel finished. All specimens were then demolded after 24 hours. All samples were covered by a plastic sheet and then stored in a plastic container for curing. The density was tested at 7, 14, and 28 days, according to ASTM C642 [37].

After the density test, the compressive strength test was carried on the cubic specimens. The compressive strength was measured by uniaxial loading in triplicates at 7, 14, and 28 days in accordance with the specifications of ASTM C109/C109M-13 [38]. The equipment used for this purpose was a Toni Technik Baustoffprüfsysteme machine with a loading rate of 100 KN/min.

The gantry concrete printer was used to fabricate flexural specimens. The nozzle used for printing was 30 mm × 15 mm ($L \times W$). The printing speed and pumping speed were 4000 mm/min and 1.8 L/min, respectively. The standoff distance was 15 mm for each layer. The printed path and side view of printed filaments are shown in Fig. 14.9. Then, printed specimens were cut into separate filaments with 350 mm in length and 300 mm in height (two layers). Afterwards, a four-point bending test with a span length of 240 mm was conducted at 7, 14, and 28 days.

14.5 RESULTS AND DISCUSSION

14.5.1 Rheological Analysis

Herein Fig. 14.10 illustrates the shear stress and shear rate relation from rheological results based on the Bingham plastic model. The raw

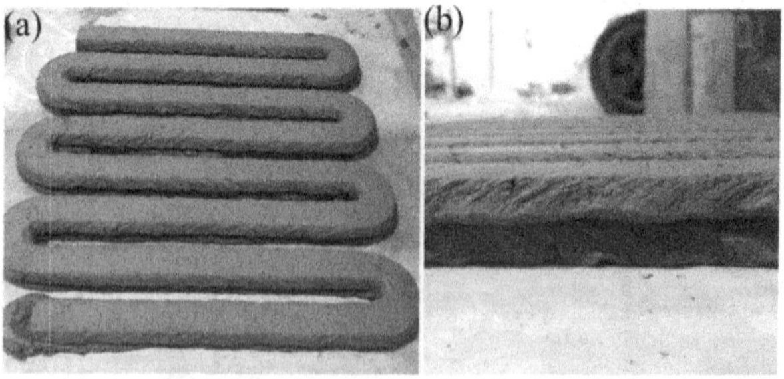

Figure 14.9 Flexural specimen preparation: (A) top view of the printed filament; and (B) side view of printed filament.

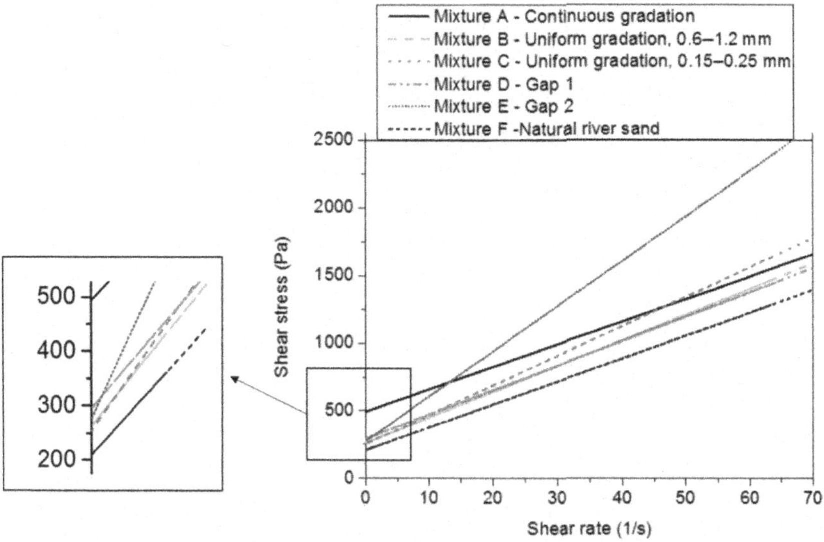

Figure 14.10 Shear stress and shear rate relation based on the Bingham plastic model.

Table 14.3 Rheological performance of different mixtures

Mixture	Static torque (N•mm)	Flow resistance (N•mm)	Torque viscosity (N · mm/rpm)	Static yield stress (Pa)	Dynamic yield stress (Pa)	Plastic viscosity (Pa · s)
A	3390	498.5	2.280	3350	492.7	16.65
B	2440	264.5	2.601	2411	261.3	19.00
C	2132	255.8	2.984	2107	252.8	21.81
D	3358	287.3	2.941	3318	290.6	18.03
E	2725	277.4	4.560	2693	274.1	33.31
F	1897	210.9	2.320	1874	208.4	16.95

data of static torque, flow resistance, and torque viscosity are shown in Table 14.3, which were then converted to static/dynamic yield stress and plastic viscosity via Eq. (14.8) and shown in the same table.

As can be seen from Fig. 14.10 and Table 14.3, mixture A has the highest static yield stress and the smallest plastic viscosity which is very desirable to ensure low pumping pressure and high buildability; mixture E possesses a highest viscosity among all the mixtures, which suggests that

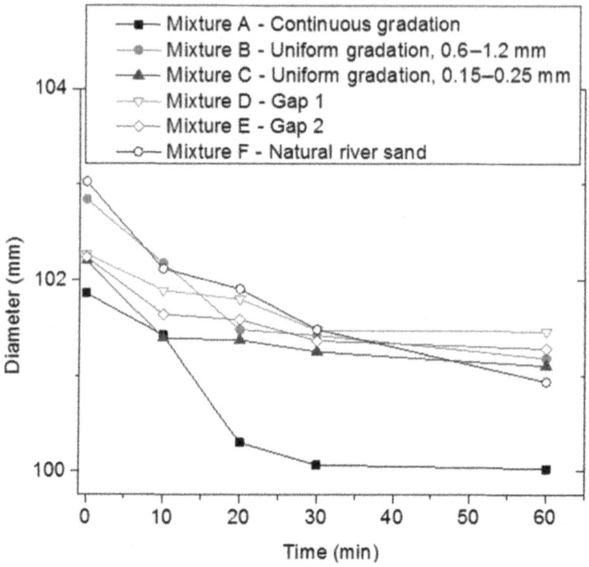

Figure 14.11 Mini-slump test result.

the pumpability of mixture E could be the worst. Fig. 14.11 reveals the results from the mini-slump test. As can be seen from Fig. 14.11, for all mixtures, the diameter tends to decline along the time. At each time point, mixture A has the lowest slump. After 20 minutes, the spread diameter of mixture A is approximately 100 mm.

Rheological and mini-slump test results indicate that the most appropriate materials for 3DCMP is mixture A with continuous gradation. With the highest yield stresses and the lowest plastic viscosity, mixture A is expected to reveal better pumpability and buildability among all mixtures investigated in this study.

14.5.2 Printing Test

A gantry printer system was used to conduct the printing test of mixtures A–F. The moving speed of the nozzle and pumping speed of were 9000 mm/min and 1.8 L/min, respectively. During the printing test, the printed part can maintain its shape initially. When the printed structure meets the failure criteria at the fresh state, which was described by Perrot et al. [17], suddenly noticeable deformation occurs. This was followed by misalignment of printing path and the collapse of the printed structure. The layer numbers with noticeable deformation and final collapse were recorded during the printing test.

Printing test results are summarized in Table 14.4, where layer number associated with noticeable deformation and final collapse are presented. Taking mixture A as an example, it maintained its shape until the 42nd layer (Fig. 14.12A), followed by large deformation (Fig. 14.12B), and finally fell down at the 43rd layer (Fig. 14.12C). For other mixtures, similar phenomena were observed during the printing test, except that they all showed large deformation and collapsed at a much earlier stage (smaller layer number). The maximum build-up height can be computed by Eq. (14.1). Nevertheless the original simple geometric factor based on

Table 14.4 Summary of layer number during which deformation and final collapse occurred

	Layer number with noticeable deformation	Layer number with final collapse
Mixture A	42	43
Mixture B	30	31
Mixture C	27	32
Mixture D	31	36
Mixture E	32	34
Mixture F	24	25

Figure 14.12 The printing test result of mixture A: (A) maintained the shape until the 42nd layer; (B) sudden deformation at the 42nd layer; and (C) fell down at 43rd layer.

Roussel and Lanos [39] is only intended for a solid column. For the hollow cylinder printed in this study, a formula for geometric factor has been derived as follows:

$$
\begin{aligned}
\alpha \quad &= \left(R_2^2 - R_1^2\right)^{-1}\left(\frac{1}{2} + \frac{C_\alpha}{R_2^2}\right)^{-1}\sqrt{\frac{3}{4} + \frac{C_\alpha^2}{R_2^4}} \\
&\cdot\left\{\frac{4}{H}\left[\frac{\left(R_2^3 - R_1^3\right)}{6} + C_\alpha(R_2 - R_1)\right]\right. \\
&+ 2C_\alpha\left(\sqrt{\frac{3R_2^4}{4C_\alpha^2} + 1} - \sqrt{\frac{3R_1^4}{4C_\alpha^2} + 1}\right) \\
&- 2C_\alpha\left[\operatorname{arcsinh}\left(\frac{2C_\alpha}{\sqrt{3}R_2^2}\right) - \operatorname{arcsinh}\left(\frac{2C_\alpha}{\sqrt{3}R_1^2}\right)\right] \\
&- 2R_1^2\left(\frac{3}{4} + \frac{C_\alpha^2}{R_1^4}\right)^{-1/2}\left(\frac{1}{4} - \frac{C_\alpha^2}{R_1^4}\right) \\
&\left.- 2R_2^2\left(\frac{3}{4} + \frac{C_\alpha^2}{R_2^4}\right)^{-1/2}\left(\frac{1}{4} - \frac{C_\alpha^2}{R_2^4}\right)\right\}
\end{aligned}
\tag{14.9}
$$

where C_α can be solved from equation

$$
\begin{aligned}
&\frac{\left(\frac{1}{2}\right) - \left(\frac{C_\alpha}{R_2^2}\right)}{\sqrt{\left(\frac{3}{4}\right) + \left(\frac{C_\alpha^2}{R_2^4}\right)}} - \frac{\left(\frac{1}{2}\right) - \left(\frac{C_\alpha}{R_1^2}\right)}{\sqrt{\left(\frac{3}{4}\right) + \left(\frac{C_\alpha^2}{R_1^4}\right)}} \\
&+ \operatorname{arcsinh}\left(\frac{2C_\alpha}{\sqrt{3}R_2^2}\right) - \operatorname{arcsinh}\left(\frac{2C_\alpha}{\sqrt{3}R_1^2}\right) = 0
\end{aligned}
\tag{14.10}
$$

The detailed derivation can be found in Appendix A. The comparison between prediction and experimental maximum built-up heights is shown in Fig. 14.13. As can be seen from Fig. 14.13, generally the model can well-predict the deformation/collapse height (layer number times 0.01). While much simplified formula for geometric factor of hollow cylinder is needed, Eq. (14.1) hollow cylinder printing may offer an opportunity for prediction and direct verification of buildability for 3DP.

Afterwards, a large printing test was conducted using mixture A. A circular nozzle with 20 mm in diameter was used in this printing. The

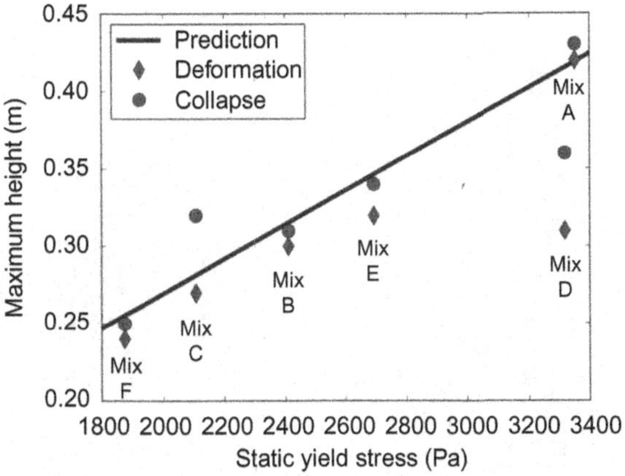

Figure 14.13 Comparison between prediction and experimental maximum built-up heights.

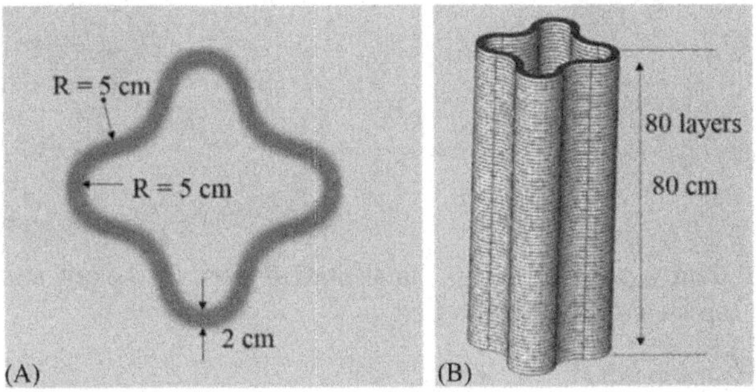

Figure 14.14 3D model for large-scale printing: (A) top view; and (B) side view.

moving speed of the nozzle and the pumping speed of the pump were 9000 mm/min and 1.8 L/min, respectively. The 3D printing model is shown in Fig. 14.14. As can be seen from the top view of the 3D model (Fig. 14.14A), it is composed of four semicircles and 4 quarter circles with 10 cm in diameter and 2 cm in thickness. Then the total layer numbers are 80 for printing and the layer height is 10 mm (Fig. 14.14B). The final printed part is shown in Fig. 14.15.

14.5.3 Density, Compressive Strength, and Flexural Strength

The density of cementitious material is related to its packing. Better packing from continuous gradation yields high density. The result of the density of all mixtures is shown in Fig. 14.16. As expected, mixture A possesses the highest density among all mixtures at 7, 14, and 28 days. Furthermore, the density of all mixtures also increases with time.

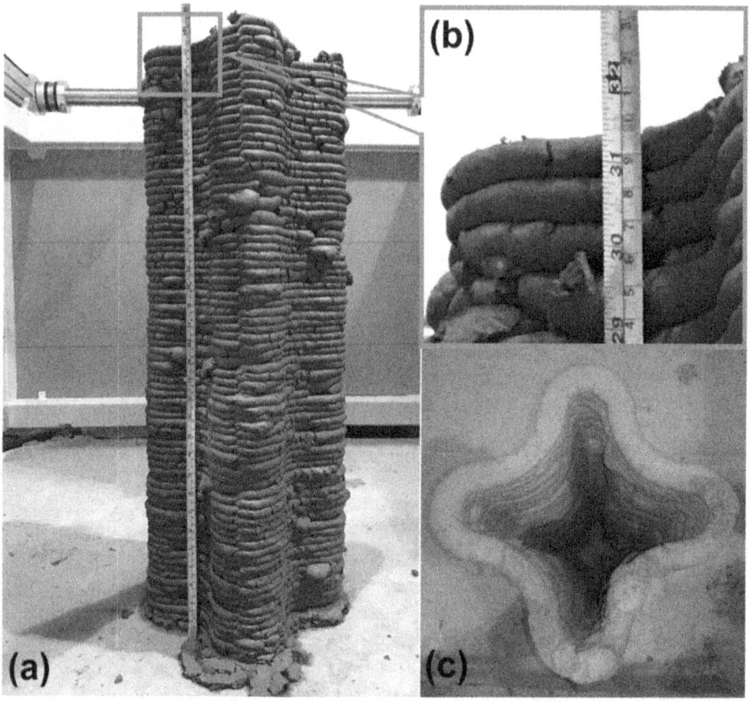

Figure 14.15 Gantry printed part: (A) final printed part; (B) height and layer thickness; and (C) top view of the printed part.

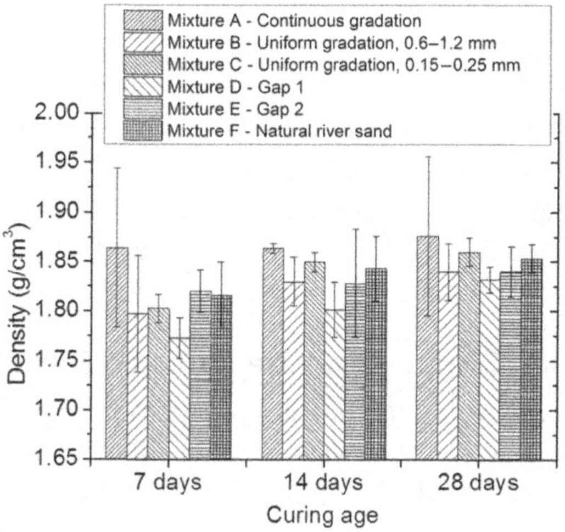

Figure 14.16 Density of all mixtures.

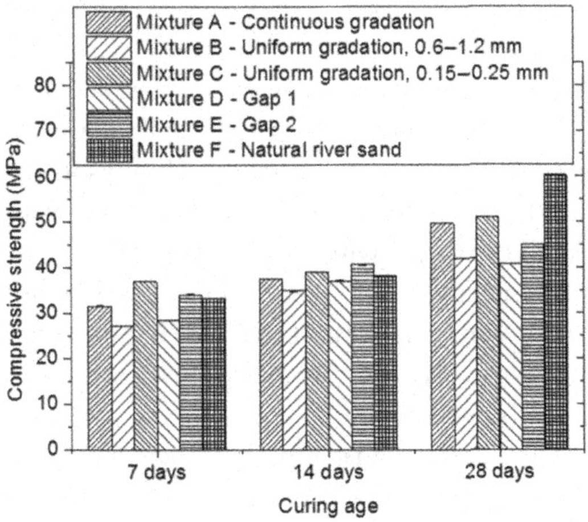

Figure 14.17 Compressive strength of all mixtures.

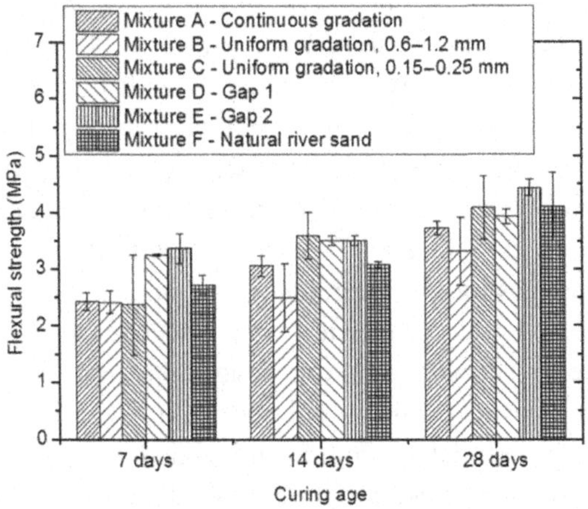

Figure 14.18 Flexural strength of different mixtures.

Figs. 14.17 and 14.18 present the results of the compressive strength and flexural strength of all the mixtures. Mechanical strength of all mixtures increases with time, as expected. Mixture A possesses appropriate compressive and flexural strength at 28 days, which are 49.7 and 3.73 MPa, respectively.

14.6 SUMMARY AND CONCLUSION

3DCMP requires the material to have low plastic viscosity and high yield stress to meet the requirements for both pumpability and buildability. The Fuller Thompson theory and Marson-Percy model were adopted as a guideline to design construction materials with proper rheological properties for 3DCMP. Six different mixtures (i.e., mixtures A–F) were prepared with various gradations using five different sands (0.6–1.2 mm, 0.25–0.6 mm, 0.15–0.25 mm, <0.15 mm, and natural river sand). The gradation of mixture A followed the Fuller Thompson theory to achieve a continuous gradation system. Mixtures B and C were two different uniform-gradation mixtures with sand particles of 0.6–1.2 mm and 0.15–0.25 mm, respectively. Mixtures D and E followed the gap-gradation method, where ranges of 0.25–0.6 mm and 0.15–0.25 mm were removed, respectively. Mixture F used river sand with particles sized bigger than 1.2 mm sieved out.

Rheological tests illustrate that mixture A has the highest yield stresses and the lowest plastic viscosity. During the printing test for buildability, it can reach as high as 40 cm (40 layers) without notable deformation. On the contrary, noticeable deformation appeared in all the other mixtures at around 30th layer. Finally, large-scale printing with mixture A was conducted and the final printed part can be as high as 80 cm without obvious deformation. Furthermore, when mixture E with highest plastic viscosity was used for the printing test, pumping pressure significantly increased compared to others. Mechanical properties, including density, and compressive and flexural strength of all mixtures were also measured.

In summary, the Fuller Thompson method and Marson-Percy model can be used as the design guideline to tailor the materials to acquire appropriate rheological properties and mechanical performance for 3DCMP. More specifically, with this design guideline, a desirable combination of plastic viscosity and yield stress can be achieved where low plastic viscosity is essential for good pumpability and high yield stress for buildability. While the direct adoption of sand gradation and mix proportion in practice is not very practical due to the scarcity of materials and other constraints, a simple design guideline by adopting a classic theory like the Fuller Thompson method and Marson-Percy model still offers meaningful insights in the further development of 3D printable material for the building and construction industry. While Eqs. (14.9 and 14.10) need to be greatly simplified for the geometric factor of a hollow cylinder, Eq. (14.1) along with hollow cylinder printing may offer an opportunity for prediction and direct verification of buildability for 3DP conveniently.

In this research, only the sand particle gradation was investigated. In the future, it is essential to extend the gradation analysis to include other filler-like materials such as silica fume and fly ash in the fresh state. These materials contribute very little to the hydration process and, therefore, mainly play the role of a filler, like sand, at least in the fresh state. Furthermore, a more quantitative link between particle size distribution/ particle interactions and rheological properties needs to be established in the context of 3D printing.

ACKNOWLEDGMENT

The authors would like to acknowledge the National Research Foundation, Prime Minister's Office, Singapore under its Medium-Sized Centre funding scheme, Singapore Centre for 3D Printing and Sembcorp Design and Construction Pty Ltd. for their funding and support in this research project.

REFERENCES

[1] C.K. Chua, K.F. Leong, C.S. Lim, Rapid Prototyping: Principles and Applications, World Scientific Publishing Company, 2010.
[2] K.V. Wong, A. Hernandez, A review of additive manufacturing, Mech. Eng. (2012) 1−10.
[3] F. Bos, R. Wolfs, Z. Ahmed, T. Salet, Additive manufacturing of concrete in construction: potentials and challenges of 3D concrete printing, Virtual Phys. Prototyping 11 (3) (2016) 209−225.
[4] S. Lim, R.A. Buswell, T.T. Le, S.A. Austin, A.G.F. Gibb, T. Thorpe, Developments in construction-scale additive manufacturing processes, Autom. Constr. 21 (2012) 262−268.
[5] Z. Malaeb, H. Hachem, A. Tourbah, et al., 3D concrete printing: machine and mix design, Int. J. Civil Eng. 6 (6) (2015) 14−22.
[6] J.B. Gardiner, S. Janssen, N. Kirchner, A realisation of a construction scale robotic system for 3D printing of complex formwork, Proc. Int. Symp. Autom. Rob. Constr., 33, Vilnius Gediminas Technical University, Department of Construction Economics & Property, 2016, p. 1.
[7] B. Khoshnevis, D. Hwang, K.T. Yao, et al., Mega-scale fabrication by contour crafting, Int. J. Ind. Syst. Eng. 1 (3) (2006) 301−320.
[8] B. Khoshnevis, Automated construction by contour crafting—related robotics and information technologies, Autom. Constr. 13 (1) (2004) 5−19.
[9] D. Hwang, B. Khoshnevis, Concrete wall fabrication by contour crafting, in: 21st International Symposium on Automation and Robotics in Construction (ISARC 2004), Jeju, South Korea, 2004.
[10] C. Gosselin, R. Duballet, P. Roux, et al., Large-scale 3D printing of ultra-high performance concrete−a new processing route for architects and builders, Mater. Des. 100 (2016) 102−109.
[11] T.H. Pham, J.H. Lim, Q.-C. Pham, Robotic 3D-printing for building and construction, in: Proceedings of the 2nd International Conference on Progress in Additive Manufacturing (Pro-AM 2016), 2016, pp. 300−305.
[12] M. Hambach, D. Volkmer, Properties of 3D-printed fiber-reinforced Portland cement paste, Cem. Concr. Compos. 79 (2017) 62−70.
[13] P. Feng, X. Meng, H. Zhang, Mechanical behavior of FRP sheets reinforced 3D elements printed with cementitious materials, Compos. Struct. 134 (2015) 331−342.
[14] P. Feng, X. Meng, J.-F. Chen, L. Ye, Mechanical properties of structures 3D printed with cementitious powders, Constr. Build. Mater. 93 (2015) 486−497.
[15] T.T. Le, S.A. Austin, S. Lim, R.A. Buswell, A. Gibb, T. Thorpe, Mix design and fresh properties for high-performance printing concrete, Mater. struct. 45 (8) (2012) 1221−1232.

[16] R.J. Flatt, P. Bowen, Yodel: a yield stress model for suspensions, J. Am. Ceram. Soc. 89 (4) (2006) 1244–1256.
[17] A. Perrot, D. Rangeard, A. Pierre, Structural built-up of cement-based materials used for 3D printing extrusion techniques, Mater. Struct. 49 (4) (2016) 1213–1220.
[18] J. Hu, A study of effects of aggregate on concrete rheology (Ph.D. thesis), Iowa State University, 2005.
[19] W.B. Ashraf, M.A. Noor, Performance-evaluation of concrete properties for different combined aggregate gradation approaches, Procedia Eng. 14 (2011) 2627–2634.
[20] C.L. Hwang, L.A.T. Bui, C.T. Chen, Application of Fuller's ideal curve and error function to making high performance concrete using rice husk ash, Comput. Concr. 10 (6) (2012) 631–647.
[21] W.B. Fuller, S.E. Thompson, The laws of proportioning concrete, Asian J. Civil Eng. Transp. 59 (1907) 67–143.
[22] Z. Li, Advanced Concrete Technology, John Wiley & Sons, 2011.
[23] M. Mangulkar, S. Jamkar, Review of particle packing theories used for concrete mix proportioning, Contributory Pap. 4 (2013) 143–148.
[24] S.A.A.M. Fennis, J.C. Walraven, Using particle packing technology for sustainable concrete mixture design, Heron 57 (2) (2012) 73–101.
[25] J. Hu, K. Wang, Effect of coarse aggregate characteristics on concrete rheology, Constr. Build. Mater. 25 (3) (2011) 1196–1204.
[26] D.M. Kalyon, S. Aktas, Factors affecting the rheology and processability of highly filled suspensions, Annu. Rev. Chem. Biomol. Eng. 5 (2014) 229–254.
[27] H.A. Barnes, J.F. Hutton, K. Walters, An Introduction to Rheology, Elsevier, 1989.
[28] G.H. Tattersall, The rheology of Portland cement pastes, Br. J. Appl.ied Phys. 6 (5) (1955) 165.
[29] C. Hu, F. De Larrard, The rheology of fresh high-performance concrete, Cem. Concr. Res. 26 (2) (1996) 283–294.
[30] Y. Weng, B. Lu, M. Tan, S. Qian, Rheology and printability of engineered cementitious composites-a literature review, in: Proceedings of the 2nd International Conference on Progress in Additive Manufacturing (Pro-AM 2016), 2016, pp. 427–432.
[31] P. Banfill, G. Starrs, G. Derruau, W. McCarter, Rheology of low carbon fibre content reinforced cement mortar, Cem. Concr. Compos. 28 (9) (2006) 773–780.
[32] P.F.G. Banfill, Rheology of fresh cement and concrete, Rheol. Rev. (2006) 61.
[33] C. Ferraris, F. De Larrard, N. Martys, S. Mindess, J. Skalny, Fresh concrete rheology: recent developments, Mater. Sci. Concr. VI (2001) 215–241.
[34] R.P. Chhabra, J.F. Richardson, Non-Newtonian Flow and Applied Rheology: Engineering Applications, Butterworth-Heinemann, 2011.
[35] R. Byron Bird, G.C. Dai, Barbara J. Yarusso, The rheology and flow of viscoplastic materials, Rev. Chem. Eng. 1 (1) (1983) 1–70.
[36] ASTM C1611/C1611M – 14 Standard Test Method for Slump Flow of Self-Consolidating Concrete.
[37] ASTM, C642, Standard Test Method for Density, Absorption, and Voids in Hardened Concrete, vol. 4, Annual Book of ASTM Standards, 2006.
[38] ASTM, C109-16A, Standard Test Method for Compressive Strength of Hydraulic Cement Mortars, American Society for Testing and Materials, Philadelphia, PA, 2016.
[39] N. Roussel, C. Lanos, Plastic fluid flow parameters identification using a simple squeezing test, Appl. Rheol. 13 (3) (2003) 132–139.

FURTHER READING

E. Yang, M. Sahmaran, Y. Yingzi, V.C. Li, Rheological control in production of engineered cementitious composites, ACI Mater. J. 106 (4) (2009) 357–366.

APPENDIX A

Considering a hollow cylinder has inner radius R_1 (m), outer radius R_2 (m), and height H (m), which is compressed by two plates with uniform velocity v (m/s), the coordinate can be built as shown in Fig. A.1.

The continuity equation under cylindrical coordinates without angular velocity can be written as:

$$\frac{1}{r}\frac{\partial}{\partial r}(ru_r) + \frac{\partial u_z}{\partial z} = 0 \tag{A.1}$$

where the velocity profiles on radial and axial directions satisfying governing equation and boundary conditions are:

$$u_r(r, z) = \frac{v}{2H}r + \frac{C}{r} \tag{A.2}$$

$$u_z(r, z) = -\frac{v}{H}z \tag{A.3}$$

where C is an undetermined constant. Thus, the strain rate tensor of this flow is:

$$D_{ij} = \begin{bmatrix} \frac{v}{2H} - \frac{C}{r^2} & 0 & 0 \\ 0 & \frac{v}{2H} + \frac{C}{r^2} & 0 \\ 0 & 0 & -\frac{v}{H} \end{bmatrix} \tag{A.4}$$

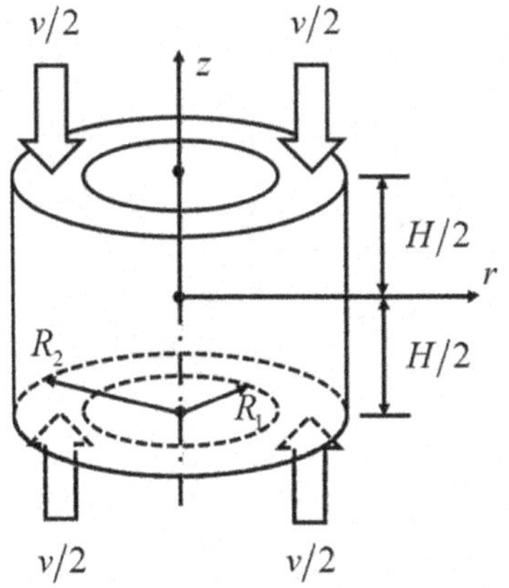

Fig. A.1 Schematic of a hollow cylinder compressed by constant velocity.

the second invariant is:

$$I_2(r,z) = \frac{3v^2}{4H^2} + \frac{C^2}{r^4} \tag{A.5}$$

and the stress tensor is:

$$\begin{aligned}\sigma_{ij}^{(d)} &= \frac{K_i}{\sqrt{I_2}} D_{ij} \\ &= K_i\left(\frac{3v^2}{4H^2} + \frac{C^2}{r^4}\right)^{-1/2} \begin{bmatrix} \frac{v}{2H} - \frac{C}{r^2} & 0 & 0 \\ 0 & \frac{v}{2H} + \frac{C}{r^2} & 0 \\ 0 & 0 & -\frac{v}{H} \end{bmatrix}\end{aligned} \tag{A.6}$$

In the absence of body forces and inertial stresses, the pressure governing equation is:

$$\frac{\partial p}{\partial r} = \frac{1}{r}\frac{\partial\left[r\sigma_r^{(d)}\right]}{\partial r} - \frac{\sigma_\theta^{(d)}}{r} + \frac{\partial \sigma_{rz}}{\partial z} \tag{A.7}$$

Integrating Eq. (A.7) on $[R_1, R_2]$:

$$\begin{aligned}\int_{R_1}^{R_2} \frac{\partial p}{\partial r} dr &= \left(\frac{v}{2H} - \frac{C}{r^2}\right)\left[\left(\frac{3v^2}{4H^2} + \frac{C^2}{r^4}\right)^{-(1/2)} K_i\right]\Bigg|_{R_1}^{R_2} \\ &+ K_i \operatorname{arcsinh}\left(\sqrt{\frac{4H^2}{3v^2}}\frac{C}{r^2}\right)\Bigg|_{R_1}^{R_2} = 0\end{aligned} \tag{A.8}$$

Let $C_\alpha = HC/v$, Eq. (A.8) can be rearranged as:

$$\begin{aligned}&\left(\frac{1}{2} - \frac{C_\alpha}{R_2^2}\right)\left(\frac{3}{4} + \frac{C_\alpha^2}{R_2^4}\right)^{-1/2} - \left(\frac{1}{2} - \frac{C_\alpha}{R_1^2}\right)\left(\frac{3}{4} + \frac{C_\alpha^2}{R_1^4}\right)^{-1/2} \\ &+ \operatorname{arcsinh}\left(\frac{2C_\alpha}{\sqrt{3}R_2^2}\right) - \operatorname{arcsinh}\left(\frac{2C_\alpha}{\sqrt{3}R_1^2}\right) = 0\end{aligned} \tag{A.9}$$

Apparently, 0 is the trivial solution of Eq. (A.9), another solution can be solved numerically from Eq. (A.9). On the plate surface, the dissipation rate is caused by the force between plates and fluids which can be computed by:

$$
\begin{aligned}
&2\int_{S_p}\sigma_{rz}\left(r,\frac{H}{2}\right)u_r\left(r,\frac{H}{2}\right)dS \\
&=\frac{4\pi\nu K_i}{H}\left[\frac{(R_2^3-R_1^3)}{6}+C_\alpha(R_2-R_1)\right]
\end{aligned}
\tag{A.10}
$$

and the dissipation rate inside of the fluid can be calculated by:

$$
\begin{aligned}
&\int_V\sigma_{ij}^{(d)}D_{ij}dV-\int_{S_R}\sigma_{ij}^{(d)}D_{ij}dS \\
&=2\pi\nu K_iC_\alpha\left(\sqrt{\frac{3R_2^4}{4C_\alpha^2}+1}-\sqrt{\frac{3R_1^4}{4C_\alpha^2}+1}\right) \\
&-2\pi\nu K_iC_\alpha\left[\operatorname{arcsinh}\left(\frac{2C_\alpha}{\sqrt{3}R_2^2}\right)-\operatorname{arcsinh}\left(\frac{2C_\alpha}{\sqrt{3}R_1^2}\right)\right] \\
&-2\pi\nu K_iR_1^2\left(\frac{3}{4}+\frac{C_\alpha^2}{R_1^4}\right)^{-1/2}\left(\frac{1}{4}-\frac{C_\alpha^2}{R_1^4}\right) \\
&-2\pi\nu K_iR_2^2\left(\frac{3}{4}+\frac{C_\alpha^2}{R_2^4}\right)^{-1/2}\left(\frac{1}{4}-\frac{C_\alpha^2}{R_2^4}\right)
\end{aligned}
\tag{A.11}
$$

Therefore, the work applied on fluid is:

$$
\begin{aligned}
Fv\;&=\frac{4\pi\nu K_i}{H}\left[\frac{(R_2^3-R_1^3)}{6}+C_\alpha(R_2-R_1)\right] \\
&+2\pi\nu K_iC_\alpha\left(\sqrt{\frac{3R_2^4}{4C_\alpha^2}+1}-\sqrt{\frac{3R_1^4}{4C_\alpha^2}+1}\right) \\
&-2\pi\nu K_iC_\alpha\left[\operatorname{arcsinh}\left(\frac{2C_\alpha}{\sqrt{3}R_2^2}\right)-\operatorname{arcsinh}\left(\frac{2C_\alpha}{\sqrt{3}R_1^2}\right)\right] \\
&-2\pi\nu K_iR_1^2\left(\frac{3}{4}+\frac{C_\alpha^2}{R_1^4}\right)^{-1/2}\left(\frac{1}{4}-\frac{C_\alpha^2}{R_1^4}\right) \\
&-2\pi\nu K_iR_2^2\left(\frac{3}{4}+\frac{C_\alpha^2}{R_2^4}\right)^{-1/2}\left(\frac{1}{4}-\frac{C_\alpha^2}{R_2^4}\right)
\end{aligned}
\tag{A.12}
$$

For a gravity driven hollow cylinder:

$$F = \rho g H \pi \left(R_2^2 - R_1^2\right) \tag{A.13}$$

when the fluid starts to flow, it must have:

$$\tau_s \alpha = \rho g H \tag{A.14}$$

where

$$\min\left[\sigma_\theta^{(d)}\right] = \sigma_\theta^{(d)}\big|_{r=R_2} = \tau_s \tag{A.15}$$

Substituting Eqs. (A.13)–(A.15) into Eq. (A.12) and rearranging:

$$\begin{aligned}
\alpha \; &= \left(R_2^2 - R_1^2\right)^{-1}\left(\frac{1}{2} + \frac{C_\alpha}{R_2^2}\right)^{-1}\sqrt{\frac{3}{4} + \frac{C_\alpha^2}{R_2^4}} \\
&\cdot\left\{\frac{4}{H}\left[\frac{\left(R_2^3 - R_1^3\right)}{6} + C_\alpha(R_2 - R_1)\right]\right. \\
&+ 2C_\alpha\left(\sqrt{\frac{3R_2^4}{4C_\alpha^2} + 1} - \sqrt{\frac{3R_1^4}{4C_\alpha^2} + 1}\right) \\
&- 2C_\alpha\left[\operatorname{arcsinh}\left(\frac{2C_\alpha}{\sqrt{3}R_2^2}\right) - \operatorname{arcsinh}\left(\frac{2C_\alpha}{\sqrt{3}R_1^2}\right)\right] \\
&- 2R_1^2\left(\frac{3}{4} + \frac{C_\alpha^2}{R_1^4}\right)^{-1/2}\left(\frac{1}{4} - \frac{C_\alpha^2}{R_1^4}\right) \\
&\left. - 2R_2^2\left(\frac{3}{4} + \frac{C_\alpha^2}{R_2^4}\right)^{-1/2}\left(\frac{1}{4} - \frac{C_\alpha^2}{R_2^4}\right)\right\}
\end{aligned} \tag{A.16}$$

where C_α can be solved from Eq. (A.9).

CHAPTER 15

Towards the Formulation of Robust and Sustainable Cementitious Binders for 3D Additive Construction by Extrusion

Dale P. Bentz, Scott Z. Jones, Isaiah R. Bentz and Max A. Peltz
Materials and Structural Systems Division, Engineering Laboratory, National Institute of Standards and Technology, Gaithersburg, MD, United States

15.1 INTRODUCTION

Additive construction by extrusion (ACE, e.g., 3-D printing of concrete) is an emerging field with numerous opportunities and challenges [1–12]. In comparison to conventional concrete construction, ACE offers potential savings of both time and money, including reduced formwork requirements and reduced labor costs. However, to date, field applications of this technology have been rather limited, as this new paradigm for construction demands pinpoint control of the rheology and setting of the composite material. In this regard, Lim et al. [1] have defined four critical characteristics of the materials employed in an ACE operation, namely:

1. *Pumpability: The ease and reliability with which material is moved through the delivery system.*
2. *Printability: The ease and reliability of depositing material through a deposition device.*
3. *Buildability: The resistance of deposited wet material to deformation under load.*
4. *Open time: The period where these three properties are consistent within acceptable tolerances. Of course, all four of these characteristics are controlled by the rheology and hydration properties of the materials and the requirements placed on these by the actual equipment being employed in an ACE operation.*

Cementitious materials (including grouts, mortars, and concretes) are typically considered to be adequately represented as Bingham-type fluids

3D Concrete Printing Technology 2019 Published by Elsevier Inc.
DOI: https://doi.org/10.1016/B978-0-12-815481-6.00015-4

in terms of their rheological properties [6,10]. With this assumption, their rheological performance is uniquely characterized by only two parameters, a yield stress (τ_0) and a plastic viscosity (η, hereafter shortened to viscosity) that define the relationship between applied strain ($\dot{\gamma}$) and resultant stress (τ):

$$\tau = \tau_0 + \eta\dot{\gamma} \tag{15.1}$$

Due to hydration reactions (including nucleation on particle surfaces) and other thixotropic effects (cement particle flocculation, etc.), these parameters develop over time following the initial mixing of water with cement. During the first several hours of curing, the hydration reactions are relatively dormant (proceeding at a slow rate in comparison to the later stages of hydration). Still, the generally observed slow, steady increase in viscosity and yield stress due to the ongoing hydration is primarily irreversible and not easily reduced by additional mixing. Cement particle flocculation also causes yield stress and viscosity to increase over time (i.e., thixotropy). However, flocculation effects are at least partially reversible, because sufficiently aggressive mixing can temporarily destroy the floc network [13–16]. The temporal evolution of these rheological properties can be further modified by chemical retarders or accelerators [17].

As shown schematically in Fig. 15.1, pumpability, printability, and buildability in an ACE application can each be represented by their own viability window on a viscosity–yield stress plot, with lower and upper bounds defining the set of acceptable values for these two parameters with respect to each of the three "operations" (pumping, printing, and building). The open time window is, therefore, defined as the intersection of these three operational windows. For some combinations of materials and 3D printers, this intersection might be nonexistent (empty set). Successful 3D printing requires operating within the open time window, so understanding how and when appropriate rheological properties are achieved is critical to creating standing and longstanding structures. Of course, in addition to simply waiting the necessary time period for a prepared mixture to enter its open time window, both chemical accelerators [17] and alternative cements [18] may be employed to alter the timing of the "open condition" relative to that expected from an Ordinary Portland Cement-based system, effectively immediately transforming a pumpable mixture into a printable and buildable one, at the printing head nozzle, for example.

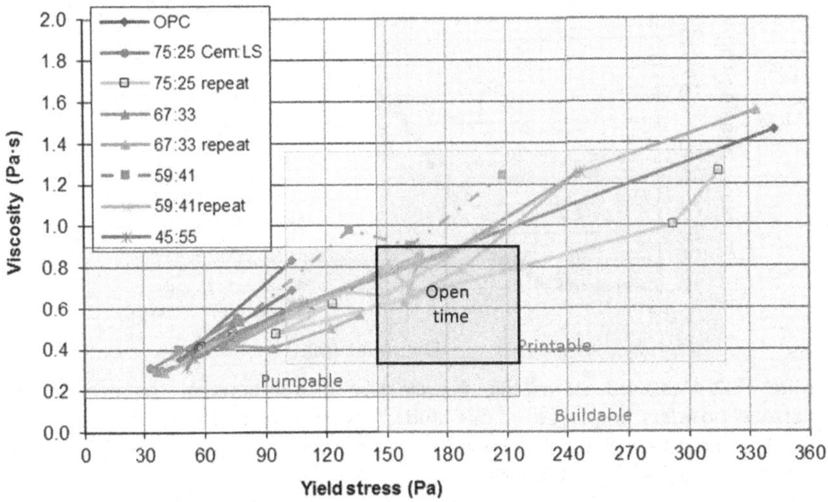

Figure 15.1 Schematic showing *hypothetical* windows for pumping, printing, and building and their intersection (open time) overlaid on viscosity versus yield stress data for a set of cement pastes prepared for this study, with various ratios of Ordinary Portland Cement to fine limestone powder as indicated in the legend. Water volume fraction was maintained constant at 53% for all the mixtures. Bingham-based yield stress and viscosity both increase with time (after mixing) for these mixtures.

The purpose of this chapter is to describe a method for formulating a set of robust and sustainable binders suitable for laboratory-scale 3D printing of cement paste, seen as a first step towards the larger-scale 3D printing of mortar (concrete). Sustainability is pursued via the replacement of a significant fraction of the cement by a limestone powder of a similar surface area (and particle size distribution), while robustness with respect to rheological performance is achieved by the incorporation of a retarder into the initial paste mixture and the subsequent injection of a conventional shotcrete accelerator.

15.2 MATERIALS AND METHODS

An ASTM C150 [19] Type I/II Ordinary Portland Cement (OPC) with about a 4% limestone addition by mass was employed in all mixtures. It has a reported Bogue-calculated phase composition of 53.1% C_3S, 14.2% C_2S, 6.5% C_3A, and 10.0% C_4AF on a mass basis. In some mixtures, a prescribed mass of the cement, having a He-pycnometry density of

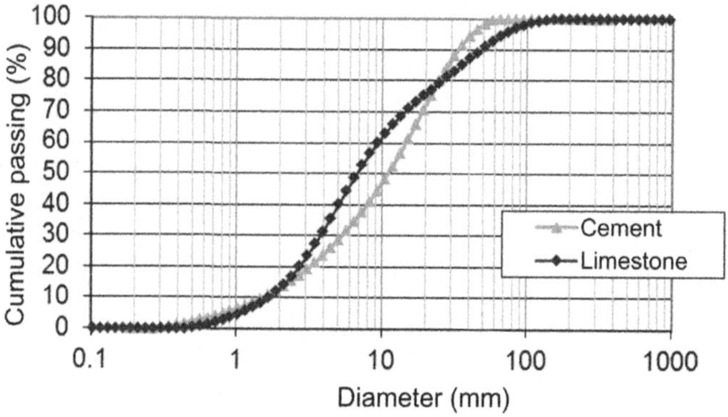

Figure 15.2 Measured cumulative particle size distributions of the cement and limestone powders employed in this study. Results shown are the average of six individual measurements and error bars (one standard deviation) would fall within the size of the symbols on the plot.

(3140 ± 10) kg/m^3, was replaced with an equivalent mass of a limestone powder, primarily composed of calcite, with a density of (2730 ± 10)[1] kg/m^3. The cumulative particle size distributions of the two powders, as measured by laser diffraction, are provided in Fig. 15.2. The measured BET surfaces areas (using nitrogen gas) of the cement and limestone powders are nearly the same, being (1.25 ± 0.01) m^2/g and (1.26 ± 0.01) m^2/g, respectively.

To keep the mixtures simple for this initial study, no high-range water reducers were employed. However, two commercially available admixtures were used to provide retardation and acceleration of the hydration, respectively. The retarder is a commercially available mixture of sodium gluconate, sucrose, water, and other minor ingredients, while the alkali-free aluminum sulfate-based accelerator is a commercial product commonly employed in shotcreting applications. Dosages were based on manufacturer recommendations.

In the first phase of this study, pastes with various proportions of cement to limestone (Table 15.1) were prepared and evaluated with respect to rheology and isothermal calorimetry. No chemical admixtures were employed in this phase and the water content was adjusted to maintain a nearly constant initial porosity (volume fraction of water), specifically within the range of 0.526 to 0.529, while

[1] In this chapter, all reported uncertainties indicate ± one standard deviation, typically for two or three replicates.

Table 15.1 Paste mixtures with various ratios of cement to limestone investigated in this study

Cement:limestone	100:0	75:25	67:33	59:41	45:55
Mass of cement (g)	300	225	200	177	135
Limestone (g)	0	75	100	123	165
Water (g)	106.6	111	112.5	112.5	115.5
Vol. frac. water	0.527	0.528	0.529	0.526	0.528
w/p	0.355	0.37	0.375	0.375	0.385
w/c	0.355	0.493	0.563	0.636	0.856
Paste density (kg/m^3)[a]	1971	1917	1899	1890	1860

[a]Assuming an air content of 2% in the prepared pastes.

the water-to-powder (w/p) mass ratio varied between 0.355 and 0.385 and the water-to-cement ratio (w/c) between 0.355 and 0.856. Mixing was performed in a temperature-controlled high shear blender using the procedures provided in ASTM Standard Practice C1738/C1738M−14, with the powder component being introduced into the blender during the first 30 seconds of mixing [20]. After mixing, specimens were prepared for either isothermal calorimetry or rheology measurements and the remainder of the paste was stored in a thermos for subsequent measurements of rheology, with a 30 s remix using a four-blade single shaft mixer rotating at 20.9 rad/s (200 rpm) prior to each measurement.

For the second phase of the study, the 67:33 cement:limestone powder paste mixture was further investigated with respect to the addition of a retarder at two dosages, as well as one dosage of a retarder/delayed accelerator combination. The retarder was added to the mixing water at dosages of 0.00133 mL/g powder and 0.002 mL/g powder, while the accelerator was injected into the lower-dosage retarded paste mixture at some fixed time after mixing at an addition rate of 0.03 g/g paste (3% by mass). For rheology measurements, the accelerator was injected into the remaining paste that was stored in a thermos and the paste remixed for 30 seconds using a single shaft mixer. For calorimetry measurements, the injection of the accelerator was performed using one of the admix modules within the calorimeter cell whose ampoule had been pre-loaded with the retarded paste. In this case, mixing the paste for 30 seconds during and following the injection was performed using the built-in mixing shaft of the admix module. Paste specimens for isothermal calorimetry were on the order of 5 g in mass and measurements were conducted at 23.00 ± 0.01°C over the course of 7 days, beginning approximately

30 minutes after the water contacted the cement. While two specimens from the same batch of paste were evaluated in some instances to provide one indication of variability, the average absolute difference between replicate specimens was previously measured to be 2.4×10^{-5} W/g (cement), for measurements conducted between 1 hours and 7 days after mixing [21].

Rheology measurements were performed using a parallel plate geometry with a set of two 35 mm diameter serrated test plates. For this study, the gap between the plates was set to 0.600 ± 0.001 mm and the temperature was controlled at $23.0 \pm 0.2°C$. Using a 3 mL capacity syringe, about 1.1 mL of cement paste was placed on the lower serrated plate for each measurement. For the shear rate loop experiments employed here, the shear rate (strain) was first increased from 0.1 to 100 s^{-1} in thirty steps equally spaced on a log scale, and subsequently decreased back down to 0.1 s^{-1}, once again in thirty steps. Eq. (15.1) is then applied over a range where the data are judged by eye to be linear to estimate values of τ_0 and η. For the measurement of stress growth, the material was sheared at a very low shear rate of 0.1 s^{-1} and the resultant stress was recorded over the course of 400 seconds. The first maximum peak value recorded was subsequently used as an approximation of the static yield stress [21,22]. Generally, the rheological parameters determined according to Eq. (15.1) or from a stress growth experiment are considered to have an uncertainty in the order of 10% [21].

15.3 RESULTS AND DISCUSSION

Insights into the hydration processes and potential property development can be gained by normalizing the isothermal calorimetry data by various mixture proportions' parameters. Here, normalizations by mass of cement, mass of powder, surface area of powder, and volume of water are presented. In Fig. 15.3, the plots with normalization by the mass of cement are presented first. Data for early ages up to 8 hours are presented in separate plots due to the particular importance of this time frame for ACE applications. Replacement of cement by limestone powder accelerates the hydration of the remaining cement, due to the tri-fold effects of dilution (increased water availability per unit cement), increased surface area per unit cement (limestone powder surfaces) available for nucleation and growth of hydration products, and the reactivity of the limestone powder itself [23–27]. Because the measured surfaces area (per gram of

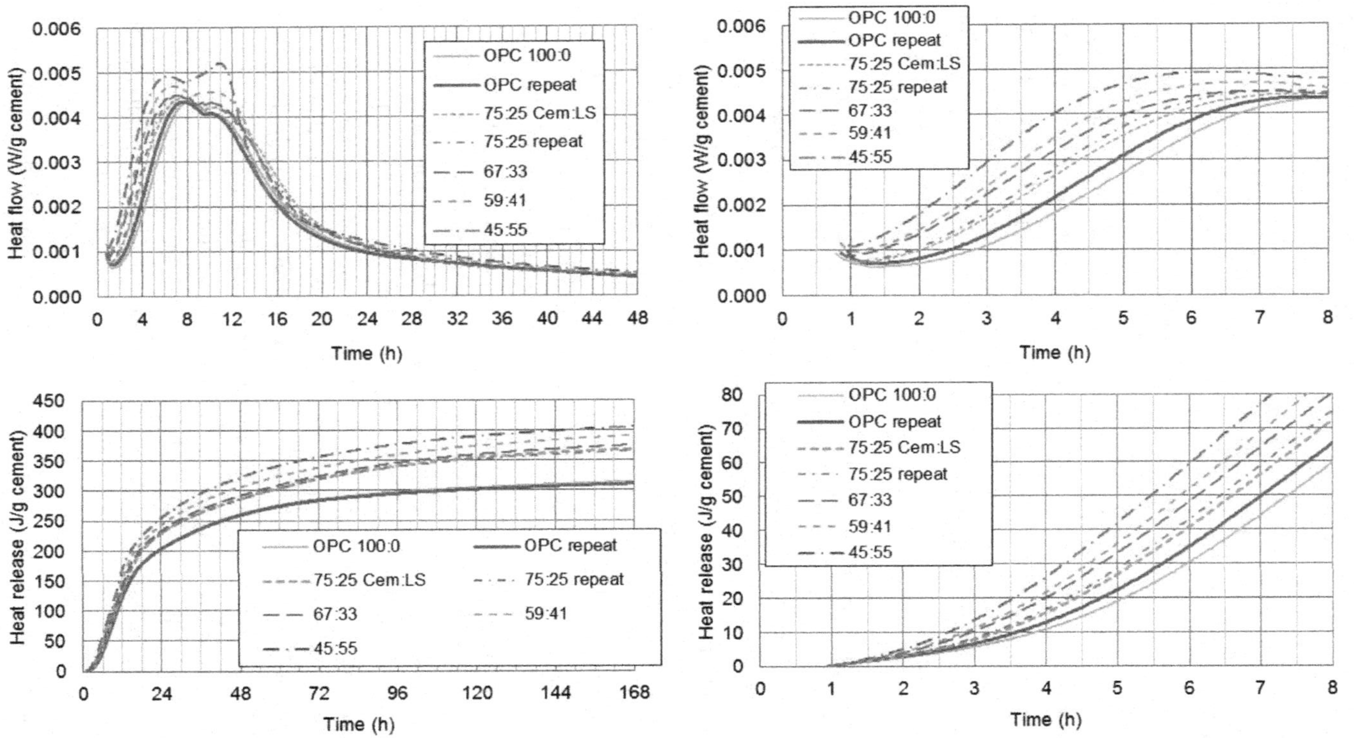

Figure 15.3 Heat flow (top) and cumulative heat release (bottom) normalized by the mass of cement in each paste mixture.

powder) of the cement and limestone powder used in this study are so similar, the total surface area available in each paste mixture is about the same. However, the ratio of total surface area to mass of cement increases dramatically as more and more cement is replaced by the limestone powder on a per mass of cement basis. This results in both an acceleration (shift to earlier times) and an amplification (increased peak heights) of the reactions of the silicates and aluminates present in the cement. Regarding the aluminates, the limestone powder can react directly with them to form carboaluminates in place of conventional sulfoaluminates and also stabilize the early-age formation of ettringite [27]; this effect is especially prominent in the 45:55 cement:limestone powder blend, where the height of the second (aluminate) peak occurring at 11 hours actually exceeds that of the first primary silicate peak at 6 hours in the top left graph of Fig. 15.3.

The roles of powder mass and surface area are further explored in Fig. 15.4 that provides the same heat flow data as in Fig. 15.3, but now normalized with respect to these two total powder characteristics. At later ages (beyond 4 hours), as would be expected, heat flow is controlled mainly by the mass of cement present in the mixture, with the heat flows ranking in direct proportion to the cement fraction in this two-component binder. During the first 4 hours or so, however, the hydration rate of the binary blend is basically controlled by the total powder mass—or more likely its surface area—because the data in Fig. 15.4 collapse to the same curve in that time range to within the measurement variability. The surface areas of the various binary mixtures prepared in this study did not vary by more than 0.6% because the surface areas of the two starting powders (cement and limestone) are so similar, but previous studies have clearly indicated that increased limestone powder surface area at a constant volume fraction accelerates early age hydration and reduces setting times [23,24]. This further implies that during this early stage of hydration, the rates are controlled not by dissolution of the cement (see right-hand side of Fig. 15.3), but rather by the availability of favorable surfaces for heterogeneous precipitation of hydration products. Limestone powder surfaces appear to be equally favorable to cement for this precipitation process, at least for limestone in its calcite, as opposed to its aragonite, form [26,28–30].

Finally, in Fig. 15.5, normalization of the calorimetry data by the initial volume of mixing water is considered. If it is assumed that the volume of generated hydration products is proportional to the heat produced, this

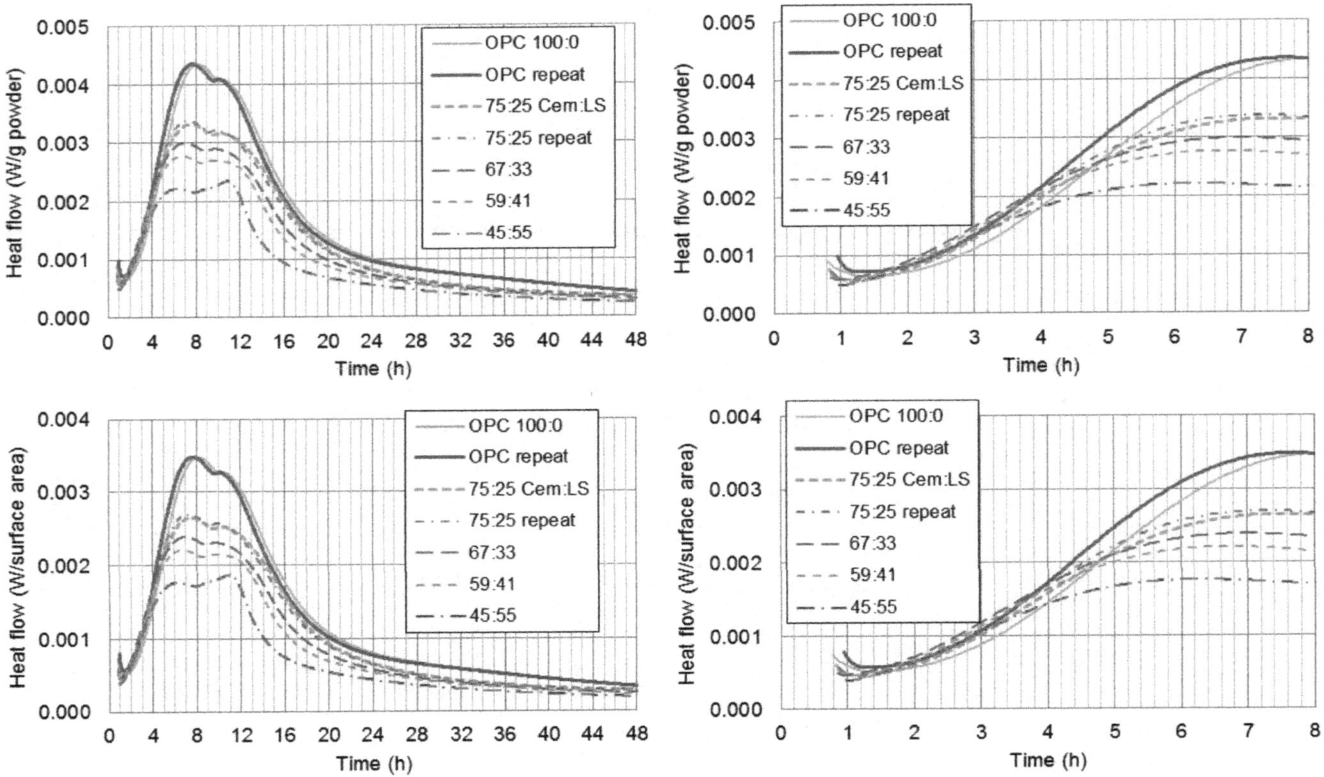

Figure 15.4 Heat flow normalized by mass of powder (top) and surface area (m^2) of powder (bottom) in each paste mixture.

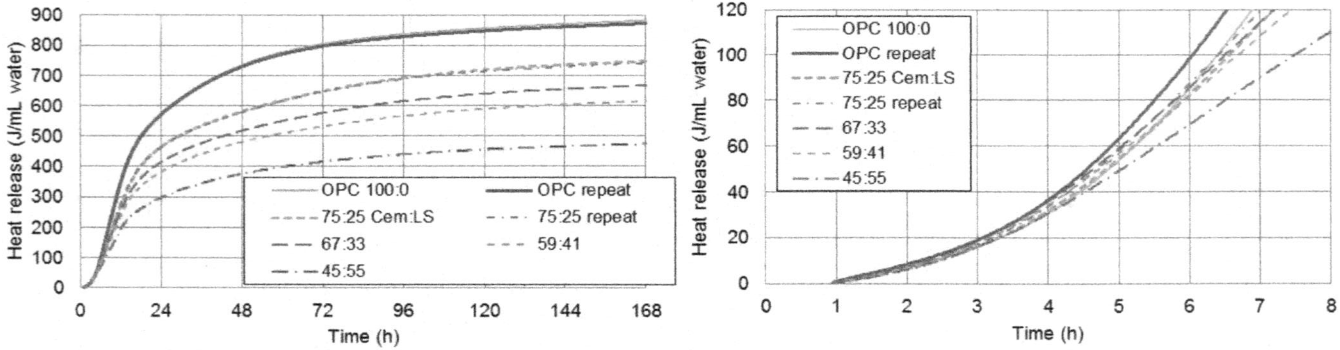

Figure 15.5 Cumulative heat release normalized by initial volume of water in each paste mixture.

normalization should be indicative of the fractional space filled by hydration. Thus, it is perhaps not surprising that this normalization generally produces a direct linear relationship between heat release and measured compressive strength, for ages from 1 to 7 days [27,31,32]. In this case, the rank order of the 7 days heat release per unit volume of water is, as expected, in direct proportion to the level of limestone replacement for cement (Fig. 15.6). The rheology results to follow will indicate that yield stress and viscosity are to a large extent controlled by the paste's solid volume fraction (or 1—water volume fraction), in agreement with previous studies [33,34]. While strength is usually engineered via changes in the *w/c* or *w/p* ratio, often with an accompanying change in solids and/or water content that would influence rheology, Fig. 15.6 suggests an alternate avenue for engineering strength in mixtures for ACE applications, in which the water content can be held constant and the replacement level of limestone for cement used instead to adjust the *w/c* and control the 7 days (and later age) strengths (basically, the approach employed in the first phase of this study). In this scenario, the rheological needs of the fresh mixture would determine its solids (water) content, while the ratio of cement to limestone powder would be used to provide a requisite level of strength (cement) without changing the water to powder (volumetric) proportions. Other benefits of replacing cement with limestone powder will generally include a cost savings [35] and a lower density paste that may provide increased buildability for ACE applications by decreasing the dead load of the upper layers on those below.

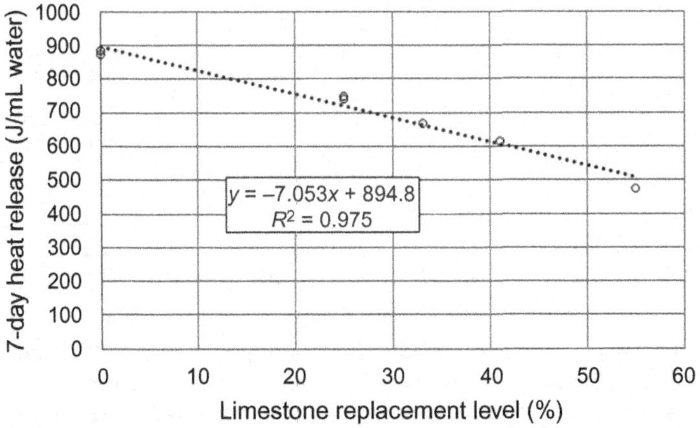

Figure 15.6 7 days heat release normalized per mL of water versus limestone powder replacement for cement level. Values for replicate specimens for the 0% and 25% limestone replacement levels are provided as an indication of variability.

Equally interesting in Fig. 15.5 is the overlap of the various curves for cumulative heat release normalized by initial volume of water for the time period between 1 hour and about 4 hours, as shown in the expanded graph on the right side of the figure. This implies that the total volume of solids being formed during this time period is nearly the same regardless of the ratio of cement to limestone powder employed in the mixture. In the blends with limestone powder, there is less cement, but it dissolves more rapidly to release about the same amount of heat. Since this overlap occurs within the time window normally attributed to initial set (typically corresponding to a cumulative heat release of about 40 J/mL water for a $w/c = 0.35$ cement paste [32]) and perhaps the beginning of final set, these mixtures with limestone powder contents ranging from 0% to 55% on a mass basis should exhibit setting behaviors that are the same within an hour or so. This provides a concrete demonstration of the potential robustness of these mixtures, as strength levels can be modified by changing the ratio of cement to limestone powder, with minimal confounding effects on critical early-age properties such as setting (as inferred from Fig. 15.5) and rheology (results to follow).

Fig. 15.7 provides an example of the data obtained from a typical shear rate loop experiment, in this case showing measurements performed at four different times after initial cement-water contact. Shear rates between

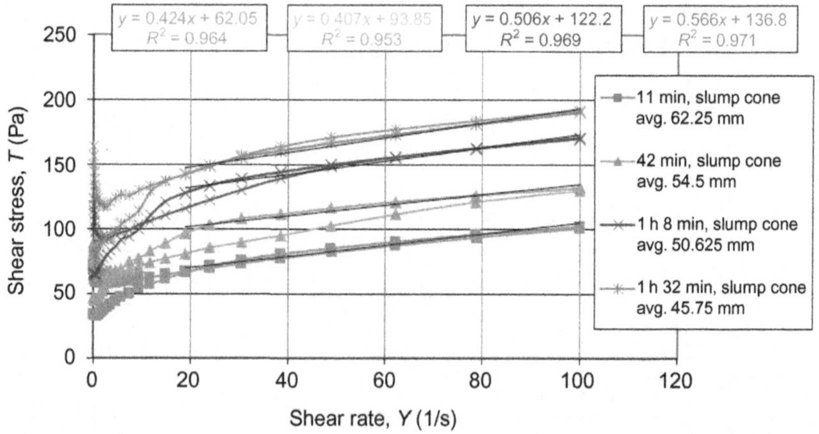

Figure 15.7 Shear stress versus shear rate for a specimen of the 67:33 cement:limestone powder paste at various times after contact between water and cement. Solid black lines illustrate best fits for an extraction of a region of the descending branch of the loop for estimation of yield stress and viscosity using Eq. (15.1) (fits shown in boxes).

100 and 18.9 s^{-1} are subsequently used for the estimation of the rheological parameters (yield stress and plastic viscosity) via Eq. (15.1). In general, by selecting this subrange of data, fits to Eq. (15.1) with an R^2 value greater than 0.95 (often >0.98) were obtained, with standard errors of less than 3%. In the worst case, among all the data presented in this chapter, the standard error in the estimated yield stress (intercept) was less than 3% and that in the estimated viscosity (slope) was less than 16%.

Fig. 15.8 provides the temporal development of the rheological parameters determined from Eq. (15.1) for the different pastes prepared in phase I of the study. Within the scatter and variability of the

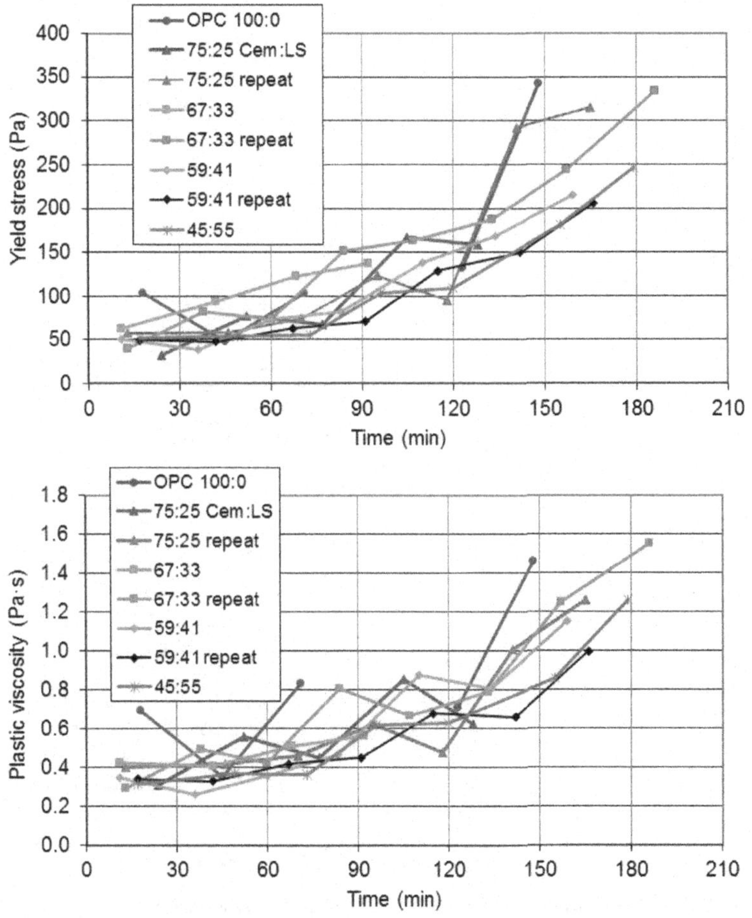

Figure 15.8 Yield stress (top) and plastic viscosity (bottom) determined via Eq. (15.1) versus time for the paste mixtures examined in phase I of the study.

measurements, the results for these pastes all prepared with nominally the same water content overlap one another to a large degree throughout the 180 minutes measurement period employed in this study. This suggests that replacing cement particles with limestone powder of a similar size/surface area has minimal impact on the resultant rheology of the system. The trends for yield stress and plastic viscosity are quite similar, with a generally slow increase at times up to about 120 minutes, followed by a more rapid increase after that. Since the early-age hydration versus time data (Fig. 15.5) are quite similar for the different mixtures, it is perhaps not surprising that the evolution of the rheological parameters is similar as well. In support of this, as shown in Fig. 15.9,

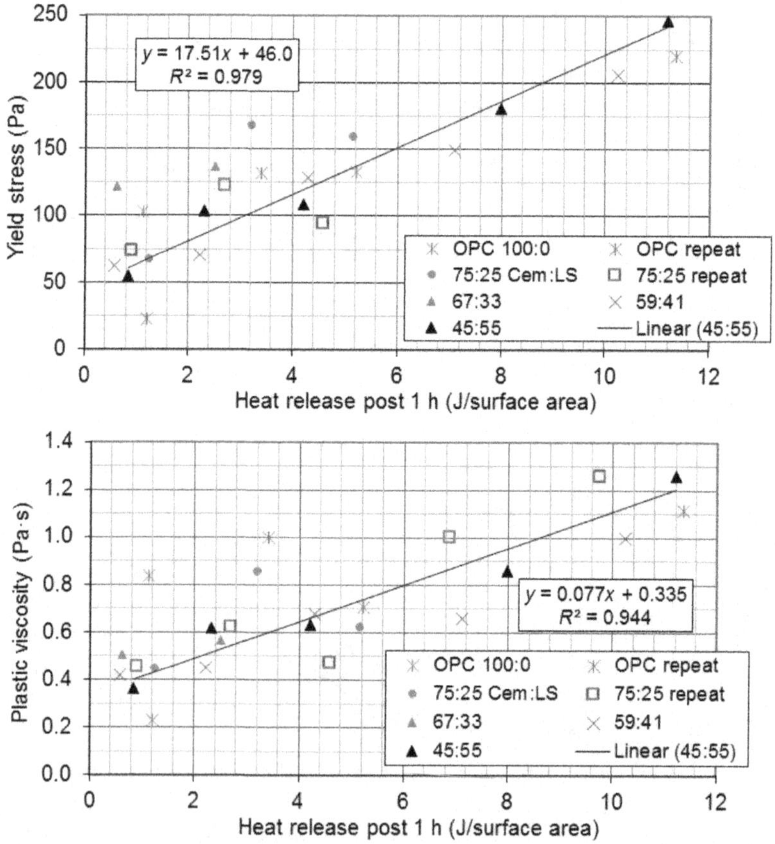

Figure 15.9 Development of yield stress (top) and plastic viscosity (bottom) versus cumulative heat release for the various pastes examined in the phase I study. In each case, the shown linear fit is only for the 45:55 cement:limestone powder paste data.

when the temporal data of Fig. 15.8 are exchanged for the equivalent cumulative heat release values, generally linear relationships between hydration (via heat release) and either yield stress or plastic viscosity are obtained.

While the phase I study has demonstrated a robust approach to mixture design in that early-age properties are generally independent of the cement:limestone powder ratio, the pastes investigated typically pose two challenges regarding their potential usage in ACE applications. First, their open time, likely limited by their viable buildability window, may be too short, as pastes of similar composition are only printable for a time period in the order of 30 minutes [36]. Additionally, their "dormant" period in the order of 2 hours may not be sufficiently long for some applications. An ideal binder for ACE applications would exhibit an extended dormant period (perhaps 6 hours or more), followed by an almost instantaneous stiffening to move seamlessly from the pumpability to the buildability window, while of course maintaining printability. Just such a performance behavior might be achieved by the judicious use of chemical retarders and accelerators, as has been demonstrated elsewhere in concrete and mortar mixtures for 3D printing applications [2,4]. Paste mixtures to demonstrate the feasibility of this approach constituted the phase II portion of this study.

The calorimetry results provided in Fig. 15.10 summarize the influence of a retarder or a retarder/injected accelerator combination on the hydration kinetics of the 67:33 cement:limestone powder mixture. A retarder dosage of 2 mL/kg powder (0.2%) produces a dormant period about 10 hours longer than that of the original paste, while a dosage of 1.33 mL/kg (0.133%) produces an extension of 5 hours. Using this retarder, the "pot" time of the mixture can be tailored easily to vary from about 2 hours (no addition) to 12 hours or more. This solves the first challenge of formulating an appropriate binder for ACE applications. As shown in Fig. 15.10, the second challenge, namely to achieve nearly instantaneous stiffening of the paste by enhanced hydration, can be addressed by the subsequent injection of an accelerator. Fig. 15.10 indicates that this accelerator shortens the "dormant" period for the silicate hydration by only about 1 hour, but in this case, stiffening instead might be supplied by a rapid hydration of a portion of the aluminate phases (e.g., ettringite and carboaluminate formation). This is indicated by the sharp peak present in the heat flow curves for the system with the injected accelerator in Fig. 15.10, although it must be recognized that a portion of

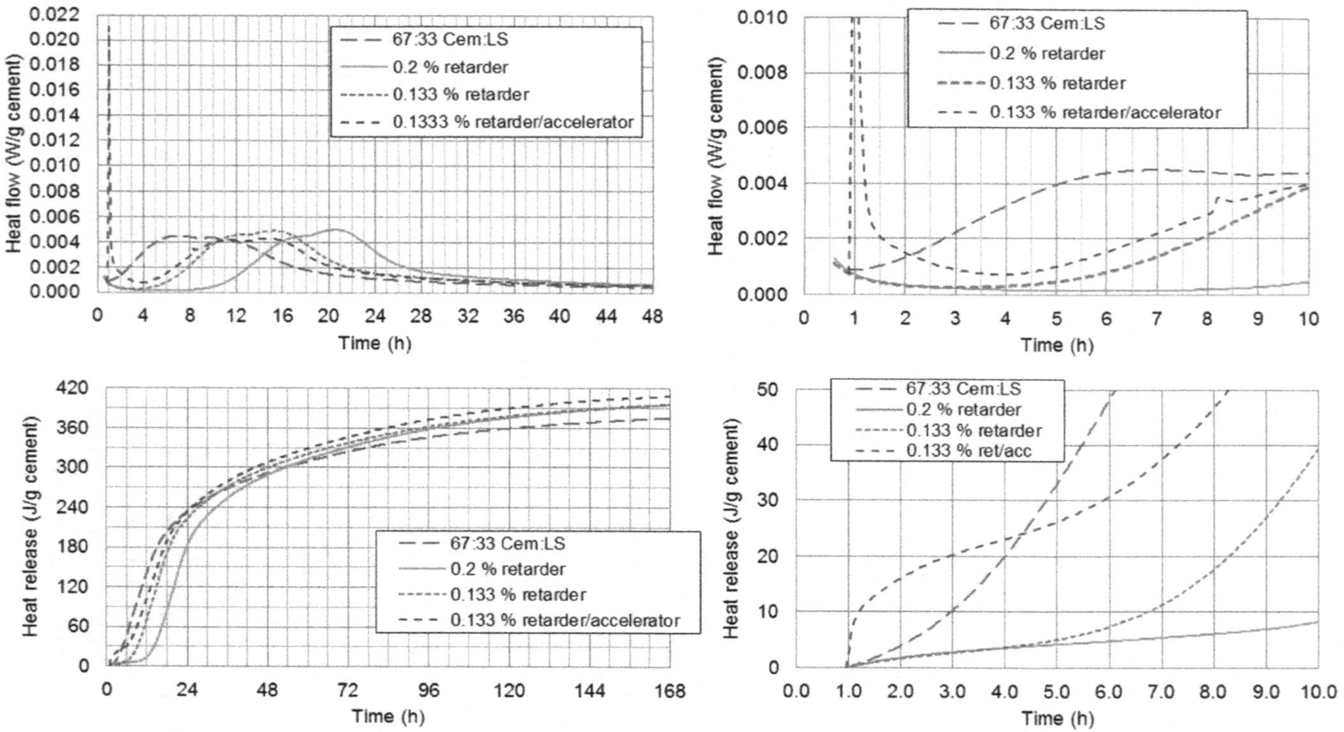

Figure 15.10 Heat flow and cumulative heat release curves for the 67:33 cement:limestone paste with and without the addition of chemical admixtures (retarder at two dosages and one combination of retarder/accelerator). 1 g cement≡0.563 mL water for these mixtures.

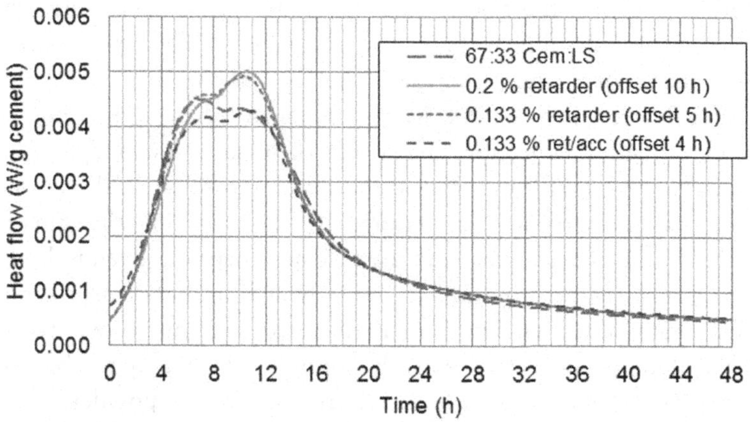

Figure 15.11 Heat flow curves each offset by a fixed time to "remove" the influence of retarder (retarder/accelerator) on the dormant period of hydration.

this heat flow is due to the in situ mixer being employed for 30 seconds to mix the accelerator into the hydrating paste in the calorimeter ampoule. Furthermore, the reaction that occurs between the aluminum sulfate in the accelerator and the fine limestone powder portion of the paste to generate carbon dioxide, calcium sulfate, and aluminum hydroxide [37] is likely endothermic and could therefore counterbalance a portion of this mixing energy input.[2] Of course, the rheology measurements to follow should corroborate that this heat peak corresponds to an actual stiffening/setting of the paste and not just an extraneous release of heat (due to the mixing energy being supplied, etc.).

Before proceeding to the rheology results for the phase II mixtures, however, it is important to note that while these chemical admixtures strongly influence the length of the dormant period in the paste mixtures, following this, the hydration proceeds in a quite similar fashion as that observed in the admixture-free control paste. This is quantitatively demonstrated in Fig. 15.11, where the heat flow curves are plotted with various offset times, each corresponding to the dormant period increase provided by that admixture (combination). With these offsets, the primary

[2] When a pure limestone powder-water paste was prepared ($w/p = 0.4$ by mass) and reacted with accelerator (3% by mass of paste), a vigorous reaction ensued, with gas generation, two color changes, and a reduction in temperature, the reduction in temperature being consistent with an endothermic reaction or at least an absorption of heat by the system. X-ray diffraction has indicated that one of the products of this reaction is calcium sulfate dihydrate (gypsum).

hydration peaks for the silicate and aluminate reactions basically overlap one another for the four cases investigated here. While the retarder does amplify the second (aluminate) peak slightly and the accelerator reduces this amplification, perhaps due to the aluminates that were rapidly consumed following its injection, the widths of these two overlapping peaks are visually the same in the four cases shown, regardless of the presence/absence of the chemical admixtures. Based only on the curves in Fig. 15.11, one would have little if any clue whether a chemical admixture had been used in the mixture once hydration starts in earnest. This suggests that for the chemical admixtures used in this study, hydration after the dormant period is primarily controlled by the system's porosity (water content) and the particle size distribution of the powder, as in the original mixture with no admixtures.

For the rheology experiments, the 67:33 cement:limestone powder paste mixture was prepared with a 0.00133 mL/g powder dosage of the retarder (added to the mixing water). In a first experiment, the rheology of this paste was monitored at discrete points over 6 hours, while in a second one, the accelerator was injected (3% dosage by mass of paste) after 3 hours and the rheology subsequently monitored over about 1 hour. For both experiments, the linear portions of the descending branch of the shear rate loop curves are provided in Fig. 15.12, along with the fits determined according to Eq. (15.1). When only the retarder is present, the paste rheology is highly stable for the first 3 hours and then the yield stress and plastic viscosity increase from about 5 hours onward. This observation is generally consistent with the hydration curves in Figs. 15.10 and 15.11 for the same paste. As suspected, the accelerator provides an "instant" stiffening of the paste mixture, but the reaction products (likely ettringite, hemicarbonate, and perhaps monocarbonate) that it generates actually produce more of a false/flash setting than a true setting of the paste. This is illustrated in the plot in Fig. 15.12, where the yield stress and plastic viscosity values achieve a maximum for the measurement performed 5 minutes after the accelerator injection and then, with subsequent remixing, fall dramatically and nearly return to their pre-injection values. Of course, in an actual ACE application, no remixing should occur once the accelerator is injected and the paste ejected from the nozzle onto the printing platen. To better simulate such a printing process, the rheological measurements were repeated for the same retarded/accelerated paste mixture, but this time with no remixing following the accelerator injection/mixing for a

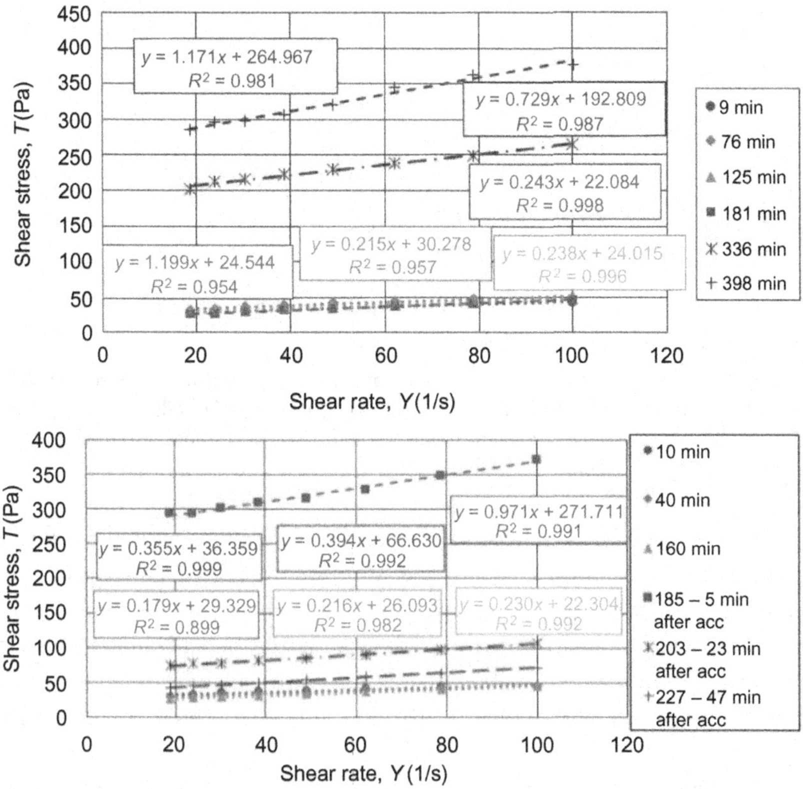

Figure 15.12 Shear stress versus shear rate for retarded 67:33 cement:limestone powder pastes without (top) and with (bottom) a subsequent injection of the accelerator at 180 min. For the fits shown in the boxes, the *y*-intercept indicates the yield stress and the slope indicates the plastic viscosity (per Eq. (15.1)).

fixed period of time, followed by the usual remixing to see if fluidity would be restored even in the case of a quite stiff paste that had been sitting/setting for 40 minutes (yield stress of about 300 Pa).

The results of this experiment with no remixing for the first 40 minutes following injection are provided in Fig. 15.13. When the accelerated paste is not remixed, it continues to stiffen as indicated by the responses recorded 20 and 40 minutes after accelerator injection. In this case, the yield stress of the retarded paste increased by less than 15 Pa during the first hour, but there was a rapid increase in yield stress of more than 200 Pa once the accelerator was introduced. After allowing the microstructure formation to proceed unabated (no mixing) for 40 minutes, a subsequent mixing did not significantly break down the developed structure, as indicated by the

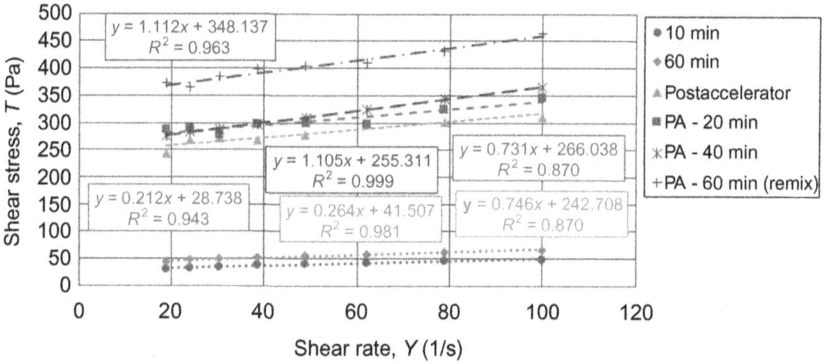

Figure 15.13 Shear stress versus shear rate for the retarded/accelerated 67:33 cement:limestone powder paste with remixing immediately following accelerator injection (at 1 h), a period of 40 min with no remixing, and finally a remixing of the accelerator-stiffened paste.

PA-60 min results in Fig. 15.13. Instead, the yield stress exhibited a further increase to about 350 Pa, while the viscosity remained at its premixing level. These results support the hypothesis that the hydration product formation provided initially by the accelerator should be sufficient to provide a buildable structure, even though this structure can be broken down by high intensity mixing applied early (within 20 minutes) in the aging process.

Based on its density provided in Table 15.1, the dead load of a 3 mm thick layer of the 67:33 cement:limestone paste can be calculated to be 55.8 Pa, implying a requisite yield stress of about 550 Pa to support 10 layers of paste in an ACE application. According to Roussel [14], thixotropy gains at early ages due to hydration of cement-based materials are typically on the order of 0.1–2 Pa/s. Relying on conventional hydration would, thus, require a wait time between successive mixtures varying between about 9 minutes and about 30 seconds depending on the magnitude of this rate of thixotropy gain. Conversely, by using the accelerator, an "instantaneous" gain of 200–600 Pa can be achieved, with a subsequent growth rate in the range of 1–2 Pa/s (see Fig. 15.14), about 10 times that exhibited by the nonaccelerated pastes examined in this study (Figs. 15.12 and 15.14). This clearly demonstrates the significant benefits of considering an injected accelerator (near the nozzle) as a necessary requisite for a viable and robust ACE production.

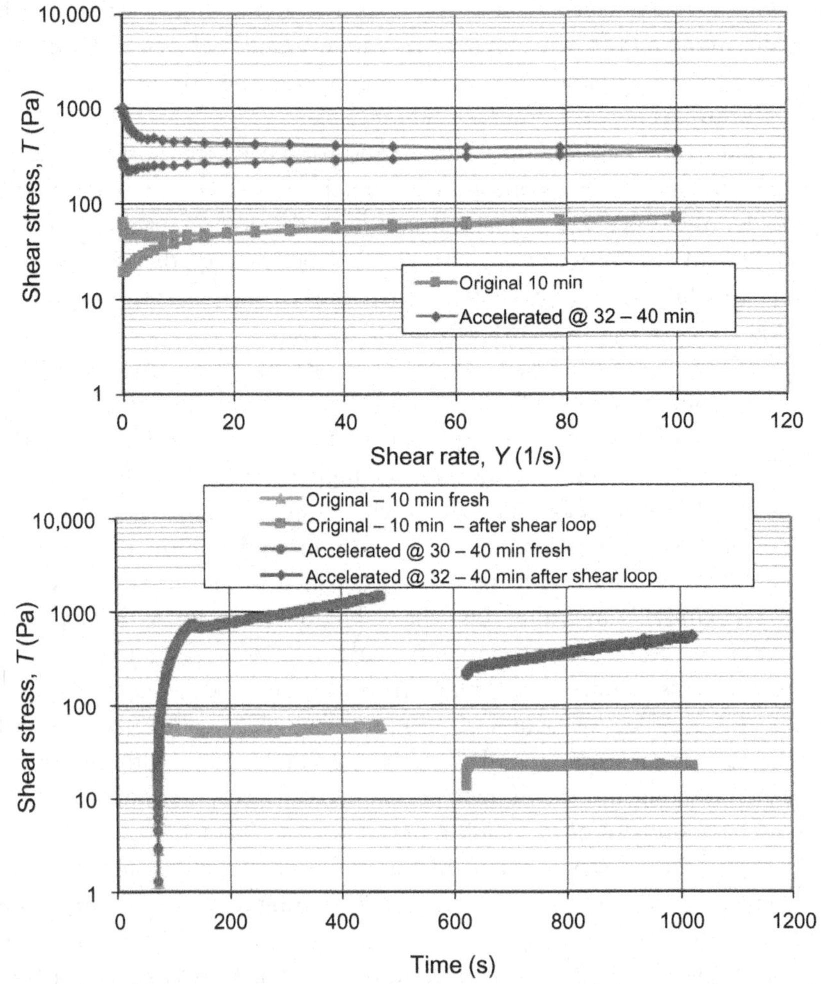

Figure 15.14 Shear rate loop (top) and stress growth (bottom) rheology measurements for a 59:41 cement:limestone paste 10 min after contact of water to cement and 8–10 min after an accelerator addition to the hydrating paste, injected at 30–32 min. No retarder was employed in this experiment. Fresh indicates that the stress growth measurement was conducted immediately after loading the specimen into the rheometer (no shear loop).

15.4 CONCLUSIONS

Two series of pastes, one with varying levels of limestone substitution for cement and one with varying dosages of chemical admixtures were prepared and evaluated with respect to early-age hydration and rheological parameters. These two analytical methods can provide valuable insights into the open time window and viability of proposed mixtures for ACE applications, as demonstrated in this chapter.

Conclusions drawn from this study include:

- When the particle size distributions and particularly the surface areas of the cement and limestone powder are similar, as was the case in this study, the early-age (to about 5 hours) rheological parameters and hydration (calorimetry) are basically independent of the replacement level of limestone powder for cement (up to the 55% investigated in this study) at a constant water volume fraction.
- A robust, sustainable paste could thus be engineered by setting the water content to provide the desired rheology and adjusting the limestone powder for cement replacement level to produce the targeted 7 days (and later age) strength. An example of this would be the 67:33 cement:limestone paste mixture employed in this study.
- For the specific retarder and accelerator used in this study, the dormant period of hydration is increased or decreased by the incorporation of the chemical admixture, but the subsequent silicate and aluminate hydration peaks are similar, particularly with respect to their overall width. This implies that setting/stiffening can be delayed in the case of the retarder or achieved nearly instantaneously in the case of the accelerator or the retarder/accelerator combination without significantly altering the subsequent primary (acceleratory period) hydration and development of mechanical properties that follows.
- Because the accelerator used in this study functions by more rapidly generating reaction products including ettringite and hemicarbonate from the available aluminates and doesn't greatly alter the course of silicate hydration, it produces more of a false/flash setting than a true setting of the paste mixtures. Still, if not followed by subsequent remixing to break down the initially created structure, the false/flash setting should be adequate to produce a buildable layer of cement paste that can support subsequent layers in the ACE production process.

ACKNOWLEDGMENTS

The authors would like to thank Carmeuse for supplying the limestone powder, GCP Applied Technologies for supplying the chemical admixtures, and Lehigh Hanson for supplying the Portland cement used in this study. The assistance of Paul Stutzman of

NIST in performing X-ray diffraction measurements on the reacting limestone powder/accelerator mixture is gratefully acknowledged, as are careful reviews of the manuscript by Dr. Jeff Bullard and Dr. Chiara Ferraris of NIST and Dr. Nathan Tregger and Dr. Ezgi Yurdakul of GCP Applied Technologies.

REFERENCES

[1] S. Lim, R.A. Buswell, T.T. Le, S.A. Austin, A.G.F. Gibb, T. Thorpe, Developments in construction-scale additive manufacturing processes, Autom. Constr. 21 (1) (2012) 262–268.

[2] T.T. Le, S.A. Austin, S. Lim, R.A. Buswell, A.G.F. Gibb, T. Thorpe, Mix design and fresh properties for high-performance printing concrete, Mater. Struct. 45 (2012) 1221–1232.

[3] F. Bos, R. Wolfs, Z. Ahmed, T. Salet, Additive manufacturing of concrete in construction: potentials and challenges of 3D concrete printing, Virtual Phys. Prototyping 11 (3) (2016) 209–225. Available from: https://doi.org/10.1080/17452759.2016.1209867.

[4] C. Gosselin, R. Duballet, Ph Roux, N. Gaudillière, J. Dirrenberger, Ph Morel, Large-scale 3D printing of ultra-high performance concrete – a new processing route for architects and builders, Mater. Des. 100 (2016) 102–109.

[5] I. Hager, A. Golonka, R. Putanowicz, 3D printing of building and building components as the future of sustainable construction? Procedia Eng. 151 (2016) 292–299. Available from: https://doi.org/10.1016/j.proeng.2016.07.357.

[6] A. Perrot, D. Rangeard, A. Pierre, Structural build-up of cement-based materials used for 3D-printing extrusion techniques, Mater. Struct. 49 (2016) 1213–1220. Available from: https://doi.org/10.1617/s11527-015-0571-0.

[7] T. Wangler, E. Lloret, L. Reiter, N. Hack, F. Gramazio, M. Kohler, et al., Digital concrete: opportunities and challenges, RILEM Tech. Lett. 1 (2016) 67–75. Available from: https://doi.org/10.21809/rilemtechlett.2016.16.

[8] M. Hamach, D. Volkmer, Properties of 3D-printed fiber-reinforced Portland cement paste, Cem. Concr. Compos. 79 (2017) 62–70. Available from: https://doi.org/10.1016/j.cemconcomp.2017.02.001.

[9] A. Kazemian, X. Yuan, E. Cochran, B. Khoshnevis, Cementitious materials for construction-scale 3D printing: laboratory testing of fresh printing mixture, Constr. Build. Mater. 145 (2017) 639–647.

[10] G. Ma, L. Wang, A critical review of preparation design and workability measurement of concrete material for largescale 3D printing, Front. Struct. Civil Eng. (2017). Available from: https://doi.org/10.1007/s11709-017-0430-x.

[11] T.S. Rushing, G. Al-Chaar, B.A. Eick, J. Burroughs, J. Shannon, L. Barna, et al., Investigation of concrete mixtures for additive construction, Rapid Prototyping J. 23 (1) (2017) 74–80. Available from: https://doi.org/10.1108/RPJ-09-2015-0124.

[12] J. Zhu, Z. Tao, R. Mansour, W. Chen, 3D printing cement based ink, and its application within the construction industry, in: MATEC Web of Conferences 120, 02003, 2017, 13 pp. https://doi.org/10.1051/matecconf/201712002003.

[13] S.P. Jiang, J.C. Mutin, A. Nonat, Studies on mechanisms and physico-chemical parameters at the origin of cement setting. 1. The fundamental processes involved during the cement setting, Cem. Concr. Res. 25 (4) (1995) 779–789.

[14] N. Roussel, A thixotropy model for fresh fluid concretes: theory, validation, and applications, Cem. Concr. Res. 36 (2006) 1797–1806.

[15] N. Roussel, A. Lemaître, R.J. Flatt, P. Coussot, Steady state flow of cement suspensions: a micromechanical state of the art, Cem. Concr. Res. 40 (2010) 77–84.

[16] N. Roussel, G. Ovarlez, S. Garrault, C. Brumaud, The origins of thixotrophy of fresh cement pastes, Cem. Concr. Res. 42 (2012) 148–157.
[17] C. Paglia, F. Wombacher, H. Böhni, The influence of alkali-free and alkaline shotcrete accelerators within cement systems I. Characterization of the setting behavior, Cem. Concr. Res. 31 (2001) 913–918.
[18] N. Khalil, G. Aouad, K. El Cheikh, S. Rémond, Use of calcium sulfoaluminate cements for setting control of 3D-printing mortars, Constr. Build. Mater. 157 (2017) 382–391.
[19] ASTM C150/C150M-17, *Standard Specification for Portland Cement*, ASTM International, West Conshohocken, PA, 2017, www.astm.org.
[20] ASTM C1738/C1738M-14, Standard Practice for High-Shear Mixing of Hydraulic Cement Pastes, ASTM International, West Conshohocken, PA, 2014, www.astm.org.
[21] D.P. Bentz, C.F. Ferraris, Rheology and setting of high volume fly ash mixtures, Cem. Concr. Compos. 32 (4) (2010) 265–270.
[22] S. Amziane, C.F. Ferraris, Cementitious paste setting using rheological and pressure measurements, ACI Mater. J. 104 (2) (2007) 137–145.
[23] L. Gurney, D.P. Bentz, T. Sato, W.J. Weiss, Using limestone to reduce set retardation in high volume fly ash mixtures: improving constructability for sustainability, Transport. Res. Record: J. Transport. Res. Board (2012) 139–146. No. 2290, Concrete Materials 2012.
[24] T. Oey, A. Kumar, J.W. Bullard, N. Neithalath, G. Sant, The filler effect: the influence of filler content and surface area of cementitious reaction rates, J. Am. Ceram. Soc. 96 (6) (2013) 1978–1990.
[25] E. Berodier, K. Scrivener, Understanding the filler effect on the nucleation and growth of C-S-H, J. Am. Ceram. Soc. 97 (12) (2014) 3764–3773.
[26] D.P. Bentz, A. Ardani, T. Barrett, S.Z. Jones, D. Lootens, M.A. Peltz, et al., Multi-scale investigation of the performance of limestone in concrete, Constr. Build. Mater. 75 (2015) 1–10.
[27] D.P. Bentz, P.E. Stutzman, F.A. Zunino, Low temperature curing strength enhancement in cement-based materials containing limestone powder, Mater. Struct. 50 (173) (2017) 14. June.
[28] J. Stark, B. Möser, F. Bellmann, Nucleation and growth of C-S-H phases on mineral admixtures, in: C.U. Grosse (Ed.), Advances in Construction Materials, Springer, Berlin, Heidelberg, 2007, pp. 531–538.
[29] D.P. Bentz, C.F. Ferraris, S.Z. Jones, D. Lootens, F. Zunino, Limestone and silica powder replacements for cement: early-age performance, Cem. Concr. Compos. 78 (2017) 43–56.
[30] G.W. Scherer, F. Bellmann, Kinetic analysis of C-S-H growth on calcite, Cem. Concr. Res. 103 (2018) 226–235.
[31] D.P. Bentz, T. Barrett, I. de la Varga, J. Weiss, Relating compressive strength to heat release in mortars, Adv. Civil Eng. Mater. 1 (1) (2012) 14. Sept.
[32] D. Lootens, D.P. Bentz, On the relation of setting and early-age strength development to porosity and hydration in cement-based materials, Cem. Concr. Compos. 68 (2016) 9–14.
[33] L. Struble, G.-K. Sun, Viscosity of Portland cement paste as a function of concentration, Adv. Cem. -Based Mater. 2 (1995) 62–69.
[34] D.P. Bentz, C.F. Ferraris, M.A. Galler, J.M. Guynn, A.S. Hansen, Influence of particle size distributions on yield stress and viscosity of cement-fly ash pastes, Cem. Concr. Res. 42 (2) (2012) 404–409.
[35] D.P. Bentz, J. Tanesi, A. Ardani, Ternary blends for controlling cost and carbon content, Concr. Int. 35 (8) (2013) 51–59.

[36] S.Z. Jones, D.P. Bentz, N.S. Martys, W.L. George, A. Thomas, Rheological control of 3D printed cement paste, 1st International Conference on Concrete and Digital Fabrication, Digital Concrete 2018 – Zurich, Switzerland, September. 10–12, 2018.
[37] T. Kuwata, T. Nakazawa, The reaction of aluminum sulfate and calcium carbonate. I. Preparation of basic aluminum sulfate sol and its hydrolysis, Bull. Chem. Soc. Jpn. 22 (4) (1949) 182–185.

CHAPTER 16

Studying the Printability of Fresh Concrete for Formwork-Free Concrete Onsite 3D Printing Technology (CONPrint3D)

Venkatesh Naidu Nerella and Viktor Mechtcherine
Technische Universität Dresden, Institute of Construction Materials, Dresden, Germany

16.1 INTRODUCTION

Concrete is the world's most widely used construction material; with a major part of it being ready-mix concrete. For placing most types of concrete, especially those of ready-mix concrete, formworks are needed. The use of formworks often leads to high material, labor, and machinery costs in addition to considerable time delays and a negative impact on the environment. Other important challenges faced by the construction industry are low productivity, scarcity of labor, limited geometrical freedom, slow speed of construction, construction in hazardous or remote areas, high cost, and low sustainability. Although modern automation and computer-aided manufacturing have revolutionized many industrial processes, they still have had a very limited influence on the construction industry. Researchers across the globe are striving to solve these limitations by bringing much-needed mechanization and automation into the construction sector with innovative digital construction (DC) technologies [1–3]. DC technologies open up a plethora of opportunities and technological advancements, such as no need of formwork (considerable time and cost reductions), high geometrical flexibility, and low dependency on skilled labor. However, the main significance and revolutionary potential of 3D concrete printing is seen in the context of Construction Industry 4.0 since it offers a logical and highly desired step from the already well-developed tool of digital design and planning (computer aided design (CAD), building information model (BIM), etc.) to the digital manufacturing, thus, making construction a fully digitalized, seamless process.

3D Concrete Printing Technology
DOI: https://doi.org/10.1016/B978-0-12-815481-6.00016-6

16.1.1 Additive Manufacturing of Construction Elements

DC approaches can be broadly subdivided into four categories: (A) methods based on extrusion [1,4]; (B) methods based on selective binding [3]; (C) vibration-based 3D concrete printing (e.g., technology implemented by Hua Shang Tengda Ltd [5]); and (D) adaptive formwork casting (e.g., smart dynamic casting [6]). This classification excludes the partial-DC technologies (e.g., variants of Total Kustom [7] and Apis Cor [8]) where a part of the structure (e.g., outer formwork shells) is 3D printed and the rest of the structure is conventionally constructed (e.g., filling self-consolidating concrete (SCC), or vibrated-concrete) (Fig. 16.1).

The most widely used approaches for DC are extrusion and selective binding. In both these approaches, a 3D CAD model is converted into a Stereo Lithography format and then divided into multiple 2D or 2.5D layers in a "slicing" process. Coordinates from these layers along with printing parameters such as velocity of printhead and rate of extrusion are then delivered to a 3D printer via its control computer program. The desired structure is then built layer-by-layer using the defined coordinates and given printing rates. In the case of extrusion-based additive construction, premixed material is extruded at the specified coordinates through a nozzle at the determined printing rates. In the case of selective binding,

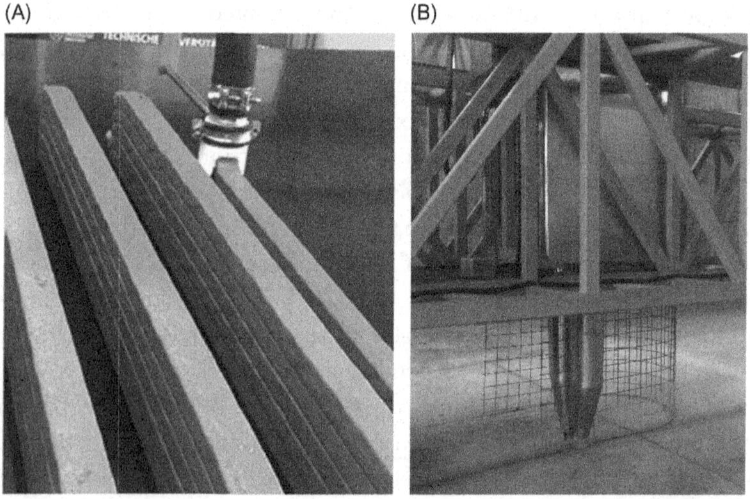

Figure 16.1 Examples from digital construction processes: (A) extruded high-performance concrete elements, from CONPrint3D [9]; and (B) printhead with forked nozzle developed by HuaShang Tengda Ltd., that simultaneously deposits concrete on both sides of the rebars. *From V. Mechtcherine.*

dry materials are first placed on a platform (bed) and binder or activator is then delivered to specified coordinates. Detailed information and example applications for all four categories of DC processes can be referred to in [9,10].

Present additive construction methods, while demonstrating many technological advantages, are subjected to some inherent limitations such as the necessity of using new, advanced machinery, small mineral aggregate sizes (generally diameter under 2 mm), and limited size of the construction elements (the size of the 3D printer must be larger than the size of the element to be printed).

16.1.2 CONPrint3D: Onsite 3D Concrete Printing

A methodology of applying 3D printing in the field of onsite concrete construction, named CONPrint3D, is being investigated at the TU Dresden. The primary objective of this project is the development of a formwork-free, monolithic construction process using 3D printing. One of the focal points of CONPrint3D is not only to develop a time-, labor-, and resource-efficient advanced construction process, but also to make the new process economically viable while achieving broader acceptance from the existing industry practitioners. This is achieved by using existing construction and production techniques as much as possible and by adopting new processes into construction site constraints. One vital aspect of the project strategy is adapting a concrete boom pump to deliver material to specific positions autonomously and accurately using a custom-developed printhead attached to the boom, see Fig. 16.2.

16.2 3D CONCRETE PRINTING: A PROCESS OF DUALITIES

3D printing of cementitious materials is a process of many dualities. For example, there is a duality of pumpability and buildability. To be pumped properly, concrete needs to be flowable and have relatively low plastic viscosity and "moderate" yield stress. At the same time, to enable the layer-by-layer printing of concrete immediately after extrusion, it must cease being flowable and exhibit much higher yield stress and plastic viscosity. Another duality results from defining the "rate of printing," which should be slow enough so that printed layers achieve sufficient buildability and fast enough to ensure both sufficient bond strength (to avoid cold joints) between individual layers [12], and to keep the construction rate economically viable [13]. From the perspective of materials science, rheological

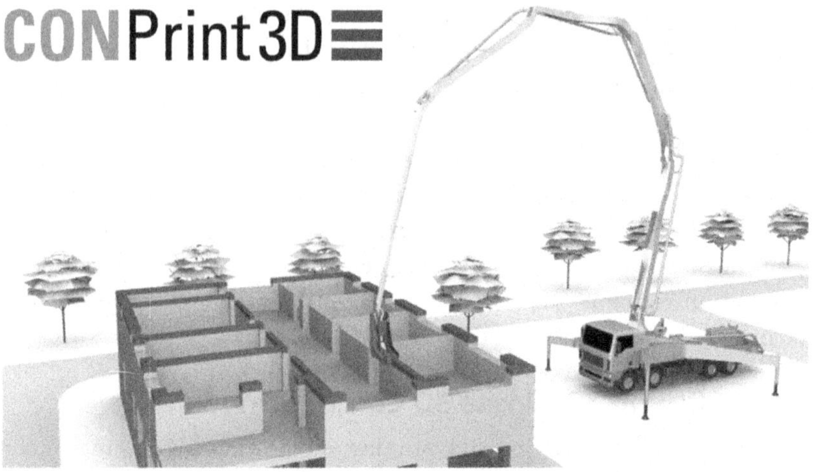

Figure 16.2 Illustration of CONPrint3D approach depicting a portable concrete pump, robotic arm, and printhead assembly. *Reproduced from [11] V.N. Nerella, M. Krause, M. Näther, V. Mechtcherine, 3D printing technology for on-site construction, Concr. Plant Int. (2016) 36–41.*

properties of fresh cementitious material are the most crucial aspect of DC, since they affect not only the extrusion process parameters, but also the properties of the final product.

To fulfill the requirements of extrusion-based DC, cementitious material should be thixotropic, quick-setting, quickly hydrated, densely packed, and possess steerable rheological properties such as static yield stress and plastic viscosity through internal or external triggers. While this general approach is understood [14,15], many fundamental questions need to be solved for its purposeful implementation into 3D concrete printing technology. The open questions include the systematic choice of rheological parameters and identification of their threshold values in the context of the processing technique and material design for achieving the desired properties, etc. In this chapter, we address how the printability of cementitious materials can be characterized with current and new test methods.

16.2.1 Important Material Aspects of Printable Concretes

To meet the physical and process requirements, cementitious compositions for DC should fulfil numerous material properties, some of the selected properties are presented in Fig. 16.3.

The holistic approach suggested and already tested at a laboratory scale at TU Dresden, followed this procedure:

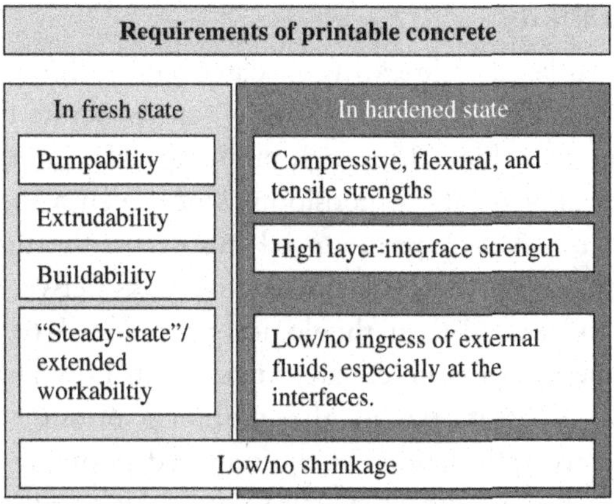

Figure 16.3 Selected requirements of printable concretes in fresh and hardened state.

1. Identify required material properties such as compressive strength based on target application.
2. Selection of raw materials and process parameters based on parametric studies and literature considering the influence of constituents and process parameters on all the material properties mentioned in step three.
3. Extensive experimental studies to develop printable concretes using investigations on:
   a. Pumpability
   b. Extrudability
   c. Buildability (ability to withhold form geometry under pressure from upper layers/"green" strength)
   d. Hydration kinetics (setting time and "extended steady state" or consistency detainment for longer time)
   e. Compressive, tensile, and flexural strengths in various combinations of loading direction with respect to the layer-interface plane.
4. Optimization of developed compositions for specific target application (e.g., load-bearing structures, nonload-bearing elements, artistic features/sculptures, etc.).

16.3 SELECTED METHODS TO CHARACTERIZE PRINTABLE CONCRETES IN FRESH STATE

In this section, methods for characterizing the primary components of printability: pumpability, extrudability, and buildability are described.

16.3.1 Pumpability

The first phase in large-scale extrusion-based DC techniques is transporting (not in the sense of "form plant to site"' the concrete to the printhead which is commonly achieved through pumping. The pumpability of concrete depends highly on its composition. Since even a slight variation in the mix design can have a pronounced impact on the behavior of fresh concrete, it is difficult to establish quantitative links between the compositions of the mixtures and their rheological properties [16]. Obviously the rheological properties of concrete are crucial to its flow characteristics and they define, to a great extent, the pumping pressure required [17]. General quantitative estimations of the required pumping pressure using, for example, nomographs and rheographs based on slump or spread values have many technical and practical limitations which have been reported in detail in the literature [16,18,19]. Various other approaches combining a broad-range of experimental tools as well as numerical tools have been introduced and successfully validated [19–22].

Among other techniques such as viscometers and tribometers, a specialized testing device—the Sliding Pipe Rheometer (Sliper; see Fig. 16.4)—has been [16,23] used to characterize pumpability. With Sliper, the pumpability is tested by filling the pipe placed in the topmost position with fresh concrete and subsequently letting the pipe slide downwards. Various speeds of the concrete in the pipe in the subsequent measurements are achieved by applying additional weights. The speed of the

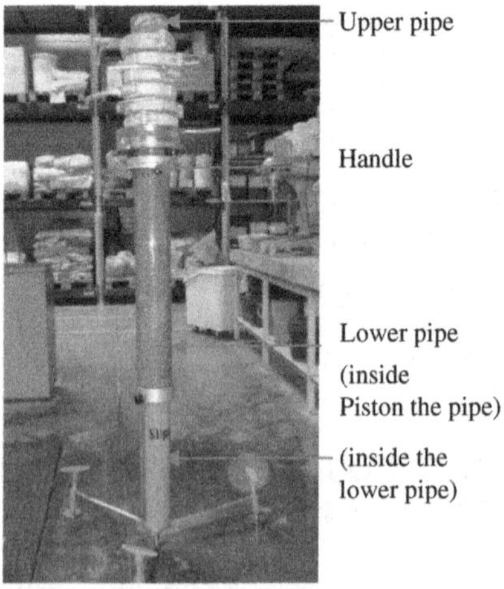

Figure 16.4 Sliding pipe rheometer (sliper).

pipe, measured by a displacement sensor, corresponds to the concrete flow rate, Q, while the pressure, P, of concrete at the piston head is associated with the pumping pressure. Eventually, the readings can be combined to plot a pressure versus flow-rate relationship P-Q. With the help of the slope and P-intercept of P-Q plot and the specifications of pumping circuit, one can estimate the discharge pressure P required for a pumping circuit under field conditions using Eq. (16.1):

$$P = \frac{4L}{D}a + \frac{16 \cdot L \cdot Q}{\pi \cdot D^3}b + \rho \cdot g \cdot H \tag{16.1}$$

where L is the length, D is the diameter of the pipeline, ρ is the density of concrete, and H is the pumping height.

For ordinary concretes it is evident from previous studies [20,24] that concrete flows as a plug in the pipelines; coarse aggregates move towards the center of the pipe forming a plug, while an easily deformable, lubricating layer is formed at the internal walls of the pipelines, leading to a considerable reduction in the required pumping pressure. Therefore, it is essential to carefully characterize and consider the properties of the lubricating layer in testing the pumpability of printable concretes. Detailed information on characterizing pumpability with experimental and numerical approaches can be referred to in [19,20,22,24–26].

16.3.2 Extrudability

Once the printable material reaches the printhead, it should possess adequate extrudability. Extrudability in DC is defined by Le et al. as "the ability of fresh concrete to pass through the small pipe and nozzles at the print-head." [4] Maleab at al. [27] described extrudability as the "capacity of the concrete to be extruded out from Nozzle." [27] Le et al. [4] used a qualitative experiment for selecting fine-grained concrete which showed no blockage and fracture in a 4.5 m extruded filament through a 9 mm nozzle [4]. A similar procedure was used to compare the length over which the cement paste can be extruded out of a nozzle without any blockage or sign of cracking and separation [27]. These methods are inadequate for DC since they do not provide any quantitative assessment. On the other hand, extrudability of some other materials (e.g., plastics) is well-investigated and quantified by using ram extruders. Using this principle, Perrot et al. [28] tested the extrudability of cement mortars quantitatively. Thus, ram extruders (see Fig. 16.5A) are, in principal, applicable for characterizing extrudability of printable concretes. Based on the

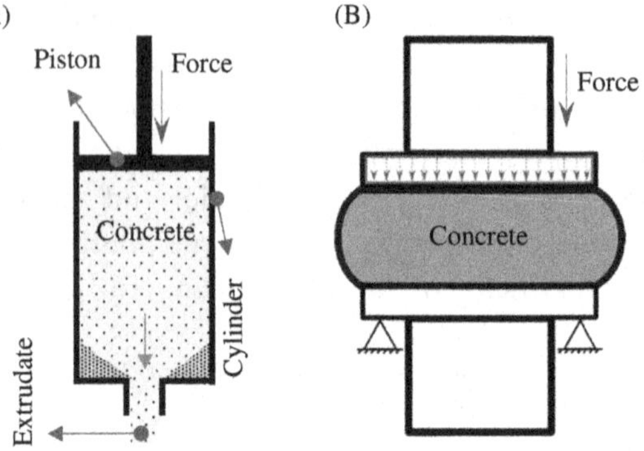

Figure 16.5 Methods for characterization of extrudability: (A) ram extruder; and (B) squeeze-flow method.

literature, two additional methods can be identified: the squeeze-flow method [29] (see Fig. 16.5B); and the penetration resistance method [30]. Detailed methodology, principles, recorded measurements, and applications of reported extrudability test methods can be referred to in [28–30].

Systematic studies on extrudability of printable concretes using ram-extrusion, the squeeze-flow method, or penetration resistance have not yet been reported. It is noteworthy that the concrete shearing in the screw-pump of a DC extruder is subjected to significantly different flow fields, shear histories, and pressures to those in the discussed approaches. Thus, new approaches are necessary to quantitatively assess the extrudability of cementitious materials.

Extrudability can be defined as the ability of a material to be extruded through the nozzle of 3D printer with minimal energy needed. Following this definition, the authors have recently developed an inline approach to quantitatively characterize extrudability of printable concretes. This is achieved with help of the 3D concrete printing test device (3DPTD) developed at TU Dresden. The principle of this approach is to measure electric power consumption, P_e, and extrudate flow rate, Q_e, at various rotational velocities of the extruder. Measured P_e and Q_e for various printable concretes are then comparatively analyzed to quantify extrudability.

16.3.3 Buildability

Buildability, that is, the resistance to deformation of a printed layer to the shear stress due to its own weight and that of subsequent layers, depends highly on the development of rheological properties of fresh concrete over time. Le et al. quantified buildability by the number of filament layers which can be built without the plastic deformation of lower layers [4]. Perrot et al. quantified buildability by the uniaxial compressive test on the extruded mortar [14].

In layer-by-layer casting, by reaching an element height of h_{max}, the lowest extrudate is considered to be under vertical stress of $\rho g h_{max}$ (full hydrostatic pressure) [14] or shear stress of $\rho g h_{max}/\sqrt{3}$ (von-Mises failure criterion) [15]. By comparing time-resolved stresses due to vertical loading with the corresponding static yield stress of the extruded material, rheology-based failure criteria were formulated. This idea was first introduced by Perrot et al. in [14], Eq. 5 and further elaborated by Wangler et al. in [10], Eq. 6:

$$\rho g h(t) \leq \alpha_g \left(\tau_{0,0} + A_{thix} t_c \left(e^{(t_{rest}/t_c)} - 1 \right) \right) \tag{16.2}$$

$$\rho g h(t) \leq \left(\alpha_g \left(\tau_{0,0} + A_{thix} t_c \left(e^{(t_{rest}/t_c)} - 1 \right) \right) \right) \sqrt{3} \tag{16.3}$$

where α_g is the geometry factor, $\tau_{0,0}$ is yield stress at the beginning of rest-time, t_c is characteristic rest-time, after which static yield stress grows exponentially.

In addition to the rheology-based failure criteria (Eqs. (16.2) and (16.3)), simple tests such as loading a concrete "slump spread" vertically with the help of a loaded plate, are also followed to estimate the buildability of concretes.

In the next sections, experimental investigations done by the authors on mechanical properties of printable concretes are described.

16.4 COMPRESSIVE AND FLEXURAL STRENGTHS OF PRINTABLE CONCRETES

Following the approach presented in the Section 16.2, appropriate raw materials were selected based on the specific requirements of the prospective application; in this case a multistory house. Aspects considered for selection of raw materials include their influence on: (1) setting and hydration times, (2) strength at very early age of concrete, (3) pumpability and workability; and (4) strength of hardened concrete. In addition to these aspects, the availability of raw materials was also taken into

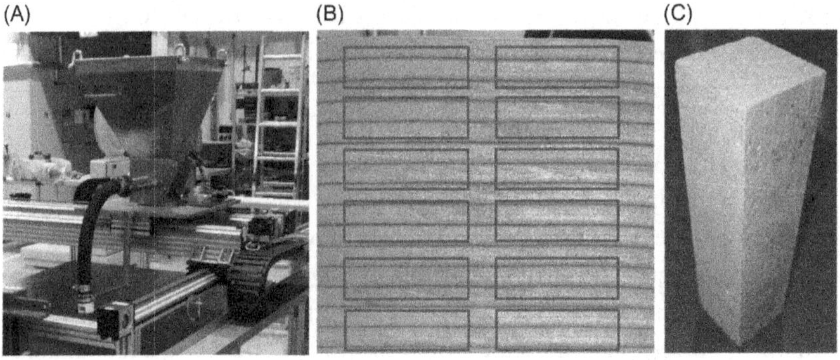

Figure 16.6 (A) Dedicated laboratory 3D concrete printing testing device; (B) a section of printed wall; and (C) saw-cut prism specimen [9].

consideration. Next, various compositions including ordinary mortar, ordinary concrete, and fiber-reinforced composites were developed. These compositions were characterized by long (up to 90 minutes) retention of consistency, high thixotropy, and controlled, rapid setting (less than 3 minutes) on addition of an accelerator inside the printhead. The "steady state" nature of concrete fresh properties, that is, workability retainment, were tested using Hägermann flow table test. Initial and final setting times are characterized using the Vicat needle test. Rheological parameters of fresh concrete, which play a vital role in defining its fresh-state flow behavior, were determined using Haake-Mars II rheometer.

Furthermore, extrudability and buildability were investigated with the new experimental method 3DPTD, see Fig. 16.6A. Prism specimens with a length of 160 mm and a square cross-section with a side dimension of 35 mm were used for compressive and flexural tests on printable concretes, see Fig. 16.6B and C. These specimens were saw-cut from a printed straight wall of length 1000 mm, height 300 mm, and breadth 38 mm, see Fig. 16.6B. All the layers were printed with a print velocity of 75 mm/s and 30 seconds time gap between printed layers.

16.4.1 Selected Results and Discussion

Table 16.1 gives the composition of a high-performance printable concrete referred to here as 3M3. This mortar was chosen for this chapter from a group of material compositions developed in the framework of the on-going CONPrint3D project as a representative one. Selected results from experimental studies are presented in Table 16.2 and in Fig. 16.7. The mechanical properties were tested at mortar ages of

Table 16.1 Material composition of printable fine-grained concrete

Constituent	Type/provider	Dosage (kg/m^3)
Cement	CEM I 52.5 R ft/OPTERRA Zement GmbH, Werk Karsdorf	430
Fly ash	Steament H 4/STEAG Power Minerals GmbH	170
Micro silica Suspension	EM SAC 500 SE/Elkem AS (solid content 50%)	180
Sand 0.06-0.2	BCS 413/Strobel Quarzsand GmbH	430
Sand 0-1	Ottendorf natural river sands from Germany	380
Sand 0-2		430
Water	Local tap water	180
Superplasticizer	MCPF 5100/MC Bauchemie	10

Table 16.2 Mechanical properties of printed and casted specimens (standard deviations are given in parenthesis)

Age (day)	Printed				Casted	
	Compressive strength (N/mm^2)		Flexural strength (N/mm^2)		Compressive (N/mm^2)	Flexural (N/mm^2)
	Vertical	Horizontal	Vertical	Horizontal	Vertical	Horizontal
3	49.7 (2.6)	45.9 (1.9)	4.3 (0.2)	4.8 (0.2)		
21	80.6 (2.1)	83.5 (4.5)	5.9 (0.4)	5.8 (0.2)	73.4 (3.5)	5.1 (0.4)

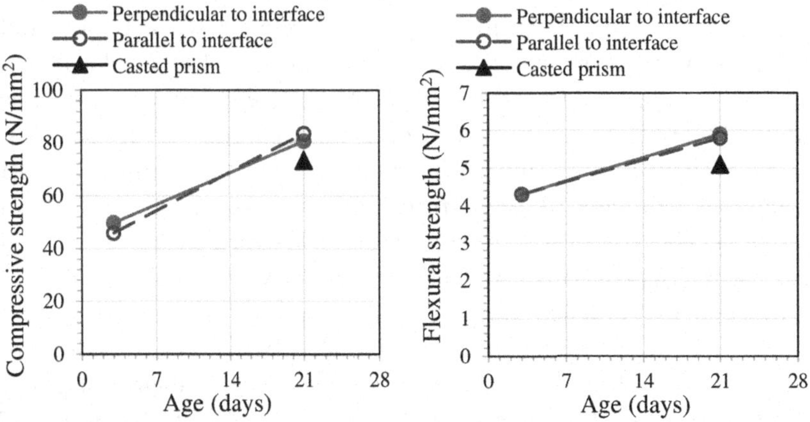

Figure 16.7 Compressive and flexural strengths of printed and casted specimens.

3 and 21 days. To investigate the bond quality between printed layers, experiments were conducted by applying force in both perpendicular and parallel directions to the layer-interface plane. In addition, conventionally casted prism specimens were tested at the age of 21 days to record any process-induced variations in material properties. These casted specimens were produced in parallel to the printing process from the same batch of printable fine-grained concrete.

Compressive and flexural strengths of the printable fine-grained concrete are presented in Table 16.2 and Fig. 16.7. The developed printable fine-grained concrete has high values of both compressive and flexural strengths already at 3 days after production. At the age of 21 days, printed specimens have a compressive strength higher than 80 MPa. Therefore, 3M3 can be termed as a high-strength, printable concrete and classified into the C80/95 compressive strength class according to DIN EN 206-1.

It is very important to study the influence of pumping and printing processes on the hardened properties of concrete. For this purpose, conventionally casted prism specimens according to DIN EN 18555 were tested at the age of 21 days. A comparison between the printed and casted specimens show that both the compressive and flexural strengths of the specimens cut off the printed walls are slightly higher than those of the casted prisms, see Fig. 16.7 and Table 16.2. The high pressure-induced compression inside the extruder's conveyer could be a reason for this.

The bond strength between different layers is generally termed the "Achilles' heel" of 3D printing with concrete. A week bond would mean a low structural stability, especially, for lateral loads. To investigate the bond quality between printed layers, experiments were conducted by applying force in both perpendicular and parallel directions to the layer-interface plane. From these tests, no significant changes in strength values were observed from both compressive as well as flexural tests. This confirms a satisfactory bond between the layers resulting in a largely isotropic material behavior. However, a further test would be needed to assess the properties of the bond quantitatively, using loading direction parallel to layer-interface plane [12]. The interested reader can find the results of tests on interface layer strengths of printed concrete, including corresponding microscopic investigations in [12] (Fig. 16.8).

Figure 16.8 Cut cross section of a printed specimen depicting apparent satisfactory bond between printed layers.

16.5 CONCLUSIONS

The significance and potential of digital construction is briefly presented in this chapter along with a novel approach of onsite 3D concrete printing, CONPrint3D. Important properties of printable concrete in fresh and hardened state were identified. Experimental methods for studying properties of printable concrete including pumpability, extrudability, and buildability are provided. Selected results of compressive and flexural strengths for printed and casted specimens were comparatively analyzed. The presented fine-grained printable concrete composition—with its high mechanical strengths and consistent printability up to 90 minutes after water addition—is a promising material for large-scale implementation. Systematic studies on quantification of extrudability and buildability of printable concretes are necessary to ensure wide-ranging implementation of digital construction in construction practice.

ACKNOWLEDGMENTS

The authors express their sincere gratitude towards the German Federal Ministry for the Environment, Nature Conservation, Building, and Nuclear Safety (BMUB) for funding this research project through research initiative Zukunft Bau of Federal Institute for Research on Building, Urban Affairs, and Spatial Development (BBSR).

We also thank our industrial partners, namely:

- Opterra Karsdorf GmbH (Werk Karsdorf), 06638 Karsdorf, Germany
- MC-Bauchemie Müller GmbH & Co. KG, 46238 Bottrop, Germany
- Putzmeister Engineering GmbH, 72631 Aichtal, Germany
- BAM Deutschland AG (NL Dresden), 01257 Dresden, Germany.

This work is based on earlier works [9,11] by the authors.

REFERENCES

[1] D. Hwang, B. Khoshnevis, Concrete wall fabrication by contour crafting, in: Proc. ISAR 2004 21st International Symposium on Automation and Robotics in Construction, 2004. https://doi.org/10.22260/ISARC2004/0057.

[2] R.A. Buswell, R.C. Soar, A.G.F. Gibb, A. Thorpe, Freeform construction: mega-scale rapid manufacturing for construction, Autom. Constr. 16 (2007) 224–231. Available from: https://doi.org/10.1016/j.autcon.2006.05.002.

[3] E. Dini, Monolite-UK-Ltd, D-Shape, 2015. <https://d-shape.com/> (accessed 23.08.15).

[4] T.T. Le, S.A. Austin, S. Lim, R.A. Buswell, A.G.F. Gibb, T. Thorpe, Mix design and fresh properties for high-performance printing concrete, Mater. Struct. 45 (2012) 1221–1232. Available from: https://doi.org/10.1617/s11527-012-9828-z.

[5] C. Scott, Chinese construction company 3D prints an entire two-story house on-site in 45 days, 3dprint.com. 17–20. <https://3dprint.com/138664/huashang-tengda-3d-print-house/>, 2016 (accessed 20.03.17).

[6] E. Lloret, A.R. Shahab, M. Linus, R.J. Flatt, F. Gramazio, M. Kohler, et al., Complex concrete structures: merging existing casting techniques with digital fabrication, Comput. Des. 60 (2015) 40–49. Available from: https://doi.org/10.1016/j.cad.2014.02.011.

[7] Totalkustom, 3D printed castle – photo gallery, 2015. <http://www.totalkustom.com/photo.html> (accessed 29.12.17).

[8] Apis-cor, Apis Cor – construction technology, 2017. <http://apis-cor.com/en/faq/texnologiya-stroitelstva/> (accessed 29.12.17).

[9] V.N. Nerella, M. Krause, M. Näther, V. Mechtcherine, Studying printability of fresh concrete for formwork free concrete on-site 3D printing technology technology (CONPrint3D), in: Proc. Rheol. Messungen an Baustoffen, Tredition GmbH, Hamburg, Regensburg, 2016, pp. 236–246.

[10] V. Mechtcherine, V.N. Nerella, 3-D-Druck mit Beton: Sachstand, Entwicklungstendenzen, Herausforderungen, Bautechnik. 95 (2018) 275–287. https://doi:10.1002/bate.201800001.

[11] V.N. Nerella, M. Krause, M. Näther, V. Mechtcherine, 3D printing technology for on-site construction, Concr. Plant Int. (2016) 36–41.

[12] V.N. Nerella, S. Hempel, V. Mechtcherine, Micro- and macroscopic investigations on the interface between layers of 3D-printed cementitious elements, in: Proceedings from the International Conference on Advances in Construction Materials and Systems, Chennai, 2017, pp. 557–565.

[13] R. Schach, M. Krause, M. Näther, V.N. Nerella, CONPrint3D: 3D-concrete-printing as an alternative for masonry, Bauingenieur (9) (2017) 355–363. http://www.bauingenieur.de/bauing/article.php?data[article_id] = 88138 (accessed October 18, 2017).

[14] A. Perrot, D. Rangeard, A. Pierre, Structural built-up of cement-based materials used for 3D-printing extrusion techniques, Mater. Struct. 49 (2016) 1213–1220. Available from: https://doi.org/10.1617/s11527-015-0571-0.

[15] T. Wangler, E. Lloret, L. Reiter, N. Hack, F. Gramazio, M. Kohler, et al., Digital concrete: opportunities and challenges, RILEM Tech. Lett. 1 (2016) 67–75. Available from: https://doi.org/10.21809/rilemtechlett.2016.16.

[16] V. Mechtcherine, V.N. Nerella, K. Kasten, Testing pumpability of concrete using sliding pipe rheometer, Constr. Build. Mater. 53 (2014) 312–323. Available from: https://doi.org/10.1016/j.conbuildmat.2013.11.037.

[17] D. Feys, K.H. Khayat, R. Khatib, How do concrete rheology, tribology, flow rate and pipe radius influence pumping pressure? Cem. Concr. Compos. 66 (2016) 38–46. Available from: https://doi.org/10.1016/j.cemconcomp.2015.11.002.

[18] O.H. Wallevik, J.E. Wallevik, Rheology as a tool in concrete science: the use of rheographs and workability boxesCem. Concr. Res. (2011) 1279–1288Elsevier Ltd. Available from: https://doi.org/10.1016/j.cemconres.2011.01.009.

[19] S.D. Jo, C.K. Park, J.H. Jeong, S.H. Lee, S.H. Kwon, A computational approach to estimating a lubricating layer in concrete pumping, Comput. Mater. Contin. 27 (2012) 189–210. Available from: https://doi.org/10.3970/cmc.2011.027.189.

[20] D. Kaplan, F. De Larrard, T. Sedran, Design of concrete pumping circuit, ACI Mater. J. (2005) 110–117. <http://www.concrete.org/PUBS/JOURNALS/AbstractDetails.asp?ID = 14304> (accessed 25.07.13).

[21] M.S. Choi, J.K. Young, K.K. Jin, Prediction of concrete pumping using various rheological models, Int. J. Concr. Struct. Mater. 8 (2014) 269–278. Available from: https://doi.org/10.1007/s40069-014-0084-1.

[22] S.H. Kwon, P.J. Kyong, J.H. Kim, P.S. Surendra, State of the art on prediction of concrete pumping, Int. J. Concr. Struct. Mater. 10 (2016) 75–85. Available from: https://doi.org/10.1007/s40069-016-0150-y.

[23] K. Kasten, Gleitrohr – Rheometer, Ein Verfahren zur Bestimmung der Fließeigenschaften von Dickstoffen in Rohrleitungen (Ph.D. thesis) [in German], Technische Universität Dresden, 2010.

[24] M. Choi, N. Roussel, Y. Kim, J. Kim, Lubrication layer properties during concrete pumping, Cem. Concr. Res. 45 (2013) 69–78. Available from: https://doi.org/10.1016/j.cemconres.2012.11.001.

[25] V.N. Nerella, E. Secrieru, V. Mechtcherine, Rheological behaviour of fresh concrete in continuous pumping and circuit-breakdown cases. Part 1: principals and experimental program, Concr. Plant Int. 5 (2015) 222–226.

[26] V.N. Nerella, V. Mechtcherine, Virtual sliding pipe rheometer for estimating pumpability of concrete, Constr. Build. Mater. 170 (2018) 366–377. Available from: https://doi:10.1016/j.conbuildmat.2018.03.003.

[27] Z. Malaeb, H. Hachem, A. Tourbah, T. Maalouf, N. El Zarwi, F. Hamzeh, 3D concrete printing: machine and mix design, Int. J. Civil Eng. Technol. 6 (2015) 14–22.

[28] A. Perrot, D. Rangeard, Y. Mélinge, P. Estellé, C. Lanos, Extrusion criterion for firm cement-based materials, in: AIP Conference Proceedings, USA, 2008, pp. 96–98.

[29] Z. Toutou, N. Roussel, C. Lanos, The squeezing test: a tool to identify firm cement-based material's rheological behaviour and evaluate their extrusion ability, Cem. Concr. Res. 35 (2005) 1891–1899. Available from: https://doi.org/10.1016/j.cemconres.2004.09.007.

[30] Y. Chen, L.J. Struble, G.H. Paulino, Extrudability of cement-based materials, Am. Ceram. Soc. Bull. 85 (2006) 9101–9105.

CHAPTER 17

Construction 3D Printing

Camille Holt, Laurie Edwards, Louise Keyte, Farzad Moghaddam and Belinda Townsend
Boral Innovation Factory, Australia

17.1 INTRODUCTION

One of the most important trends in manufacturing over the past decade has been the rise of additive manufacture, or 3D printing; 3D printing allows direct manufacture of finished articles from computer models, unlike traditional mass production techniques which require expensive tools or molds. 3D printing, therefore, allows mass customization where it is no more expensive to produce unique items than multiples.

3D printing techniques have been gaining traction in various industries as the technology has evolved. Some industries have embraced 3D printing much more rapidly than others. For example, direct manufacture of titanium parts through 3D printing has generated significant interest as it minimizes waste of the costly metal and allows highly complex shapes to be created. 3D printing has also generated interest in objects as diverse as bicycle frames, firearms, and chocolates [1–3].

This chapter examines the drivers behind 3D printing and explores how additive manufacturing could revolutionize construction.

17.2 BACKGROUND

3D printing, or additive manufacture as it is also known, refers to a body of technologies that began to be developed in the 1980s [4]. While these technologies are varied in both approach and material, they share the unique feature of constructing an object through the layer-by-layer addition of material. This is fundamentally different from traditional approaches of manufacture such as casting into a mold, die casting, or removing material by machining.

3D Concrete Printing Technology. DOI: https://doi.org/10.1016/B978-0-12-815481-6.00017-8

Adapted with permission from Construction 3D Printing (Vol 42, No 3, Sept 2016, pp. 30–35) by Concrete Institute of Australia.

All forms of additive manufacture follow the same basic process. First, a three-dimensional blueprint of an object is created and is sliced into cross sections using computer software. This is commonly referred to as computer aided design (CAD) or computer aided manufacturing [5]. Next, the design is converted to a set of manufacturing instructions which specify deposition of material at the required locations. Finally, material is deposited in successive laminations to build the object. Deposition of material can be achieved through a number of techniques that can mostly be divided into three key groups: Stereolithography (SLA), fused deposition modeling (FDM), or selective laser sintering (SLS).

In 1986, the first 3D printing-related patent was issued to Charles (Chuck) Hull for the method of SLA [6]. He had invented a technique that could bypass the several-week wait normally expected when creating a plastic prototype and instead have it ready to be handled within a day. Hull's invention takes digital cross sections which are then communicated to a Stereolithography machine consisting of a build platform suspended in a UV-sensitive liquid photopolymer resin-filled basin and a guidable UV laser. This laser traces the path defined by the cross section, applying energy to trigger solidification of the resin. Once a layer has been completed, the build platform drops slightly and the laser begins an adjacent cross-section, thus, building the object one cross section at a time. For some applications, stereolithography is the superior choice due to its ability to create highly detailed objects with thin walls and a clean surface finish. However, there are also some challenges specific to this technique; as the object is being created in fluid, any objects with overhang require support structures. The necessity of support structures increases material wastage and labor time as support materials are frequently difficult to remove. Also, the phase-change of the material can sometimes cause shrinkage, warping, or curling. Finally, due to the reliance on photosensitive liquids, materials are limited.

Concurrently, Carl Deckard was working on an entirely different method of additive manufacture at the University of Texas. By 1989 he was granted a US patent for the process of SLS [7]. SLS enables three-dimensional models to be rendered in a similar fashion to SLA, however, the basin is filled with a powdered material rather than a liquid. In this case, a laser traces the cross section and applies thermal energy to sinter the material. The build platform within the basin is then lowered and a roller or scraper system deposits a fresh layer of powder so the process can be repeated; this time the laser simultaneously creates a new layer and fuses it to the layer below. Unlike stereolithography, the raw material provides support to the model as it is being created, meaning additional supports are rarely required. However selective laser sintering is also subject to limited suitable materials.

Perhaps the most similar technology to traditional two-dimensional printing is FDM. FDM was first patented in 1992 by Scott Crump [8] who partnered with his wife Lisa and founded a company called Stratasys. In early fused deposition modeling, a thermoplastic polymer or metal filament spool is fed to the print head in a similar way to thread in a sewing machine. The print head then melts the material and follows the path specified by the digital cross section, depositing a layer of molten material. The successive layer is then deposited before the previous layer has fully solidified, enabling lamination. Depending on the machine being used, the build platform may be lowered as layers are printed or the print head may be raised. As FDM technology has advanced, the range of materials able to be printed in this way has broadened. Almost any substance that can start in a liquid state before solidifying can be used in this technique.

17.3 APPLICATIONS IN CONSTRUCTION

The construction industry is particularly well-suited to take advantage of the benefits of 3D printing. 3D printing could bring improvements in safety, reductions in labor and time, and advances in customization and form.

Perhaps the largest drawcard for 3D printing in construction is the reduction in labor requirements, as this can translate to a saving in both cost and time. 3D printers would allow a house to be built by a skeleton crew, rather than a full team spanning multiple trades [9]. This reduction in labor would result in both decreased cost and an increased level of site safety, particularly in harsh or dangerous environments. Automated construction could also minimize costly errors and defects.

As well as the improved cost, timeline, and safety, 3D printing also removes many design limitations. Rectilinear forms are known to be structurally weaker than curvilinear forms [10]. However, the creation of curvilinear forms in construction requires specialty formwork or engineering. This usually comes at a dramatic increase in expense and time. The use of 3D printing would enable curvilinear designs to be executed as easily as more traditional angular structures. This offers a structural advantage as well as an aesthetic one. Likewise, elements that are precast are limited to being solid whereas those which are printed are able to be created with cavities, saving on material and also creating channels for essential utilities [11].

The construction industry is well-positioned to capitalize on the benefits of 3D printing as the use of modeling is already commonplace. In fact, the majority of information needed to create a 3D blueprint is generated during the design of a building. In building information modeling, which is rapidly growing in popularity, it is standard procedure to create three-dimensional CAD models of buildings [12]. It is a relatively small step to move from this type of model to instructions for a 3D printer.

17.4 CURRENT PROGRESS

Early versions of 3D concrete printers operate by extruding very low slump mortar, designed to be self-supporting, from a deposition head mounted on a gantry [13]. This technique is analogous to FDM. Although FDM is not the only technique with potential application in the construction industry, it utilizes the cheapest materials and is the favored choice so far.

Since 2007, a team at Loughborough University in the United Kingdom, led by Dr. Richard Buswell and Professor Simon Austin, have been focused on 3D printing applications within the construction industry. They are working to build and commercialize a 3D concrete printing robot, as well as helping to develop a supply chain for the required materials [14]. Construction company Skanska are collaborating with Loughborough University with the aim of facilitating the transition of 3D printing techniques to a commercially viable form of construction.

In 2014, the Chinese company WinSun demonstrated how 3D printing could be used in practical terms by printing 10 basic dwellings in 24 hours [15]. The structures are claimed to have cost less than US$5000 each to manufacture and were created from mostly recycled materials. WinSun used a large gantry-style printer and high-performance mortar to manufacture the house components offsite. These completed parts were then transported and assembled in a similar fashion to traditional precast construction. The houses were produced rapidly and cheaply and while the finish on the elements was poor compared to typical precast concrete, WinSun proved that 3D printing could one day be a viable construction technique. However, the true benefits of 3D printing were not realized, as the elements still required transport and assembly onsite.

WinSun have shown that 3D printing can be applied to the construction of taller structures, building a 5-story apartment building and a 1100 m^2 concrete mansion, complete with interior fittings. The mansion was printed at a cost of approximately US$160,000 [16]. WinSun have since announced receiving an order for 10 of these mansions from the Taiwanese real estate company Tomson Group [17]. WinSun have added 80 and 130 m^2 homes with a Chinese style courtyard to the list of buildings they have printed with assembly taking two days for each house. It appears that despite architectural improvements in the newer WinSun buildings compared to the earliest creations, there has been some improvement of the esthetic of the printed elements with the company developing facades which can be affixed to walls with screws. There are plans to license their technology so that it can be used outside China, but uptake in China has been slow with WinSun constructing mostly bus stops and public toilets 3 years after unveiling its first 3D printed buildings [18].

Rival Chinese company, HuaShang Tengda, has also printed a two-story villa but using a different approach. A frame was erected around which the 400 m^3 structure was printed. Overall, the villa took 45 days to construct and due to the grade and thickness of the printing material used. the building is estimated to be able to withstand an earthquake measuring 8 on the Richter scale [19].

In 2013, DUS Architects announced plans to build a full-scale 3D printed canal house in Amsterdam using their 3D printer dubbed the KamerMaker (which translates as Room Maker in English) [20]. The team behind the project aimed to learn as they built the full-sized house and were able to increase the printing speed threefold during the build which began in 2014. The building is made from an environmentally friendly plastic containing linseed oil which can be shredded and recycled once the building has reached the end of its useful life. In August 2016, the same group created an urban cabin with a footprint of 8 m^2 and a total volume of 25 m^3. The cabin only has room for a bed but boasts a 3D printed bathtub which is located outside the cabin. Such structures, while small, can be printed quickly and cheaply to provide emergency shelter which can be easily removed and recycled once no longer needed [21,22].

The Dubai government unveiled a 250 m^2 3D printed office space in May 2016 located within Dubai's Emirates Towers complex [23]. WinSun's gantry style printer (6 m × 36 m × 12 m) along with smaller printers were used to create the offices over a period of 17 days and required one person to oversee the printing process. The individual pieces were printed in China in WinSun's factory and assembled in Dubai. The cost of building the offices was reported to be approximately US$140,000 and the cost of labor approximately half of that required for buildings of a similar size [24,25]. After construction was complete, internal and external finishes were added. Just prior to opening the 3D printed office, the Dubai government announced that they will strive towards having 25% of Dubai's buildings constructed using 3D printing by 2030. The government foresees 3D printing one day permeating all aspects of life to the benefit of humanity and aims to become a hub for research into 3D printing. Utilization of 3D printing technology in Dubai is set to increase by approximately 2% in 2019 and continue to increase further over time depending on the further growth and reliability of 3D printed buildings [26].

Dubai's interest in 3D printing has attracted companies such as Cazza to the area who were involved in the Dubai Futures Accelerators program and have since moved their headquarters to the UAE. Cazza released their own line of robotic construction 3D printers in June 2017 that they hope will increase the speed of construction and safety while decreasing costs and harm to the environment. The printers are capable of printing a 100 m^2 dwelling within 24 hours and requires one person to place steel reinforcements between the layers [27,28]. Cazza has also announced plans to construct a skyscraper in Dubai built using their 3D printing technology with construction planned to start by 2023 [29].

US-based Emerging Objects are employing entirely different technology to develop building components with a focus on the aesthetic advances that 3D printing has enabled. Using a proprietary cement-based material, the components produced are claimed to be lightweight and are toughened by the addition of fiber reinforcement. Emerging Objects offer a number of postprinting finishes such as sand blasted, glossy, or satin. With the increased shape capabilities of 3D printing they have been able to incorporate functionality in many products such as their interlocking seismically resistant blocks and acoustic dampening walls [30].

Outside of engineering labs, Andrey Rudenko built and fine-tuned an FDM printer in his own backyard. By August 2014, Rudenko had successfully printed a castle in layers of concrete 10 mm high by 30 mm wide [31]. Two years after starting work on his 3D printed castle, Andrey Rudenko finished it. He overcame challenges with the programming, the mortar mix design, as well as problems with the physical deposition of the cement, such as the clogging of the extruder. He was able to print layers of varying heights and widths. After completing the castle, the printer was redesigned. Rudneko is now looking for partners to build with [32]. He teamed up with the Lewis Grand Hotel in the Philippines to build a 130 m^2 villa in the hotel with his 3D printing technology which was completed in September 2015, including a 3D printed spa bath. Local materials (sand and volcanic ash) were used which imparted favorable qualities to the printed elements. While more difficult to extrude, it resulted in good bonding between the printed layers. Rudneko estimated that with the equipment in place and tuned, a medium-sized house could be printed in approximately a week including the installation of plumbing and electrical elements. The owner of the hotel, Lewis Yakich, was interested in applying the technology to the development of cheap homes for people with low incomes [33]. Unfortunately, Yakich went missing after a business meeting in 2015 [34].

World's Advanced Saving Project (WASP) aims to use 3D printing to encourage sustainable development and build housing for the world's disadvantaged people. Their research has seen the development of DeltaWASP printers with the largest of this series, the BigDelta (12 m high and 7 m wide), being used to build the Shamballa village in Massa Lombarda, Italy using clay. The village will be the first to be created using this construction method. Structures are built using locally available mud and clay as well as plant fibers, allowing for environmentally friendly construction. The cost of the first house, completed in 2016, was €48 [19,35]. They were able to increase the height of the structure at a rate of approximately 60 cm a day. On hot days, this could possibly be increased to 1 m as the clay and straw mixture hardened more quickly in the heat [36]. They used local materials (clay, straw, water) to fabricate structures which can be recycled when they have reached the end of their useful life leading to a minimization of waste and environmental impact. The clay and dirt used as the main building material has favorable properties for housing. It is a thermal insulator, keeping the structure cool in summer and warm in winter. It also possesses good acoustic insulation properties, fire resistance, and is able to regulate humidity due to its breathable nature. Lime and sand can also be added to improve the stability of the structure. During the design process they considered thermal and acoustic insulation, location of cables and plumbing, where to place internal wood structures, and how to make the structure resistant to earthquakes [37].

The group has also had success in applying their sustainable philosophy to printing with cement, producing structural beams which contain less cement than traditional concrete beams using software to eliminate "redundant material" [38]. They have experimented with many types of materials including clay and mud, lime, cement, geopolymers, as well as polystyrene and popcorn to decrease the density of the material [39].

Another construction-related application has been demonstrated by multinational engineering firm Arup, who have demonstrated steel components printed using selective laser sintering of powdered metal. The use of maraging steel has resulted in a low carbon, low weight, ultra-high-strength prototype. Arup are hoping further refining will result in a market-ready product in the near future. While these components are currently more expensive to create than their traditional counterparts, Arup believe this method of manufacturing will, in time, significantly reduce both cost and waste [40]. A notable example of construction 3D printing using metals is MX3D who are currently constructing a canal bridge that is 8 m long and 4 m wide in Amsterdam using robots fitted with a welding attachment. Printing was originally planned to take place onsite, but was deemed too dangerous [41].

A binder jetting method for 3D printing large structures with a stone-like finish called D-Shape uses sand mixed with a metal oxide catalyst as the powder bed onto which a low viscosity solution containing inorganic compounds (in particular, chlorides) is printed [42]. Binder jetting is, in some ways, analogous to an inkjet printer wherein the printer head moves across the page depositing ink in predetermined locations. Once a layer has been printed a hopper deposits the next layer of sand which is then compacted prior to printing the next layer. The binder for this system, a metal oxide combined with a solution of inorganic salts, is reminiscent of magnesium oxychloride (Sorel) cement which is made by mixing a solution of magnesium chloride with powdered magnesium oxide [43] and boasts strengths greater than those of traditional cement. A main advantage of this method in comparison to other examples of 3D printing in construction is that the whole structure is supported through the printing process by the excess sand surrounding the printed element. The D-Shape method also includes printing a wall around the structure as it is printed to hold the sand. Once the printing process is complete, this wall is broken and the excess sand removed to reveal the printed structure. Also, the printer itself is transportable and can be set up onsite. This method is currently being used in the construction of a four-bedroom house in

New York as well as military bunkers [44]. Collaboration between the Institute of Advanced Architecture of Catalonia and Enrico Dini resulted in the construction of a bridge (12 m long and 1.75 m wide) in Madrid, Spain which was completed in 2017 [19]. There is also talk of buildings in Perth being constructed from elements made using this technique [45]. There is currently no start date for this project.

Contour crafting (CC) [11] is a 3D printing system in which clay is extruded from a nozzle in a manner similar to that of the FDM technique. Attached to the nozzle is a pivoting trowel which smooths the sides of the structure as they are printed. The fact that the trowel can pivot means that curved structures such as cones or vases can also be produced with smooth surfaces. Individually printed layers are visible in the structures produced by both WinSun and Andrey Rudenko and need an extra finishing step to improve their aesthetic appeal. Given that the CC system is compatible with gantry systems which could be used onsite, a combination of the CC technique with cementitious materials could result in a higher quality 3D printing system for construction. In June 2017, CC started production on a series of robotic printers which have been designed to be lightweight and easy to transport [46,47].

San Francisco-based Apis Cor printed a small house onsite in Russia using their 3D printer in less than 24 hours in late 2016. The printer can reach a maximum height of 3.1 m and 5 m width. The house is 38 m^2 in size with a bathroom, kitchen, and living area which converts into a bedroom. The roof, insulation, and windows, as well as interior furnishings were not printed but added after printing was complete. The total price of the home was US$10,134 which didn't include the cost of furnishings or appliances [48]. Apis Cor claim the building can last up to 175 years and needed only two people onsite to operate the printer and place reinforcement between the printed layers [49].

Researchers at MIT have taken a more holistic approach to automated construction with their digital construction platform (DCP). They believe that no one method or material will be sufficient in all situations and construction methods and materials should be tailored to a specific site. They have focused on creating an extrusion-based automated system as there is scope for a high-level of customization in building design. For example, increasing the complexity of structures does not necessarily come with an increase in time or cost and structures need not be a single thickness throughout, allowing for added functionality. The DCP was tested through the construction of an open dome which was 14.6 m in diameter, 3.7 m in height and made from polyurethane foam which could be used as formwork for concrete and left in place to serve as insulation. Print time was 13.5 hours over the course of 2 days and the final structure, which was not backfilled with concrete or other material, was able to support the weight of a person. The authors noted that their methods are similar to, and compatible with, currently available techniques possibly making implementation of the DCP easier. An analysis of the costs of construction using the DCP and existing construction methods showed that the 3D printing technology resulted in lower costs, due largely to a decrease in labor, consistent with other forms of 3D printing technology. [50] The authors aim to build an entirely self-sufficient system that collects its own energy through the use of solar panels, is able build with locally available materials, and have equipped the DCP with the capability to dig and mill material [51].

In addition to these projects and achievements, there are a range of other contributions that have been made to this field as well as projects yet to be started such as the Landscape House which was inspired by the Mobius strip to be printed using D-Shape technology [52], CyBe have printed street furniture and a laboratory [53], the US Army are trialing 3D printing technology for building barracks and temporary shelters [54,55], the Additive Manufacturing Integrated Energy (AMIE) project at Oak Ridge National Laboratory [56], the DFAB House in Dübendorf, Switzerland which will be designed, planned, and built using predominantly digital processes [57], and Renca, LafargeHolcim, and Siam Cement Group who are producing binder systems to be used with 3D printers [58–61].

17.5 KEY CHALLENGES

Although some clear benefits of 3D printing are already apparent in the construction industry, there are numerous challenges to be overcome before the technique can be widely accepted. Unfortunately, concrete, as a material, is susceptible to the effects of weather, ambient temperature, and deficiencies in mix design. These factors can make concrete unpredictable, even when utilizing current technology. With the additional complications of extruding concrete through a finely tuned print head, there is a lot of opportunity for errors to occur. [11] The setting time and rheology of concrete are also not ideal for layer-by-layer construction; however, these properties can be controlled to a certain extent through the use of admixtures. The problems associated with using hydraulic binders as 3D printing materials have, in part, begun to be addressed through the advances in large-scale 3D printing. While concrete itself is not amenable to extrusion due to the size of the aggregates it contains, cement paste and mortar have been successfully utilized in the construction of buildings. Alternative binders, such as that used in the D-Shape system and geopolymer systems, are other options to explore further.

Another consideration is the finish produced when concrete is 3D printed. The clearly ridged texture is apparent; this may be irrelevant in applications where speed or large cost savings are the driving forces, such as emergency housing. However, structures like WinSun's mansion beg the question of how this finish will be received by consumers considering higher-end printed dwellings [13]. If the completed surface is deemed to require further finishing, the cost and effort may outweigh the initial savings. A simple answer to this problem has already been reported by Behrokh Khoshnevis, inventor of CC. Through the addition of a trowel to the system which smooths each layer as it is printed Khoshnevis is able to produce structures in a variety of geometries, including both curves and right angles, with a smooth outer wall [11]. Combination of such a system with a printer that extrudes cement paste may go some way in improving the appearance of objects printed in this manner. It is likely, however, that the final step in constructing a 3D printed building will be tidying up its appearance. This seems to have been the case for the offices recently opened in Dubai [23], the small building constructed by Apis Cor [49], and WinSun who have produced cladding which can be affixed to the outside of the buildings [18].

A further challenge is the incorporation of reinforcement. Extrusion in layers makes conventional reinforcing difficult to achieve. Fiber reinforcement does not solve the issue, as fibers will be aligned with the direction of printing and, therefore, provide no inter-laminar bonding. Reinforcement is a key issue to overcome if structural concrete is to be 3D printed. Incorporation of reinforcement appears to be conversant with the CC methodology with printing able to take place over the top of an embedded metal coil due to the pressure with which the clay paste is extruded. It is not clear how the metal coil is incorporated into the system, however, nor have other materials such as cement pastes been tested. Another interesting enhancement to the CC system is the ability to produce hollow layers. It was suggested that a similar method could be used to extrude two materials simultaneously with the aim of having a reinforcing material surrounded by clay [11]. The D-Shape system claims to be able to create structures so strong that traditional iron reinforcement is unnecessary. However, reinforcing fibers such as metal shavings are being investigated for incorporation into the house to be built in New York to ensure it can support a roof [62]. Another possible method is to manually insert steel bars between the printed layers like Apis Cor has done [49].

Large-scale printers face additional challenges created by transportation. Generally, a printer remains in situ and can be expected to produce the same results with each task. 3D printers capable of constructing buildings must be sufficiently durable to withstand the rigors of transport and operation at construction sites. A gantry system could be employed which would allow a building to be printed as a whole, saving time and money as the need for assembling individual parts would be eliminated. Static systems such as gantry-based 3D printing systems are well-suited to pre-fabrication due to their large size and the difficulty involved in moving and setting up these types of systems. These systems are able to construct simple designs with ease, but can struggle with more complex designs such as those with overhanging sections and are limited by the footprint of the gantry. Researchers at MIT who developed the DCP believe that mobile autonomous systems will provide extra benefit compared to static systems as they are easier to transport and quicker to set up onsite. Systems which use a robotic arm are more flexible and can, therefore, be used to build more complex designs and are often not constrained by the size of a gantry, but have the disadvantage of lower load capacities and greater sensitivity to disturbances [50].

Industry pushback could delay the rise of 3D printing applications in the industry. Mechanization is almost always met with skepticism and any new technology that has the potential to automate industry jobs will encounter resistance [63]. Phillips & James Construction Consulting Firm estimates that construction of a typical family home can employ up to 63 construction workers over a 4-month period [64]. 3D printing techniques could see this number drastically slashed; a prospect that may see sectors of the construction industry meet the new technology with some resistance. Retraining current workers could provide a solution to this resistance while also addressing the lack of workers skilled in the use of this technology.

The types of materials that have already proven themselves compatible with current 3D printing technology mostly consist of special preparations which are expensive to purchase. In addition, such materials are not fit for use in the construction of buildings. For 3D printing to prove itself as a viable construction method there needs to be a cheap, failsafe, and readily available system that can be prepared easily onsite. WinSun used recycled building materials along with glass fibers, steel, and cement in their projects and the likes of Rudenko and Apis Cor used cement-based mixes made to a strict recipe. However, DUS Architects use a plastic based on linseed oil and the buildings constructed in Shamballa village by WASP were built with clay and straw. While it is likely that cement will be the main material from which 3D printed buildings would be constructed, there are still a lot of variables that need to be considered when designing a system to be used in a 3D printer, such as the setting time of the cement. Once cement has started to set it will be harder to pump or extrude making printing impossible. In this case, a long setting-time is desired. However, once a layer of an object has been printed it would be advantageous to be able to complete the next layer in a short amount of time so the printed cement needs to set quickly so it can support the weight of subsequent layers. Alternatively, the D-Shape system utilizes a quick setting two-part cement which can be incorporated into the sand from which the structure is made [42]. Research and testing of potential cement mixtures for use in a 3D printing system would need to take place, along with long-term stability testing to ensure the final structure is satisfactory. The foam dome that was printed by the DCP could serve as formwork into which traditional concrete could be poured with the foam remaining in place as insulation [50]. Incorporating the use of traditional construction materials into structures built using 3D printing technology is another way to help ease the introduction of 3D printing into the industry.

Another unknown is whether 3D printed structures will comply with building codes. This is a new method of construction which does not necessarily use standard materials and so no regulations exist to aid those designing and constructing buildings using 3D printing techniques. The first residential structures built using 3D printing have been constructed in the past few years so there is no information regarding long-term performance. For 3D printing to become an acceptable method of construction it will need to be further investigated by regulatory bodies so that guidelines and standards can be written. Given the current difficulties in including traditional reinforcement in 3D printed structures and the fact that nonstandard materials will likely be used, it is conceivable that elements produced using 3D printing may not be as durable or last as long as traditionally produced elements. Such elements may also have different requirements for maintenance.

Depending on the material or equipment used to make an object, some level of postprocessing may also be required. For example, removal of rough edges or an additional treatment to improve the strength of the finished article may be necessary. While the houses constructed by WinSun [9] did not require further treatment after printing to improve strength since they were made of cement, the overall finish was lacking. It appears that in most complete projects to date, some level of finishing has been required.

17.6 LIMITATIONS OF 3D PRINTING

3D printing has the potential to make a significant and positive contribution to the construction industry. However, to make the best use of this technology, not only should what it can do be investigated but the limits of 3D printing should also be recognized. Arguably, the most important limitation of current 3D printing technology is the range of compatible building materials. The structures made by WinSun and Rodenko use a cement paste which, in the case of WinSun, also contained building waste materials as reinforcement. It is unlikely that these methods could be altered in such a way so as to allow traditional concrete to be used in place of cement paste. However, the increase in popularity of 3D printing in the construction industry has seen an increase in the number of binder systems available specifically for use with 3D printing technology. D-Shape uses sand to form its structures which gives the system flexibility in that it can use sand from any source which can reduce the cost of materials. Similarly, the use of locally available clay or dirt and straw by WASP is another way to reduce raw materials and transportation costs.

There is currently no "one size fits all" 3D printer for fabrication of structural elements. While there are similarities between the reported examples of 3D printing in construction, each is different and specific to an individual situation or set-up. It is envisioned that over time as the popularity of 3D printing increases and the price of the technology decreases, further exploration into the use of 3D printing in construction will yield modifications to the current technology that will allow for increased flexibility and ease of set-up. For example, the creators of the DCP want to reimagine how buildings are constructed with complete individualization based on the nature of the site. For example, printing thicker load-bearing walls, including curved elements to help withstand wind, and site-specific choice of where supporting pillars are placed [65].

Printing elements with overhanging areas is a challenge for 3D printing as support structures are needed to hold up overhanging areas until they have developed enough strength to support their own weight. SLS [7] and other powder bed methods, in which the unfinished object is supported by a bed of powder during fabrication, present a solution to this problem, but require extra time and effort to remove the powder after production is complete. As far as examples such as the WinSun, Rudenko, and Apis Cor structures are concerned, this may mean that printing an entire building, including the roof, all at once is beyond the capabilities of current 3D printing technology. The D-Shape system, being a powder bed method, allows the object being printed to be entirely supported throughout printing, but the current size of the printer means that the size of the object is limited to what can be printed in a 6 × 6 m space. This size cannot accommodate an entire house. The future of construction 3D printing may lie with mobile robots such as the DCP which are not constrained by the footprint of a gantry or the reach of an arm.

17.7 COST-BENEFIT ANALYSIS OF 3D PRINTING IN CONSTRUCTION

It is hard to evaluate exactly what impact 3D printing will have on construction. What is known, however, is that there is potential for savings using 3D printing. In its current state, the largest immediate impact that 3D printing can have is in producing models from technical drawings and designs. Often, a three-dimensional model can more easily communicate an idea or concern than technical drawings or computer models. In addition, a 3D printed model can cost less than a model made in a traditional manner. A recent study suggested that a sensible estimate of the cost of such a model is approximately 0.1% of the total project cost, not including costs relating to purchase of equipment, materials and software, or labor associated with building the model. Depending on the level of complexity of the model, it can be made in a short space of time ranging from hours to a number of days which compares favorably with current methods [66].

Table 17.1 Cost estimates for constructing a wall from 40 MPa concrete in a multistory building in the Sydney CBD using formwork

	Cost	Amount	Price
Supply of concrete	$200/m^3$	150 m^3	$30,000
Pumping	$20/m^3$	150 m^3	$3000
Labor	$20/m^3$	150 m^3	$3000
Formwork	$100/m^2$	1500 m^2	$150,000
Total			$186,000

Table 17.2 Estimate of cost for constructing a wall from 40 MPa concrete using 3D printing

	Cost	Amount (m^3)	Price
Supply of concrete	$250/m^3$	150	$37,500
Pumping	$20/m^3$	150	$3000
Labor	$20/m^3$	—	—
Formwork	$100/m^2$	—	—
Total			$40,500

A major cost associated with concrete construction is formwork. Cost estimates for construction in the Sydney CBD show that when the cost of building a concrete wall in a multistory building is broken down, formwork contributed approximately 80% of the total costs (Table 17.1). By eliminating the need for traditional formwork during construction, there is potential for considerable savings to be made (Table 17.2). The cost of the printer is currently unknown, and this is not included in Table 17.2. Given the magnitude of the gap, there is significant scope for printer costs to be absorbed and still provide a cost-effective solution. Without the need for installation and removal of formwork, a saving of approximately 1 day per floor could also be achieved. The WinSun, Rudenko, or Apis Cor examples did not require the use of formwork to construct their elements, but were unable to cope with overhanging sections without the incorporation of supports. Even so, buildings can be made piecewise without the need for formwork. Powder bed methods (such as D-Shape) allow a great deal of flexibility in the types of shapes which can be produced as the unreacted powder provides support to an object as it is being printed eliminating the need for formwork altogether. Another option is to use the 3D printer to print the formwork and fill using concrete, such as what could be achieved using the DCP.

Overall, the cost of labor would also be expected to fall as the number of workers required to construct a building would be decreased. Only those people required to oversee the printing activities, assembly, and installation of utilities would be onsite resulting in a massive reduction of labor costs. With less people working onsite, the potential for injury is also decreased. With the reduction in traditional roles, an increase in new roles would also be seen. Staff with expertise in design and the ability to create commands for a printer, technicians conversant with the technology, as well as those skilled in the assembling, dismantling, and transportation of the printer will all be required. There will also be costs associated with training and development of staff as new skills are acquired.

It is likely that the cost of building materials consumed during a project could be less for a 3D printing project. More exact placing of material means less waste is generated. There is also the possibility of incorporating construction waste materials into the printing media, as was demonstrated by WinSun. [8] The inventors of D-Shape, WASP, and the DCP can use material from any source, such as the building site itself, reducing the costs of raw materials and impact on the environment caused by production and transportation of these materials. Another way in which an automated system can reduce waste is through lessening the impact of human error on the building process.

There are also cost savings to be realized through the speed with which building can be constructed with 3D printing versus traditional building methods. 3D printing boasts improved speed and efficiency and also greater transportability with systems that are able to be taken apart and rebuilt at the construction site. The completed elements can then immediately be placed without the need for transportation from a factory or warehouse to the site.

Another cost associated with the implementation of 3D printing is the cost of research and development. The current methods, while providing an exciting glimpse into the future, can be improved upon and scaled up to more effectively service the needs of the industry. There are also issues surrounding the incorporation of reinforcement; currently there are no systems which are compatible with traditional steel reinforcement and there is a reliance on fibers and metal shavings for additional strength. New binder systems may also improve the quality and strength of 3D printed systems. In addition, efforts into making more user-friendly programs for creating digital models for use with 3D printers could increase the popularity of 3D printing techniques and make them a more cost-effective option.

The nature of 3D printing is such that it costs no more in equipment to print a once-off design than it does to mass produce that same design. Also, curved and irregularly shaped structures will likely cost no more to build with a 3D printer than a structure with a regular shape, giving architects and designers a wider degree of freedom in creating the shape of a building without inflating costs. However, this could result in buildings for which there are no standards by which to evaluate structural integrity. Therefore, a cost to governments and authorities will likely result as they introduce new standards specific for 3D printed structures. Intellectual property issues must also be investigated to avoid plagiarism of technology and designs.

17.8 CONCLUSION

Recent activity indicates that there is growing interest in construction-scale 3D printing. There are clear benefits in safety, speed, design flexibility, and cost, but also significant practical hurdles to overcome. The key benefits of 3D printing align well with the continual drivers of the construction industry: improved safety, reduced cost, and increased design freedom. Although in its infancy, 3D printing is likely to become an important part of construction in the relatively near future.

REFERENCES

[1] Empire Cycles Ltd., Empire's 3D printed MX6-R, 2014. [Online]. Available from: <http://empire-cycles.com/article.php?xArt = 31> (accessed 05.06.15).

[2] Defense Distributed, Defense distributed, 2015. [Online]. Available from: <https://defdist.org/> (accessed 05.06.15).

[3] T. Edwards, 3D systems unveils CocoJet Chocolate 3D Printer At 2015 CES, January 6, 2015. [Online]. Available from: <http://3dprint.com/35081/culinary-printing-3d-systems/> (accessed 05.06.15).

[4] RedOrbit, The history of 3D printing, 2014. [Online]. Available from: <http://www.redorbit.com/education/reference_library/general-2/history-of/1112953506/the-history-of-3d-printing/>. (accessed 11.02.15).

[5] J. Kim, D. Robb, 3D printing: a revolution in the making, Univ. Auckland Business Rev. 17 (1) (2014) 16–25.

[6] C. Hull, Method and apparatus for production of three-dimensional objects by stereolithography. European Patent Patent 0171069, 1986.

[7] C. Deckard, Method and apparatus for producing parts by selective sintering. United States Patent Patent 4863538, 1989.

[8] S. Crump, Apparatus and method for creating three-dimensional objects. US Patent 5121329A, 1992.
[9] M. Fulcher, Chinese firm 3D prints villa and apartment block, Architects' J. (2015).
[10] M. Abrams, 3D printing houses, 2014. [Online]. Available from: <https://www.asme.org/engineering-topics/articles/construction-and-building/3d-printing-house-s> (accessed 11.02.15).
[11] B. Khoshnevis, Automated construction by contour crafting – related robotics and information technologies, J. Autom. Constr. 13 (1) (2004) 5–19.
[12] C. Eastman, P. Teicholz, R. Sacks, K. Liston, IM Handbook: A Guide to Building Information Modeling for Owners, Managers, Designers, Engineers and Contractors, second ed., John Wiley & Sons, Hoboken, NJ, 2011.
[13] G. Gibbons, R. Williams, P. Purnell, E. Farahi, 3D Printing of cement composites, Adv. Appl. Ceram.: Struct., Funct. Bioceram. 109 (5) (2010) 287–290.
[14] L. University, Partnership aims to develop 3D concrete printing in construction, 2014. [Online]. Available from: <http://www.lboro.ac.uk/news-events/news/2014/november/204-skanska.html> (accessed 11.02.15).
[15] L. Hock, 3D printing builds up architecture, Prod. Des. Dev. (2014).
[16] M. Starr, World's first 3D-printed apartment building constructed in China, January 19, 2015. [Online]. Available from: <https://www.cnet.com/news/worlds-first-3d-printed-apartment-building-constructed-in-china/> (accessed 15.11.17).
[17] M. Russon, Chinese man creates world's tallest 3D-printed building and a villa in just 10 months, 2015. [Online]. Available from: <http://www.ibtimes.co.uk/chinese-man-creates-worlds-tallest-3d-printed-building-villa-just-10-months-1485354> (accessed 11.02.15).
[18] Z. Aldama, 'We could 3D-print Trump's wall': China construction visionaries set to revolutionise an industry rife with graft and old thinking, May 13, 2017. [Online]. Available from: <http://www.scmp.com/magazines/post-magazine/long-reads/article/2093914/we-could-3d-print-trumps-wall-china-construction> (accessed 13.11.17).
[19] T. Koslow, 3D printed house – world's 35 greatest 3D printed structures, June 22, 2017. [Online]. Available from: <https://all3dp.com/1/3d-printed-house-homes-buildings-3d-printing-construction/> (accessed 15.11.17).
[20] E. Chalcraft, Amsterdam architects plan 3D-printed canal house, March 9, 2013. [Online]. Available from: <https://www.dezeen.com/2013/03/09/amsterdam-architects-plan-3d-printed-house/> (accessed 15.11.17).
[21] V. Anusci, Going to Amsterdam? Visit the 3D printed house, April 11, 2015. [Online]. Available from: <https://all3dp.com/going-amsterdam-visit-3d-printed-house/> (accessed 15.11.17).
[22] DUS Architects 3D prints an 8 sqm Urban Cabin and Accompanying Bathtub in Amsterdam, August 30, 2016. [Online]. Available from: <http://www.3ders.org/articles/20160830-dus-architects-3d-prints-an-8-sqm-urban-cabin-and-accompanying-bathtub-in-amsterdam.html> (accessed 15.11.17).
[23] A. Williams, World's first 3D-printed office building completed in Dubai, May 25, 2016. [Online]. Available from: <http://www.gizmag.com/3d-printed-office-dubai-completed/43522/> (accessed 01.06.16).
[24] M. Starr, Dubai unveils world's first 3D-printed office building, May 25, 2016. [Online]. Available from: <http://www.cnet.com/news/dubai-unveils-worlds-first-3d-printed-office-building/> (accessed 01.06.16).
[25] L. Alter, Office of the future is 3D printed in Dubai, May 31, 2016. [Online]. Available from: <https://www.treehugger.com/green-architecture/office-future-3d-printed-dubai.html> (accessed 14.11.17).

[26] UAE aims to be global hub for 3D printing, Gulf News, April 27, 2016. [Online]. Available from: <http://gulfnews.com/news/uae/government/uae-aims-to-be-global-hub-for-3d-printing-1.1813400> (accessed 14.11.17).
[27] C. Scott, Cazza construction technologies is ready to populate the world with 3D printed smart cities, November 14, 2016. [Online]. Available from: <https://3dprint.com/155335/cazza-construction-technologies/> (accessed 14.11.17).
[28] S. Saudners, Cazza construction technologies introduces 3D printing construction robots and sees high demand, September 28, 2017. [Online]. Available from: <https://3dprint.com/189228/cazza-3d-print-construction-robots/> (accessed 14.11.17).
[29] Cazza to begin work on world's first 3D-printed skyscraper in Dubai by 2023, October 22, 2017. [Online]. Available from: <http://www.designmena.com/thoughts/cazza-to-begin-work-on-worlds-first-3d-printed-skyscraper-in-dubai-by-2023> (accessed 14.11.17).
[30] E. Objects, Emerging Objects, 2015. [Online]. Available from: <http://www.emergingobjects.com/> (accessed 11.02.15).
[31] A. Rudenko, 3D printed concrete castle is complete, 2014. [Online]. Available from: <http://www.totalkustom.com/3d-castle-completed.html> (accessed 11.02.15).
[32] A. Rudenko, 3D concrete house printer, [Online]. Available from: <http://www.totalkustom.com/> (accessed 14.11.17).
[33] Lewis Grand Hotel teams with Andrey Rudenko to develop world's first 3D printed hotel, planning 3D printed homes, September 8, 2015. [Online]. Available from: <http://www.3ders.org/articles/20150909-lewis-grand-hotel-andrey-rudenko-to-develop-worlds-first-3d-printed-hotel.html> (accessed 14.11.17).
[34] S.J. Grunewald, Where is Lewis Yakich? The man behind the world's first 3D printed hotel suite vanished a year ago without a trace, September16, 2016. [Online]. Available from: <https://3dprint.com/149554/where-is-lewis-yakich/> (accessed 14.11.17).
[35] WASP, WASP's BigDelta returns, March 23, 2016. [Online]. Available from: <http://www.wasproject.it/w/en/wasps-bigdelta-returns/> (accessed 01.06.16).
[36] The clay and straw wall by the 3 meters, August 10, 2016. [Online]. Available from: <http://www.wasproject.it/w/en/il-muro-di-terra-e-paglia-alle-soglie-dei-3-metri/> (accessed 15.11.17).
[37] 3D printed houses for a renewed balance between environment and technology, January 30, 2017. [Online]. Available from: <http://www.wasproject.it/w/en/3d-printed-houses-for-a-renewed-balance-between-environment-and-technology/> (accessed 15.11.17).
[38] K. Campbell-Dollaghan, This Bizarre concrete beam is the smartest use of 3D printing in architecture yet, August 11, 2015. [Online]. Available from: <http://gizmodo.com/this-bizarre-concrete-beam-is-the-smartest-use-of-3d-pr-1723340656> (accessed 01.06.16).
[39] BigDeltaWASP 12m, [Online]. Available from: <http://www.wasproject.it/w/en/3d-printing/bigdeltawasp-12m/> (accessed 15.11.17).
[40] S. Robarts, Arup uses 3D printing to create structural steel component, 2014. [Online]. Available from: <http://www.gizmag.com/arup-laser-sintering-construction/32457/> (accessed 11.02.15).
[41] M. Fairs, Joris Laarman's canal bridge in Amsterdam could take 3D printing "to a higher level", October 19, 2015. [Online]. Available from: <https://www.dezeen.com/2015/10/19/joris-laarman-3d-printed-canal-bridge-amsterdam/> (accessed 15.11.17).
[42] E. Dini. United States of America Patent US 8,337,736 B2, 2012.
[43] P.C. Hewlett, Lea's Chemistry of Cement and Concrete, fourth ed., Elsevier, Oxford, 2008, pp. 817–818.

[44] E. Krassenstien, D-Shape looks to 3D print bridges, a military bunker, and concrete/metal mixture, December 18, 2014. [Online]. Available from: <http://3dprint.com/27229/d-shape-3d-printed-military/> (accessed 29.05.15).
[45] E. Krassenstien, D-Shape intern unveils plans to 3D print unique buildings in Australia & beyond, May 22, 2015. [Online]. Available from: <http://3dprint.com/64469/3d-printed-buildings-australia/> (accessed 29.05.15).
[46] C. Scott, Contour crafting prepares for series production of robotic construction 3D printers, June 15, 2017. [Online]. Available from: <https://3dprint.com/178100/contour-crafting-series-production/> (accessed 15.11.17).
[47] Contour Crafting Corporation, New Press Release, June 8, 2017. [Online]. Available from: <http://contourcrafting.com/press-release/> (accessed 15.11.17).
[48] A. Williams, Portable 3D printer builds a tiny house for a tiny price, March 4, 2017. [Online]. Available from: <https://newatlas.com/apis-cor-3d-printed-tiny-house/48231/> (accessed 15.11.17).
[49] T. Petyhova, Features and perspectives of 3D-printing, August 18, 2017. [Online]. Available from: <http://apis-cor.com/en/about/blog/features-and-perspectives-of-3d-printing> (accessed 15.11.17).
[50] S.J. Keating, J.C. Leland, L. Cai, N. Oxman, Toward site-specific and self sufficient robotic fabrication on architectural scales, Sci. Robot. 2 (5) (2017).
[51] B. Heater, MIT's giant mobile 3D printer can build a building in 14 hours, and some day it may be headed to Mars, April 27, 2017. [Online]. Available from: <https://techcrunch.com/2017/04/27/mits-giant-mobile-3d-printer-can-build-a-building-in-14-hours-and-some-day-it-may-be-headed-to-mars/> (accessed 15.11.17).
[52] V. Woollaston, I. Steadman, The race to build the first 3D-printed building, March 6, 2017. [Online]. Available from: <http://www.wired.co.uk/article/architecture-and-3d-printing> (accessed 15.11.17).
[53] CyBe Construction, Projects, [Online]. Available from: <https://www.cybe.eu/projects/> (accessed 15.11.17).
[54] B. Jackson, U.S. army seeks comercialization of 3D printed cement barracks, August 24, 2017. [Online]. Available from: <https://3dprintingindustry.com/news/u-s-army-seeks-commercialization-3d-printed-cement-barracks-120491/> (accessed 15.11.17).
[55] M. Jazdyk, 3D printing a building, August 23, 2017. [Online]. Available from: <https://www.rdmag.com/article/2017/08/3d-printing-building> (accessed 15.11.17).
[56] Oak Ridge National Laboratory, AMIE demonstration project, [Online]. Available from: <http://web.ornl.gov/sci/eere/amie/> (accessed 15.11.17).
[57] Building with robots and 3D printers, June 29, 2017. [Online]. Available from: <https://www.ethz.ch/en/news-and-events/eth-news/news/2017/06/building-with-robots.html> (accessed 15.11.17).
[58] M. Mensley, Startup renca creates eco-cement fit for 3D printing, March 23, 2017. [Online]. Available from: <https://all3dp.com/renca-geoconcrete/> (accessed 15.11.17).
[59] Dubai-based Renca develops 'green' 3D printing cement made from industrial waste, March 23, 2017. [Online]. Available from: <http://www.3ders.org/articles/20170323-dubai-based-renca-develops-green-3d-printing-cement-made-from-industrial-waste.html> (accessed 15.11.17).
[60] LafargeHolcim, LafargeHolcim innovates with 3D concrete printing, September 20, 2016. [Online]. Available from: <http://www.lafargeholcim.com/lafargeholcim-innovates-with-3D-concrete-printing> (accessed 15.11.17).
[61] S.J. Grunewald, Thai company SCG develops custom 3D printable cement for 3D printing houses and structures, April 27, 2016. [Online]. Available from: <https://3dprint.com/131560/scg-3d-printable-cement/> (accessed 15.11.17).

[62] M. Brassfield, Home, sweet 3D printed home, October 4, 2014. [Online]. Available from: <http://www.notimpossiblenow.com/the-latest/3d-printed-estate> (accessed 29.05.15).
[63] F. Ishengoma, A. Mtaho, 3D printing: developing countries perspectives, Int. J. Comput. Appl. 104 (11) (2014) 30–34.
[64] D. Rivero, Could 3D printers soon kill construction, 2014. [Online]. Available from: <http://fusion.net/story/4565/could-3d-printers-soon-kill-construction/> (accessed 11.02.15).
[65] D.L. Chandler, 3-D printing offers new approach to making buildings, April 26, 2017. [Online]. Available from: <http://news.mit.edu/2017/3-d-printing-buildings-0426> (accessed 15.11.17).
[66] D. Foy, F. Shahbodaghlou, 3D printing for general contractors: an analysis of potential benefits, in 51st ASC Annual International Conference Proceedings, Texas, 2015.

FURTHER READING

D. Sher, Texas student builds concrete 3D printer with hopes of house printing, July 30, 2015. [Online]. Available from: <http://3dprintingindustry.com/news/texas-student-builds-backyard-concrete-3d-printer-everyone-54448/> (accessed 01.06.16).

CHAPTER 18

Properties of Extrusion-Based 3D Printable Geopolymers for Digital Construction Applications

Behzad Nematollahi[1], Ming Xia[1], Praful Vijay[2] and Jay G. Sanjayan[1]
[1]Centre for Sustainable Infrastructure, Faculty of Science, Engineering and Technology, Swinburne University of Technology, Hawthorn, VIC, Australia
[2]Institute of Construction Materials, Faculty of Civil Engineering, TU Dresden, Dresden, Germany

18.1 INTRODUCTION

Additive manufacturing (AM), commonly known as three-dimensional (3D) printing is rapidly transforming manufacturing, aerospace, and medical industries. It has "the potential to revolutionize the way we make almost everything" as said by the former US President, Barack Obama in his state of the union address. If the 3D printing can be used with concrete, it can potentially change the way we build concrete structures, saving huge costs in formworks and allow freeform construction. Freeform construction would enhance architectural expression, where the cost of producing a structural component will be independent of the shape, providing the much-needed freedom from the rectilinear design [1,2].

In the past few years, several emerging 3D concrete printing (3DCP) technologies have been explored in the construction industry. These 3DCP technologies are principally based on two techniques, namely: (1) powder-based; and (2) extrusion-based techniques. The powder-based 3DCP technique is an offsite process, which has the potential to make building components with complex geometries such as panels, permanent formworks and interior structures which can be later assembled onsite [1,2]. There is a demand in the construction industry for such components. Based on the available construction systems, the use of expensive formworks are essential to build such components. The powder-based 3DCP technique can satisfy this industrial demand as it is capable of making robust and durable building components at a reasonable speed without the use of expensive formworks. However, the commercially available powder-based 3D printers usually use proprietary printing materials, which are not suitable for the construction

3D Concrete Printing Technology
DOI: https://doi.org/10.1016/B978-0-12-815481-6.00018-X

applications. To tackle this limitation, the authors of this study developed an innovative methodology to adopt geopolymer materials in commercially available powder-based 3D printers for construction applications [2–9].

The extrusion-based 3DCP technique is similar to fused deposition modeling method that extrudes cementitious material from a nozzle mounted on a gantry, crane, or a robotic arm to build a concrete component layer-by-layer without the use of expensive formworks. This technique has been aimed at onsite construction applications such as large-scale concrete components with complex geometries. One of the limitations of the extrusion-based 3DCP technique is the limited range of printable concretes. Conventional concrete in its current form is not suitable for the extrusion-based 3DCP process. Setting characteristics of Ordinary Portland cement (OPC) limit its use for 3DCP. In addition, it is well-established that production of OPC is highly energy and emissions intensive. The emissions due to manufacture of OPC are the fourth largest source of carbon emissions after the petroleum, coal and natural gas and estimated to account for 5%–7 % of all anthropogenic emissions [10]. Therefore, it is essential to develop concretes without OPC which are suitable for extrusion-based 3DCP [2].

Geopolymers are an emerging sustainable alternative to OPC and can be synthesized by alkaline activation industrial byproducts such as fly ash and/or slag [11]. Fourteen million tons of fly ash are produced annually in Australia, of which only two to three million tons are utilized in some form mainly as partial replacement of OPC in concrete production [12]. In comparison to OPC, geopolymers offer excellent advantages, including: (1) a geopolymer has adjustable setting characteristics and can develop higher strengths in a short period [11,13], which is vital for layer-by-layer build-up capacity in 3DCP; (2) geopolymers demonstrate significantly superior resistance to fire, sulfate, and acid attacks [12], (3) CO_2 emissions due to manufacture of geopolymer is about 80% less than OPC [14]; and (4) as the disposition of fly ash as an industrial byproduct has always been a global issue, so the utilization of fly ash in geopolymers is considered to be particularly beneficial [2].

Although some researchers have tried to utilize geopolymers as the base material for extrusion-based 3DCP, there has not been any systematic work to optimize mixture proportions of 3D printable geopolymers. Most of the works have been based on trial and error to achieve a geopolymer mix which is extrudable and buildable [15]. To tackle this problem, the authors of this chapter recently investigated the effect of various mixture parameters such as type of activator, modulus of sodium silicate, and water-to-geopolymer-solids ratio on the printability and mechanical properties of geopolymer mixtures [16]. Based on the results, an optimum 3D printed geopolymer mixture exhibiting desirable properties was developed [16]. The first part of this book chapter (Part I) reports the effects of testing direction on the compressive and flexural strengths of the optimum 3D printed geopolymer mix. The interlayer bond strength of the optimum 3D printed geopolymer mix was also investigated. The second part of this book chapter (Part II) reports the effects of type of fiber on the flexural and interlayer bond strengths of the optimum 3D printed geopolymer mix [2].

Recently, Panda et al. [17] investigated the effect of incorporation of chopped glass fibers with different lengths (3, 6, and 8 mm) and volume fractions (0.25%, 0.50%, 0.75%, and 1.0%) on the mechanical properties of their 3D printed geopolymer. They concluded that inclusion of fiber resulted in negligible increase of compressive strength of printed geopolymer. Nevertheless, the flexural and tensile strengths of 3D printed geopolymers considerably increased with increase of fiber content up to 1.0% [2]. However, Panda et al. [17] did not compare the effect of the inclusion of fibers on the properties with respect to the 3D printed geopolymer without any fiber (i.e., they did not report the results of 3D printed geopolymer without fibers). In addition, the effect of the type of fiber on the properties of printed geopolymers has not been investigated in their study. Therefore, the first objective of Part II of this book chapter is to investigate the effect of the type of fiber on the flexural strength of the optimum 3D printed geopolymer mixture developed in the previous study by the authors [2,16]. The properties of the fiber-reinforced printed geopolymer were also compared with those of the optimum 3D printed geopolymer with no fiber [2].

One of the main challenges of the extrusion-based 3DCP technique is the weak bond strength between the printed layers. The low interlayer bond strength is a major weakness of 3D printed concrete, as potential flaws can be created between extruded layers, which induce stress concentration [2,18]. The interlayer bond strength depends on the adhesion between extruded layers, which is a function of print-time interval between layers, referred to as delay time. Recently, the authors of this study investigated the effect of delay time on the interlayer strength of extrusion-based 3D printed OPC-based concrete [19,20]. Panda et al. [21] recently studied the effect of printing time gap between layers, nozzle speed, and nozzle standoff distance on the interlayer bond strength of 3D printed geopolymer. However, the effect of the type of fiber on the interlayer bond strength of 3D printed geopolymer has not been investigated in their study. Therefore, the second objective of Part II of this chapter is to investigate the effect of type of fiber on the interlayer bond strength of the optimum 3D printed geopolymer mixture developed in the previous study by the authors [2,16]. The interlayer bond strength of the fiber-reinforced printed geopolymer was also compared with that of the optimum 3D printed geopolymer with no fiber [2].

18.2 MATERIALS

The low calcium (Class F) fly ash used in this study was supplied from Gladstone power station in Queensland, Australia. Table 18.1 presents the chemical composition and loss on ignition (LOI) of the fly ash determined by X-ray fluorescence. The total does not add up to 100% because of rounding errors. Fig. 18.1 presents the particle size distribution of fly ash measured by using a CILAS 1190 laser diffraction particle analyzer [2].

Two types of silica sands with different particle sizes were used in this study. The relatively coarser sand denoted as "CS" supplied by Dingo Cement Pty Ltd had average and maximum particle sizes of 330 and

Table 18.1 Chemical composition of fly ash

Chemical	Component (wt%)
Al_2O_3	25.56
SiO_2	51.11
CaO	4.3
Fe_2O_3	12.48
K_2O	0.7
MgO	1.45
Na_2O	0.77
P_2O_5	0.885
TiO_2	1.32
MnO	0.15
SO_3	0.24
LOI[a]	0.57

[a]Loss on ignition.

Data from B. Nematollahi, M. Xia, J. Sanjayan, P. Vijay, Influence of type of fiber on flexural and inter-layer bond strength of extrusion-based 3D printed geopolymer concrete, Mater. Sci. Forum 939 (2018) 155–162.

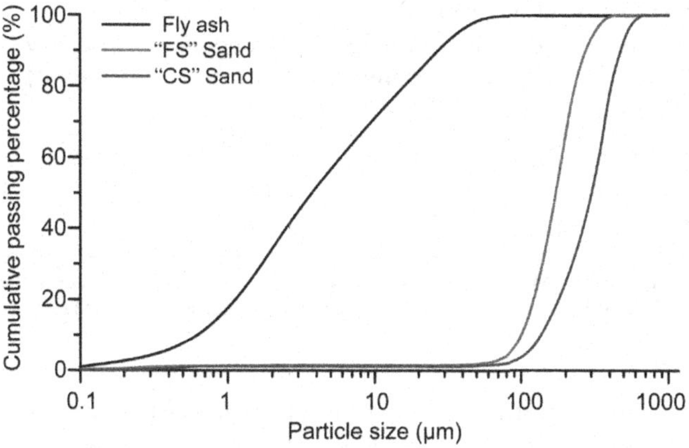

Figure 18.1 Particle size distributions of fly ash and silica sands.

Table 18.2 Specifications of N grade Na_2SiO_3 solution

Na_2SiO_3 solution	SiO_2 (wt%)	Na_2O (wt%)	H_2O (wt%)	Modules ratio $M = SiO_2/N_2O$	Viscosity at 20°C (cps)	Unit weight (g/cm^3)
N Grade	28.7	8.9	62.4	3.22	100-300	1.38

Data from B. Nematollahi, M. Xia, J. Sanjayan, P. Vijay, Influence of type of fiber on flexural and inter-layer bond strength of extrusion-based 3D printed geopolymer concrete, Mater. Sci. Forum 939 (2018) 155–162.

465 μm, respectively. The relatively finer sand denoted as "FS" supplied by TGS Industrial Sand Ltd had an average and maximum particle sizes of 172 and 271 μm, respectively. The particle size distributions of both silica sands measured by using a CILAS 1190 laser diffraction particle analyzer are also presented in Fig. 18.1 [2].

A sodium-based (Na-based) activator composed of 8.0 M NaOH and N Grade Na_2SiO_3 solutions with Na_2SiO_3/NaOH mass ratio of 2.5 was used in this study. The selection of this activator was based on the authors' previous research, which indicated that the 3D printed geopolymer made by using this activator provided the optimum properties [2,16]. The NaOH solution was prepared using NaOH beads of 97% purity supplied by Sigma-Aldrich and tap water. Table 18.2 presents the specifications of N grade Na_2SiO_3 solution supplied by PQ Australia. Sodium carboxymethyl cellulose (CMC) powder supplied by DKS Co. Ltd, Japan was used as a viscosity modifying agent in this study [2].

Three types of polymeric fibers including polyvinyl alcohol (PVA), polypropylene (PP), and polyphenylene benzobisoxazole (PBO) fibers

Table 18.3 Properties of the fibers

Fiber type	Diameter (μm)	Length (mm)	Young's modulus (GPa)	Elongation (%)	Density (kg/m^3)	Nominal strength (MPa)
PVA	26	6	37	6.0	1300	1600
PP	11.2	6	13.2	17.6	900	880
PBO	12	6	270	2.5	1560	5800

Data from B. Nematollahi, M. Xia, J. Sanjayan, P. Vijay, Influence of type of fiber on flexural and inter-layer bond strength of extrusion-based 3D printed geopolymer concrete, Mater. Sci. Forum 939 (2018) 155–162.

were used in this study. Properties of the fibers are presented in Table 18.3. It should be noted that while the length of fiber was kept constant as 6 mm, the Young's modulus and tensile strength of the fibers were significantly different, depending on the type of fiber. The PVA, PP, and PBO fibers were supplied by Kuraray Co. Ltd., Japan, Redco NV, Belgium and Toyobo Co. Ltd., Japan, respectively [2].

18.3 EXPERIMENTAL PROCEDURES

The experimental program conducted in this study was divided into two parts:

- *Part I*

  The first part investigates the effect of testing directions on the compressive and flexural strengths, and measure the interlayer bond strength of the optimum 3D printed geopolymer mix developed in the previous study by the authors [16].
- *Part II*

  The second part investigates the effects of the type of fiber on the flexural and interlayer bond strengths of the optimum 3D printed geopolymer mix developed in the previous study by the authors [2,16].

18.3.1 Mix Proportions

Table 18.4 presents the mix proportions of 3D printable geopolymer mixtures investigated in this study. The "GP-Plain" mixture represents the optimum 3D printed geopolymer mixture developed in the previous study by the authors [16]. The effect of the type of fiber was evaluated in "GP-PVA," "GP-PBO," and "GP-PP." All proportions were kept constant in these mixes, except the type of fiber and the amount of CMC

Table 18.4 Mix proportions of the 3D printable geopolymer

Mix ID	Fly ash	Activator[a]	"CS" sand	"FS" sand	CMC[b]	Fiber
GP-Plain	1.0	0.38	1.135	0.365	0.0120	–
GP-PP	1.0	0.38	1.135	0.365	0.0024	0.0025
GP-PVA	1.0	0.38	1.135	0.365	0.0052	0.0025
GP-PBO	1.0	0.38	1.135	0.365	0.0010	0.0025

Note: All numbers are mass ratios of fly ash weight except fiber content (volume fraction).
[a]Composed of composed of 8 M NaOH and N Grade Na_2SiO_3 solutions with Na_2SiO_3/NaOH mass ratio of 2.5.
[b]Sodium carboxymethyl cellulose powder as viscosity modifying agent.
Data from B. Nematollahi, M. Xia, J. Sanjayan, P. Vijay, Influence of type of fiber on flexural and inter-layer bond strength of extrusion-based 3D printed geopolymer concrete, Mater. Sci. Forum 939 (2018) 155–162.

powder was adjusted to achieve the desired rheological properties necessary for extrusion-based 3DCP process [2].

18.3.2 Mixing, Printing, and Curing

All mixtures were prepared in a Hobart mixer. The fly ash and silica sands were dry mixed for about 1 minute at low speed. Then, the alkaline solution was gradually added to the dry mix and the mixing was continued for about 4 minutes. After the mixture ingredients were thoroughly mixed, the fibers (if any) with a volume fraction of 0.25% were gradually added and mixed, taking care to ensure uniform fiber dispersion. Subsequently, the CMC powder was added and the mixing was continued for about 2 minutes to achieve the appropriate rheology for the extrusion process (visually assessed). The fresh mixtures were gradually loaded into the extruder of the printer. Moderate external vibration was applied to the extruder to ensure adequate compaction of the mixture inside the extruder [2].

A custom-made 3D printer was designed and manufactured to simulate the extrusion-based 3DCP process. A piston-type extruder was developed for this 3D printer to extrude the fresh material through a metallic cylinder measuring 50 mm × 600 mm (diameter × length). As can be seen in Fig. 18.2A, a 45 degrees nozzle with a 25 mm × 15 mm opening was designed and attached to the end of the extruder. The specimens were extruded by moving the extruder in the horizontal direction with a constant speed. The printed specimens consisted of two extruded layers. In this study the delay time (i.e., the print-time interval between layers) was chosen to be 15 minutes. The first layer was extruded with the

Figure 18.2 (A) 45 degrees nozzle with a 25 mm (W) × 15 mm (H) opening attached to the extruder; and (B) a typical 3D printed geopolymer sample measuring 250 mm (L) × 25 mm (W) × 30 mm (H). *Figure from B. Nematollahi, M. Xia, J. Sanjayan, P. Vijay, Influence of type of fiber on flexural and inter-layer bond strength of extrusion-based 3D printed geopolymer concrete, Mater. Sci. Forum 939 (2018) 155–162.*

dimensions of 250 mm (L) × 25 mm (W) × 15 mm (H). Then, after the chosen delay time of 15 minutes, the second layer was extruded on top of the first layer. Fig. 18.2B shows a typical 3D printed geopolymer composed of two layers with the dimensions of 250 mm (L) × 25 mm (W) × 30 mm (H) [2].

Heat curing was adopted in this study, for which all specimens were placed in a sealed container to minimize excessive moisture loss and cured in an oven at 60°C for 24 hours. At the end of heat curing period, the specimens were removed from the oven, kept undisturbed until being cool, and left in the laboratory at ambient temperature until the day of mechanical tests. Previous studies reported that strength of fly ash-based geopolymer after completion of the heat curing does not change significantly over time [22], typically reaching more than 90% of its long-term strength after 3 days [23–26], in a similar fashion to OPC under room temperature curing which reaches more than 90% of its long-term strength after 28 days which has become the standard age for comparison purposes. Thus, in this study all specimens were tested 3 days after printing [2].

18.3.3 Testing

In Part I of this study to measure the compressive strength of the optimum 3D printed geopolymer mix, 50 × 25 × 30 mm specimens were sawn from the 250 × 25 × 30 mm printed filaments and tested in one of the three directions, namely the perpendicular, the lateral and the longitudinal directions (Fig. 18.3). In this part, at least 10 printed specimens were tested for each direction. All specimens were tested in uniaxial compression under load control at the rate of 20 MPa/min [16].

In Part I of this study to measure the flexural strength of the optimum 3D printed geopolymer mix at least three 250 × 25 × 30 mm specimens were tested in one of the two directions, namely the perpendicular and the lateral directions (Fig. 18.4) [16]. However, in Part II of this study to measure the flexural strength of 3D printed fiber-reinforced geopolymer mixtures at least three 250 × 25 × 30 mm specimens were loaded in perpendicular direction only. All specimens were tested in a three-point bending test setup with a span of 200 mm under displacement control at the rate of 1.0 mm/min using a MTS testing machine [2,16].

In both Part I and Part II of this study to measure the interlayer bond strength of the optimum 3D printed geopolymer mix and the 3D printed fiber-reinforced geopolymer mixtures, 50 × 25 × 30 mm specimens were sawn from the 250 × 25 × 30 mm printed samples and loaded in uniaxial tension. The test set-up and specimen for the interlayer bond strength measurement are shown in Fig. 18.5. At the interface of layers, a small notch with an approximate depth of 5 mm was made on both cross sections of the specimens to ensure failure of the specimen at the interface. Two metallic brackets were epoxy glued on the top and bottom of the 3D printed specimen. The interlayer bond strength test was conducted using a MTS testing machine under displacement control at the rate

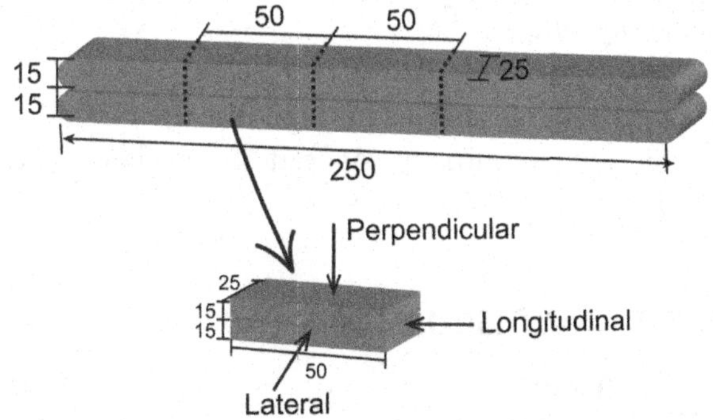

Figure 18.3 Cutting diagram and loading directions for compression testing of the optimum 3D printed geopolymer mix. *Figure from B. Nematollahi, V. Praful, J. Sanjayan, M. Xia, V. Nerella, V. Mechtcherine, Systematic Approach to develop geopolymers for 3D concrete printing applications, Under review in Archives of Civil and Mechanical Engineering, 2018.*

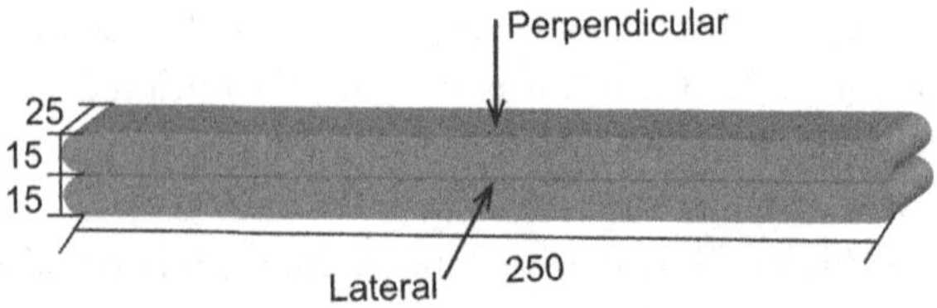

Figure 18.4 Testing directions for flexural testing of 3D printed geopolymers. *Figure from B. Nematollahi, V. Praful, J. Sanjayan, M. Xia, V. Nerella, V. Mechtcherine, Systematic Approach to develop geopolymers for 3D concrete printing applications, Under review in Archives of Civil and Mechanical Engineering, 2018.*

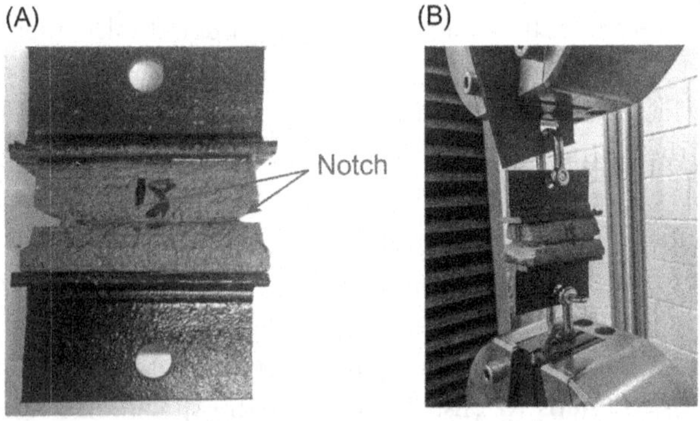

Figure 18.5 Interlayer bond strength test of 3D printed geopolymer: (A) test specimen; and (B) test set-up. *Figure from B. Nematollahi, M. Xia, J. Sanjayan, P. Vijay, Influence of type of fiber on flexural and inter-layer bond strength of extrusion-based 3D printed geopolymer concrete, Mater. Sci. Forum 939 (2018) 155–162; B. Nematollahi, V. Praful, J. Sanjayan, M. Xia, V. Nerella, V. Mechtcherine, Systematic Approach to develop geopolymers for 3D concrete printing applications, Under review in Archives of Civil and Mechanical Engineering, 2018.*

of 1.0 mm/min. Care was taken to align the specimen in the testing machine to avoid any eccentricity. At least six specimens were tested for each mix [2,16].

18.4 RESULTS AND DISCUSSIONS

18.4.1 Results of Part I

18.4.1.1 Compressive Strength

Fig. 18.6 presents the compressive strength of the optimum 3D printed geopolymer mix with respect to different testing directions. The compressive strength of the printed geopolymer exhibited anisotropic

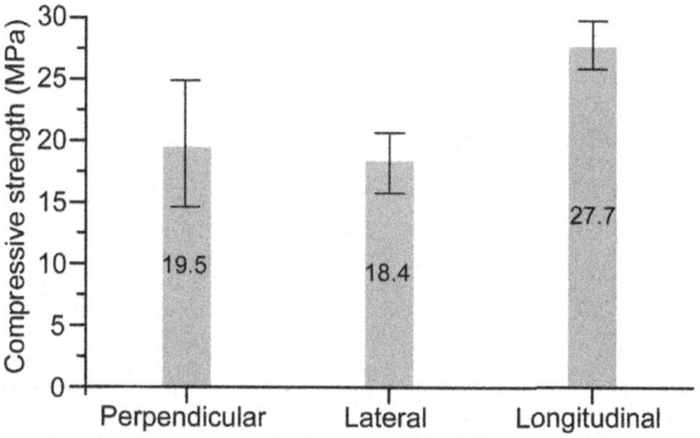

Figure 18.6 Compressive strength of the optimum 3D printed geopolymer mix in different directions. *Figure from B. Nematollahi, V. Praful, J. Sanjayan, M. Xia, V. Nerella, V. Mechtcherine, Systematic Approach to develop geopolymers for 3D concrete printing applications, Under review in Archives of Civil and Mechanical Engineering, 2018.*

phenomenon depending on the orientation of the loading relative to the printed layers. It can be argued that the difference in the compressive strengths of the optimum 3D printed geopolymer mix between the perpendicular and lateral directions are within the scatter of the results. However, Le et al. [18], Sanjayan et al. [20], and Panda et al. [15] also reported anisotropic phenomenon for the compressive strength of printed OPC concrete and geopolymer concrete. The highest mean compressive strength was obtained in the longitudinal direction. This could be attributed to high pressure exerted on the material in this direction during the extrusion process [20]. The lowest mean compressive strength was obtained in the lateral direction. The lateral direction has the least amount of pressure during the setting process. The fresh geopolymer is free to settle and expand in the lateral direction, as there is no mold or formwork to prevent the lateral flow and settling of the material. This, in turn, results in weakness in the lateral direction. The mean compressive strength in the perpendicular direction was in between the longitudinal (extrusion direction) and the lateral direction (free to expand and settle). This is because there is some level of pressure in the perpendicular direction during the setting process owing to the self-weight of the material. In summary, it can be concluded that the highest mean compressive strength was obtained in the longitudinal direction followed by the perpendicular and lateral directions. The same trend was reported by Sanjayan et al. [20] and Panda et al. [15] for 3D printed OPC mortar and geopolymer mortar, respectively [16].

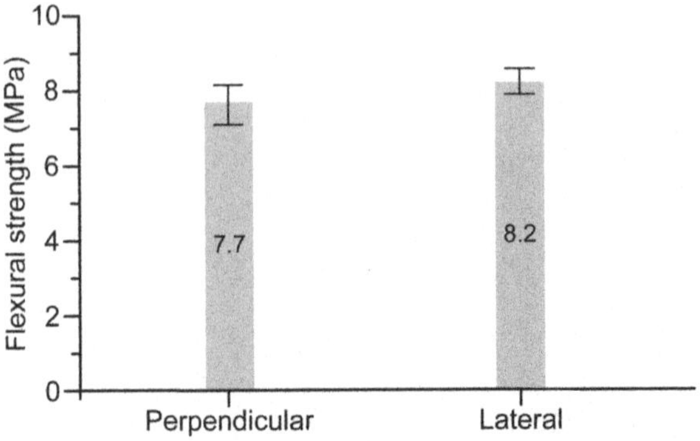

Figure 18.7 Flexural strength of the optimum 3D printed geopolymer mix in different directions. *Figure from B. Nematollahi, V. Praful, J. Sanjayan, M. Xia, V. Nerella, V. Mechtcherine, Systematic Approach to develop geopolymers for 3D concrete printing applications, Under review in Archives of Civil and Mechanical Engineering, 2018.*

18.4.1.2 Flexural Strength

Fig. 18.7 presents the flexural strength of the optimum 3D printed geopolymer mix with respect to different testing directions. Similar to the compressive strength, the flexural strength of the printed geopolymer also exhibited an anisotropic phenomenon depending on the testing direction. Similar to the compressive strength results (Fig. 18.6), it can be argued that the difference in the flexural strengths of the optimum 3D printed geopolymer mix between the perpendicular and lateral directions are within the scatter of the results. However, Le et al. [18] and Panda et al. [15] also reported anisotropic phenomenon for the flexural strength of the printed concrete. According to Oxman et al. [27] the directional dependency of the compressive and flexural strengths of the printed concrete is an inherent characteristic of the layer-by-layer 3D printing process.

18.4.1.3 Interlayer Bond Strength

The optimum 3D printed geopolymer mix exhibited a high interlayer bond strength of 3.03 ± 0.02 MPa, which was sufficiently high to prevent interfacial shear failure. This is supported by the mode of failure during the flexural tests, where the flexural failures of the specimens tested in the perpendicular direction were governed by the tensile strength of the bottom layer rather than the interlayer shear strength. As can be seen in Fig. 18.8A, no visible interface between the printed layers could be seen in the printed specimen where the surfaces in the lateral and longitudinal directions were ground. This indicates the layers were bonded together

(A) (B)

Figure 18.8 Interlayer bond strength of the optimum 3D printed geopolymer mix: (A) no visible interface can be seen in the printed specimen with ground surfaces; and (B) failure mode of interlayer bond specimens. *Figure from B. Nematollahi, V. Praful, J. Sanjayan, M. Xia, V. Nerella, V. Mechtcherine, Systematic Approach to develop geopolymers for 3D concrete printing applications, Under review in Archives of Civil and Mechanical Engineering, 2018.*

seamlessly, which is further evidence for the high interlayer bond strength obtained in this study. It is interesting to note that the interlayer bond strength of 3D printed geopolymer developed in this study was significantly higher than the ones reported previously [17] where the strength was around 0.5 MPa compared to 3.0 MPa in this study. This highlights the importance of optimizing the mixture parameters across multiple performance criteria. As shown in Fig. 18.5, the bond test specimens were loaded in uniaxial tension to measure the bond strength between the printed layers. The failure mechanism is tensile failure (Mode I fracture) according to the fracture mechanics theory. Once the tensile stress applied perpendicular to the interface of the printed layers exceeded the interlayer strength of the samples, the specimens failed at the interface between the first and second layers (as shown in Fig. 18.8B) [16].

18.4.2 Results of Part II

18.4.2.1 Flexural Strength

The average flexural strength of each mix in perpendicular direction is presented in Table 18.5. As expected, the flexural strengths of GP-PBO, GP-PVA, and GP-PP containing different types of fibers were 34%, 23%, and 17%, respectively, higher than that of GP-Plain (plain geopolymer mix without fiber). This is attributed to the incorporation of 0.25% (v/v) randomly oriented short fibers, which bridged the cracks and contributed to the flexural strength of the 3D printed geopolymer. This result is consistent with numerous published literature on the effect of incorporation of fibers on flexural and tensile strength of conventionally mold-cast geopolymer [23–25]. Among the fibers investigated, the PBO fiber was the

Table 18.5 Flexural strength of 3D printed geopolymers in the perpendicular direction

Mix ID	Flexural strength (MPa)
GP-Plain	7.7 ± 0.69
GP-PP	9.0 ± 0.56
GP-PVA	9.5 ± 0.22
GP-PBO	10.3 ± 0.37

Data from B. Nematollahi, M. Xia, J. Sanjayan, P. Vijay, Influence of type of fiber on flexural and inter-layer bond strength of extrusion-based 3D printed geopolymer concrete, Mater. Sci. Forum 939 (2018) 155–162.

Table 18.6 Interlayer bond strength of 3D printed geopolymers

Mix ID	Interlayer bond strength (MPa)
GP-Plain	3.03 ± 0.02
GP-PP	2.43 ± 0.15
GP-PVA	2.58 ± 0.31
GP-PBO	2.33 ± 0.13

Data from B. Nematollahi, M. Xia, J. Sanjayan, P. Vijay, Influence of type of fiber on flexural and inter-layer bond strength of extrusion-based 3D printed geopolymer concrete, Mater. Sci. Forum 939 (2018) 155–162.

most effective in increasing the flexural strength of the 3D printed geopolymer, followed by the PVA and PP fibers. This is attributed to the significantly different tensile strength and elastic modulus of the fibers. As shown in Table 18.3, the PBO fiber has the highest tensile strength and elastic modulus, followed by the PVA and PP fibers. It should be noted that the failure mode during flexural tests was due to the tensile failure of the bottom layer rather than the shear failure of the interface [2].

It is interesting to note that the flexural strength of all 3D printed plain and fiber-reinforced geopolymer mixes investigated in this study were significantly higher than that of the mix developed by Panda et al. [17]. As compared to the flexural strength of 7.7–10.3 MPa obtained in this study, Panda et al. [17] reported a flexural strength of only 4.8 MPa for their 3D printed geopolymer mix reinforced by chopped glass fibers for the same volume fraction of 0.25% and the same length of 6 mm [2].

18.4.2.2 Interlayer Bond Strength

The interlayer bond strength of each mix is presented in Table 18.6. The plain geopolymer mix without fiber (GP-Plain) exhibited the highest interlayer bond strength. The interlayer bond strengths of GP-PBO, GP-PVA, and GP-PP containing different types of fibers were 23%, 15%, and 20%, respectively, lower than that of GP-Plain. It is hypothesized that the reduction in the interlayer bond strength of the 3D printed fiber-reinforced geopolymer could be attributed to the increase in stiffness of the mix due to the incorporation of the fibers. Visual observations confirmed that the mixtures containing fibers were relatively stiffer as compared to the mix without fiber. The increased stiffness of the fresh mix reduced the ability of the freshly placed layers to deform and form a seamless interface. Thus, in the mixtures containing fibers the interface of printed layers may be more porous than that of the mix without fiber, leading to reduction in the interlayer bond strength. Further research is needed to confirm this hypothesis. It should be pointed out that all 3D printed geopolymer mixes (with or without fiber) developed in this study exhibited sufficiently high interlayer bond strengths to prevent interfacial shear failure. This is supported by the mode of failure during flexural tests, where the flexural failures of the specimens were governed by the tensile strength of the bottom layer rather than the interlayer shear strength. It is interesting to note that the interlayer bond strength of GP-Plain (plain mix without fiber) developed in this study was significantly higher than that of the mix developed by Panda et al. [21], where the interlayer bond strength was 0.47 MPa for the same 15 minutes delay time, compared to 3.03 MPa obtained in this study [2].

18.5 CONCLUSIONS

This chapter reports the results of a study on an extrusion-based 3D printable geopolymer for digital construction applications. The interlayer bond strength and the effects of testing direction on the compressive and flexural strengths of the optimum 3D printed geopolymer mix recently developed in the previous study by the authors were experimentally measured. The effect of incorporation of three types of short fibers on the interlayer bond and flexural strengths of the optimum 3D printed geopolymer mix were also evaluated. The following specific conclusions are drawn:

1. The compressive strength of the optimum 3D printed geopolymer exhibited an anisotropic behavior depending on the testing directions. The highest mean compressive strength was achieved in the longitudinal direction followed by the perpendicular and lateral directions. This

is consistent with the results reported in the literature for 3D printed OPC mortar and geopolymer mortar.

2. The flexural strength of the optimum 3D printed geopolymer also exhibited an isotropic behavior depending on the testing direction. However, the anisotropic phenomenon was less pronounced compared to the compressive strength results.
3. The interlayer bond strength of 3D printed fiber-reinforced geopolymers was 15%–23% lower than that of the 3D printed geopolymer with no fiber. Among the fibers investigated, the PVA fiber resulted in the lowest reduction in the interlayer bond strength of 3D printed fiber-reinforced geopolymer.
4. The incorporation of only 0.25% (v/v) fibers considerably (17% to 34%) increased the flexural strength of the 3D printed fly ash-based geopolymer. Among the fibers investigated, the PBO fiber was the most effective type of fiber in increasing the flexural strength of the 3D printed geopolymer.
5. All 3D printed geopolymer mixes developed in this study (with or without fiber) exhibited sufficiently high interlayer bond strengths to prevent interfacial shear failure. This is supported by the mode of failure during flexural tests where the flexural failures of the specimens were governed by the tensile strength of the bottom layer rather than the interlayer shear strength.

REFERENCES

[1] B. Nematollahi, M. Xia, J. Sanjayan, Current progress of 3D concrete printing technologies, in: ISARC, Proceedings of the International Symposium on Automation and Robotics in Construction, Vilnius Gediminas Technical University, Department of Construction Economics & Property, Taipei, 2017.

[2] B. Nematollahi, M. Xia, J. Sanjayan, P. Vijay, Influence of type of fiber on flexural and inter-layer bond strength of extrusion-based 3D printed geopolymer concrete, Mater. Sci. Forum 939 (2018) 155–162.

[3] M. Xia, B. Nematollahi, J. Sanjayan, Printability, accuracy and strength of fly ash/slag geopolymer made using powder-based 3D printing for construction applications, Under review in Automation in Construction, 2018.

[4] M. Xia, B. Nematollahi, J. Sanjayan, Compressive strength and dimensional accuracy of Portland cement mortar made using powder-based 3D printing for construction applications, in: The 1st RILEM International Conference on Concrete and Digital Fabrication (Digital Concrete 2018), Zurich, Switzerland, 2018.

[5] M. Xia, B. Nematollahi, J. Sanjayan, Influence of binder saturation level on green strength and dimensional accuracy of powder-based 3D printed geopolymer, Mater. Sci. Forum 939 (2018) 177–183.

[6] M. Xia, J. Sanjayan, Method of formulating geopolymer for 3D printing for construction applications, Mater. Des. 110 (2016) 382–390.
[7] M. Xia, J. Sanjayan, Post-processing methods for improving strength of geopolymer produced using 3D printing technique, in: ICACMS, International Conference on Advances in Construction Materials and Systems, Chennai, 2017.
[8] M. Xia, J. Sanjayan, Methods of enhancing strength of geopolymer produced from powder-based 3D printing process, Mater. Lett. 227 (2018) 281–283.
[9] B. Nematollahi, M. XIA, J. Sanjayan, Enhancing strength of powder-based 3D printed geopolymers via post-processing methods. In: Proceedings of the First International Conference on 3D Construction Printing, (3DcP), Melbourne, Australia, 26–28 November, 2018.
[10] D.N. Huntzinger, T.D. Eatmon, A life-cycle assessment of Portland cement manufacturing: comparing the traditional process with alternative technologies, J. Clean. Prod. 17 (7) (2009) 668–675.
[11] B. Nematollahi, J. Sanjayan, F.U.A. Shaikh, Synthesis of heat and ambient cured one-part geopolymer mixes with different grades of sodium silicate, Ceram. Int. 41 (4) (2015) 5696–5704.
[12] J. Sanjayan, Materials technology research to structural design of geopolymer concrete, in: Mechanics of Structures and Materials XXIV: Proceedings of the 24th Australian Conference on the Mechanics of Structures and Materials (ACMSM24, Perth, Australia, 6-9 December 2016), CRC Press, 2016, p. 31.
[13] B. Nematollahi, J. Sanjayan, Effect of different superplasticizers and activator combinations on workability and strength of fly ash based geopolymer, Mater. Des. 57 (0) (2014) 667–672.
[14] P. Duxson, J.L. Provis, G.C. Lukey, J.S. Van Deventer, The role of inorganic polymer technology in the development of 'green concrete', Cem. Concr. Res. 37 (12) (2007) 1590–1597.
[15] B. Panda, S.C. Paul, L.J. Hui, Y.W.D. Tay, M.J. Tan, Additive manufacturing of geopolymer for sustainable built environment, J. Clean. Prod. 167 (2017) 281–288.
[16] B. Nematollahi, V. Praful, J. Sanjayan, M. Xia, V. Nerella, V. Mechtcherine, Systematic Approach to develop geopolymers for 3D concrete printing applications, Under review in Archives of Civil and Mechanical Engineering, 2018.
[17] B. Panda, S.C. Paul, M.J. Tan, Anisotropic mechanical performance of 3D printed fiber reinforced sustainable construction material, Mater. Lett. 209 (2017) 146–149.
[18] T.T. Le, S.A. Austin, S. Lim, R.A. Buswell, R. Law, A.G.F. Gibb, et al., Hardened properties of high-performance printing concrete, Cem. Concr. Res. 42 (3) (2012) 558–566.
[19] T. Marchment, M. Xia, E. Dodd, J. Sanjayan, B. Nematollahi, Effect of delay time on the mechanical properties of extrusion-based 3D printed concrete, in: ISARC, Proceedings of the International Symposium on Automation and Robotics in Construction, Vilnius Gediminas Technical University, Department of Construction Economics & Property, 2017.
[20] J.G. Sanjayan, B. Nematollahi, M. Xia, T. Marchment, Effect of surface moisture on inter-layer strength of 3D printed concrete, Constr. Build. Mater. 172 (2018) 468–475.
[21] B. Panda, S.C. Paul, N.A.N. Mohamed, Y.W.D. Tay, M.J. Tan, Measurement of tensile bond strength of 3D printed geopolymer mortar, Measurement 113 (2018) 108–116.
[22] D. Hardjito, S.E. Wallah, D.M.J. Sumajouw, A.B.V. Rangan, On the development of fly ash-based geopolymer concrete, ACI Mater. J. 101 (6) (2004) 467–472.
[23] B. Nematollahi, R. Ranade, J. Sanjayan, S. Ramakrishnan, Thermal and mechanical properties of sustainable lightweight strain hardening geopolymer composites, Arch. Civil Mech. Eng. 17 (1) (2017) 55–64.

[24] B. Nematollahi, J. Sanjayan, J. Qiu, E.-H. Yang, Micromechanics-based investigation of a sustainable ambient temperature cured one-part strain hardening geopolymer composite, Constr. Build. Mater. 131 (2017) 552–563.
[25] B. Nematollahi, J. Sanjayan, J. Qiu, E.-H. Yang, High ductile behavior of a polyethylene fiber-reinforced one-part geopolymer composite: a micromechanics-based investigation, Arch. Civil Mech. Eng. 17 (3) (2017) 555–563.
[26] B. Nematollahi, J. Qiu, E.-H. Yang, J. Sanjayan, Micromechanics constitutive modelling and optimization of strain hardening geopolymer composite, Ceram. Int. 43 (8) (2017) 5999–6007.
[27] N. Oxman, E. Tsai, M. Firstenberg, Digital anisotropy: a variable elasticity rapid prototyping platform, Virtual Phys. Prototyping 7 (4) (2012) 261–274.

CHAPTER 19

Industrial Adoption of 3D Concrete Printing in the Australian Market: Potentials and Challenges

Daniel Avrutis, Ali Nazari and Jay G. Sanjayan
Centre for Sustainable Infrastructure, Faculty of Science, Engineering and Technology, Swinburne University of Technology, Hawthorn, VIC, Australia

19.1 INTRODUCTION

The construction industry in Australia is one of the main sectors of national employment. This industry employs approximately 1.1 million people, nearing 10% of the entire national employment workforce. Employment in the Australian construction industry has seen an increase of over 130 thousand, equating to 14% over the past 5 years and is expected to increase to over 1.134 million people by 2020. This industry is essential in maintaining Australia's economic stability and future growth as it is comprised of various subdivisions such as builders, architects, engineers, and certifiers. Therefore, due to the high dependence in this industry for economic and financial stability, any changes to the current building platform may cause major ripples throughout the nation. However, just like any other industry, it is essential that modern technologies and innovations be constantly adopted to ensure the industry progresses in a positive direction.

3D printing is a fairly new technique which is being adopted in various industries due to its automated innovativeness. However, only recently has this technologically advanced method of fabrication been used for construction purposes. The basis of this technology allows for digitally designed 3D objects to be autonomously printed via robotic control. Currently there are only three major techniques which have been developed for 3D printing of construction components. These techniques are contour crafting (CC) [1], D-Shape [2], and 3D concrete printing [3], each presenting their own benefits and limitations. CC [1] is based around

3D Concrete Printing Technology
DOI: https://doi.org/10.1016/B978-0-12-815481-6.00019-1

layering manufacturing, which builds the component layer by layer, as the printer extrudes material according to the outline of the designed digital model. D-shape [2] is a process of extruding a binder throughout a powder deposition and as the binder sets the hardened component is then extracted from the remaining powder material. Concrete printing [3,4] is similar to CC and based on extrusion; however, this method of printing allows for a much smaller resolution of filaments and can develop finer details. Benefits from 3D printing include lower overall costs form labor savings, cleaner construction process with less material wastage and lower embodied energy, safer method of construction by removing dangerous construction works, and time savings via faster building erection. Limitations include cost of machinery, minimal materials, no regulations or standards to comply with, and almost no reference points from previous cases due to the innovativeness of this technique [5].

The application of 3D printing of construction components is a completely new field which has not been explored in great depth. In 2014, Dus Architects from Amsterdam decided to construct a house only using components manufactured from a 3D printer, the first project of its kind in Europe [6]. Named Canal House, these architects wanted to prove that not only is this method of construction possible, but also by forming these components onsite it would eliminate building waste and significantly reduce any transportation costs which are associated with constructing modular components offsite [7]. Simultaneously, the WinSun project in China managed to construct a 5-story apartment building with the aid of 3D printing of the construction components. D-Shape is currently constructing a 220 m^2 house, swimming pool, Jacuzzi, and a carport using the 3DP (3D Powder) process in upstate New York. It will use materials found onsite and bind them using a magnesium-based cement to build the structures [2]. Another application of this 3DP technology is being applied in a bridge-strengthening project [8], where the bridge pier sheaths are printed using 3DP technology to match the old piers which were damaged due to scour. Krassenstein [8] also reported that military bunkers are being constructed using the 3DP technology under contract from the UK government.

The Fishbone Diagram, also known as the Ishikawa diagram or the cause and effect diagram, was developed in the 1960s and is used as an analytical tool to discover the causes and effects of a particular problem [9]. These causes are then further broken down to subcategories which are used to assess and evaluate the problem at hand. This model identifies and categorizes the correlation between an event and its predecessors via its unique fishbone-like structure which makes it an easier method of problem solving. A benefit of this assessment is that it assists in identifying the foundation of a particular problem in a structured manner while identifying areas which may need further consideration [10].

The aim of this chapter is to investigate potentials and challenges of adopting concrete 3D printing in the Australian construction industry. A fishbone problem diagnosis diagram was developed to investigate the effect of each parameter on the matter studied in this chapter. Initially, a focused problem statement is developed and placed at the head of the fish. The more refined the statement, the easier for the root causes to be identified. A backbone is then drawn from the head to allow for segmentation of major causes. The major types of causes are generally categorized depending on whether the process is for manufacturing or a service. Thus, for manufacturing these categories are people, materials, machinery, methods, environment, and measurement while for services these are people, policies, procedures, location, and measurement. Each major cause is then segmented into subcauses, allowing for the root of each cause to be determined and assessed [11].

19.2 INDUSTRIAL ADOPTION OF 3D CONCRETE PRINTING

19.2.1 Innovation in the Australian Construction Industry

Innovation is taking new imaginative ideas, which are advantageous to a particular industry, and transforming them from conceptual to tangible. This idea should preferably delve into the unknown as it is these new ideas which attracts attention and creates new direction towards an unimagined future. Such creations are essential for business growth, stability, and competitiveness in any well-established industry. It provides the opportunity for businesses to assert their value and importance to the industry and places them on the high-end of the value ladder. However, if innovation is not undertaken, businesses have a higher risk of jeopardizing their position in the industry and may result in the need for cost-cutting or even closure. Additionally, corporations which attempt to innovate have a higher chance of precedence as this process forces companies to analyses, understand, and anticipate future trends regardless of the outlining impact on the industry. By remaining aware of industry trends, businesses will have the opportunity to capitalize on any niches or openings which may arise [12].

According to the Business Council of Australia, Australia has seen 23 years of economic growth; however, in recent years statistics has shown that national productivity has slowed from 2.2% to a concerning 1.5%. Thus, proof that national complacency is beginning to slow the economy and business innovativeness has slowed in comparison to its international competitors. These figures are a worrying sign of future instability and misdirection which may cause a massive dent in its economy. Furthermore, due to the lengthy period of economic growth, many Australians have not experienced the consequences which come with economic downturn and industry collapse, thus, a possible reason for the lack of innovation and drive. This is a major cause for concern as implications for this lack of preparation will be far more dramatic and unexpected if the downturn does arise. Additionally, another cause of concern of national complacency is that this has allowed international construction companies to increase their activity in the Australian market. This international activity results in higher acquisition of local jobs and ultimately takes away from our local businesses. As these international companies drown our market with their innovative methods and cost efficiencies, Australian construction companies will have no choice but to follow the trend or this low innovation and productivity mindset will have major penalties to local business and the national economy [13].

19.2.2 Impact of Emerging Technology on Society

Robotic reliance in the business industry is an inevitable scenario. The use of robotic equipment to replace human involvement is an expanding area within the developed world. It is this innovative area which has seen the most significant action as it is proven that robotic replacements reduce the cost of human labor, eliminates human error, increases productivity, and ultimately achieves the ideal scenario of simultaneously increasing affordability and efficiency. The global increase of operational robotic equipment has risen from 1.2 million in 2013 to 1.9 million in 2017. This shows a dramatic increase of approximately 700,000 machines over a 4-year period which is a staggering value since this technological advancement of robotics is still in its early stages. By further breakdown the countries which have invested and increased their robotic reliance are Japan, the United States, China, South Korea, and Germany, that is, the countries which have the greatest global presence and economic influence. Thus, due to these main countries taking the innovative path of robotic reliance, this sector is expected to be worth $67 billion by 2025, four times the estimated current market worth [14].

Moreover, this dramatic increase in technological advancements is due to positive business impact; however, it is not without an adverse effect on the workforce. Many of these technologically reliant businesses have economic stability without the need for a standard number of employees. This shows the innovativeness of robotics is reducing the need for traditional labor, eliminating jobs, and having consequences for the working class. Although evidence has shown that jobs are being taken away due to automation operation, simultaneously, new jobs are created. However, as the rate of job elimination is greater than job creation there is a concern that the use of robotics within our workforce may ultimately send the national economic stability into a downward spiral [14].

19.2.3 Potential Impact of 3D Printing on the Construction Industry and Economy

As 3D printing within the construction industry is still in its early stages it is difficult to accurately predict the impact of such an innovative technology. However, during the establishment of such technology and tests undertaken there have been predicted advantages and disadvantages to construction businesses and national economy.

When focusing on the business impact of 3D printing such known benefits include less wastage of materials as the printer creates the exact digital design, reduced transportation of materials as additional construction materials are not required, efficiency in design allowing for new and different designs to be developed, reduced labor costs as the number of laborers requires is minimal, reduced safety hazards as human error is eliminated, and faster construction as the machine is operable nonstop. On the other hand, known disadvantages include it being more expensive than conventional construction due to set-up and unfamiliarity, the limitation of applicable materials, requiring clear access on site due to the large framework of printers, and requires specialized digital modeling to be developed [15].

Moreover, when focusing on how 3D printing may affect the economic sector various scenarios have been considered. Due to the versatility of such machinery and elimination of stick build construction such technology will allow for a level playing field and may reduce any economic imbalances. Also, as this technology is quickly spreading, this will require people to learn and manage this area, creating new and unique positions which previously were not required. These newly created positions can be associated with, but not limited to, printer production, engineering, design, material supply, and software development. On the other hand, as 3D printing is computer-controlled it eliminates the need for onsite labor leading to significant reduction in construction jobs, opening the doors for possible economic imbalance [16].

19.3 METHODOLOGY

19.3.1 Research Strategy

The strategy of this research was based on previously analyzed data and literature. Due to the wide range of information which is currently available this allows for an in-depth research report to be undertaken and rational conclusions to be developed. Although a well-established area of research this research report has the distinguishable objective of assessing both the local and global effects of 3D printing of construction components to evaluate the overall impact on the Australian construction industry and economy.

19.3.2 Research Method

The method of this research is to undertake both a qualitative and quantitative analysis in order to generate a comprehensive evaluation and to produce the most accurate conclusion. This concurrent method of research allows for both textual and numerical data to be used simultaneously for a more in-depth research.

19.3.3 Research Approach

Due to the innovativeness of this topic a limited amount of information is currently available on this technology in the Australian market. Thus, the approach of this research was to use a broad spectrum of available literature and data to develop rational conclusions on local and global effects.

19.3.4 Data Collection Method and Tools

The basis of this report is to perform an in-depth literature review and data collection approach in order to be able to adequately analyze the information and rationally summarize the results and observations. Post this review and analysis, the Fishbone Diagram tool was used to undertake the comprehensive viability evaluation.

19.3.5 Research Process

The process of this research is to develop six appropriate categories and their respective subsections, to analyses the observed information via the Fishbone Diagram tool. In turn, each subsection is given a positive or negative impact rating which is used to graphically illustrate the viability assessment.

19.3.6 Data Analysis

The impact rating is graphically summarized in order to determine the viability percentage for each category. Once calculated, the average percentage of the six categorical results will be used to determine the overall viability percentage for the Australian market.

19.3.7 Research Limitations

Various assumptions had to be made in order to obtain certain conclusions, due to the limited amount of information which is currently available for the Australian industry. Thus, the overall viability percentage cannot be guaranteed.

19.3.8 Research Conclusions

The conclusion of this research report will determine the potential viability of 3D printing of construction components for the Australian market.

19.4 RESULTS AND OBSERVATIONS

Post reviewing the available literature, it is clear that robotic dependency and autonomous innovation is becoming increasingly popular across many business sectors throughout the world. A subcategory of robotic automation is 3D printing. Although still in its early stages this innovative technology is taking over many business sectors and is beginning to make traction throughout construction industry. 3D printing of construction components is an innovative advancement to modular construction. It is a process where sections of a building are digitally designed, autonomously printed via a 3D printer and then assembled onsite. Benefits include scheduling efficiency due to parallel production and weather protected fabrication, increase in quality due to the manufacture of smaller modules, and labor safety due to simple onsite assembly. Conversely constraints include major preplanning and preparation to ensure each module and section has been developed precisely to ensure smooth onsite assembly, increase in project coordination for quality control for offsite fabrication, and onsite construction. However, the underlining differences between modular stick build construction and 3D printing of construction components are; eliminated transportation costs as the 3D printing process can be done onsite, less material wastage as 3D printing prints the exact material as required, the assurance of top quality for every print due to precision of 3D printing, and the flexibility to alter the digitally designed model and print the updated models with less complications.

Currently, there are only three major techniques which have been developed as acceptable methods for 3D printing of construction components each presenting their own benefits and limitations. Benefits of 3D printing of construction components are lower overall cost including construction, transportation and storage, cleaner construction process with less material wastage and lower embodied energy, safer construction method by removing dangerous construction works, and time savings via faster building erection. On the other hand, limitations include the restricted number of materials which can be used for the printing process, component compressive strength reduction in comparison to in situ construction, no developed standards or compliance requirements which can be followed, unattractive finishes, and no tangible examples which demonstrate the effect of longevity. Although on the local scale construction via 3D printing of components has many advantageous accompaniments for builders, the global scale also requires to be observed to identify how this method of construction may affect employees and the national economy. Although difficult to identify the exact impact which 3D printing may have on the industry, the literature observed has allowed for an analytical prediction to be developed.

In the last quarter century Australia has seen an economic growth, however in recent years the nations productivity and advancements have slowed to a concerning low of 1.5%. This decline is assumed to be caused by the nation's complacent mentality and unwillingness to adopt the mind set to innovate and revolutionize their respective industry. This is a major concern, as international competitors which have the innovative and progressive mentality now have a competitive edge against Australian business. This edge ultimately allows these businesses to enter our market and acquire local jobs due to their ability to offer new and exciting products in conjunction with cost efficiencies. These job losses are a worrying sign for the already declining Australian economy, as businesses will be unable to sustain work, force closure and ultimately unemployment levels will rise.

Although 3D printing will help Australian construction businesses develop an innovative mindset and introduce new and exciting prospects into the industry there are both positive and negative impacts which will result from this technology. Global advantages for businesses include innovative progression and industry leadership resulting in sustained industry employment and business succession, maintain industry competitiveness through innovation, and decrease the risk of business collapse resulting in a positive effect on national employment. On the other hand, disadvantages include industry employee losses as the need for laborers will decrease, hefty investment costs, and the daunting realm of the unknown factor. As a result, 3D printing in the construction industry is a double-edged sword as this innovative technology allows for businesses to be competitive; however, it will almost certainly replace laborers and result in job losses. On the other hand, if businesses do not innovate and maintain a competitive edge, they are at risk of being left behind in this aggressive industry which may force closure.

As a result, it is essential for Australian businesses to be proactive in observing innovation in order to sustain a competitive edge and to succeed in the industry. Although 3D printing and robotic reliance is becoming an increasingly popular advancement across the globe, this technology will result in both positive and negative impacts locally and globally. Thus, before implementation analytical investigations are required to be carried out prior to such a hefty investment and risk. Regardless of these potential impacts, corporations which attempt to innovate have a greater chance of survival within the construction industry as the innovation process forces companies to change their mindset by analyzing, understanding and anticipating future trends. Accordingly, when taking the declining Australian productivity into account it is the innovative mindset which will help Australian businesses remain a chance against international competitors, otherwise the current complacent mentality will have major implications to local business and national economy.

Ultimately, 3D printing is an innovative process which offers endless possibilities. However due to this being such an advanced technology with limited understanding or tangible case studies, this process of construction will require a detailed analysis from all angles to develop an understanding of the implications which this technology might have on the industry and the nation. This analysis must review the cause and effect of such an industry changing method of construction, in order to determine the probable benefits and concerns prior to being released into the industry. Such review will be essential in determining how 3D printing could impact such an important industry and will form the basis of determining whether or not it is a viable option for use within Australia. Therefore, in order to determine whether or not 3D printing of construction components is a viable method for the Australian construction industry, a managerial method of analysis has been carried out via Fishbone Diagram. This system of analysis examines the cause and effect of a particular problem, where the cause is distinguished by various categories depending on the problem. For example, for manufacturing the causes can be categorized by people, materials, machinery, methods, environment, and measurement and for services these can be categorized by people, policies, procedures, location, and measurement. Since 3D printing of construction materials is a combination of service and manufacturing the causes have been categorized by environment, policies, procedures, materials, people, method, and machine, as shown in Fig. 19.1.

19.4.1 Evaluation for Viability Analysis

For the Fishbone Diagram evaluation six critical subquestions for each category have been established. These questions will be answered upon the combination of information gathered from the literature review and my personal judgement. Each question will be given a yes or no/positive or negative rating, permitting the expression of results in a statistical format and will be the basis of my viability evaluation.

19.4.1.1 Environment

- *Can 3D printing combat climate change?*

  No, although 3D printing has a lower embodied energy by eliminating the need for transportation and reducing the amount of material wastage, this method of construction still relies on fossil fuel to fabricate construction components.
- *Do these machines meet the conditions of the Environmental Protection Agency?*

  Yes, since the Environmental Protection Agency covers segments such as air, climate, noise, water, odor, and waste it is difficult to determine whether or not 3D printing of components meets all these conditions as this process has not been tested nationally. Although from the derived literature and observation it can be assumed that

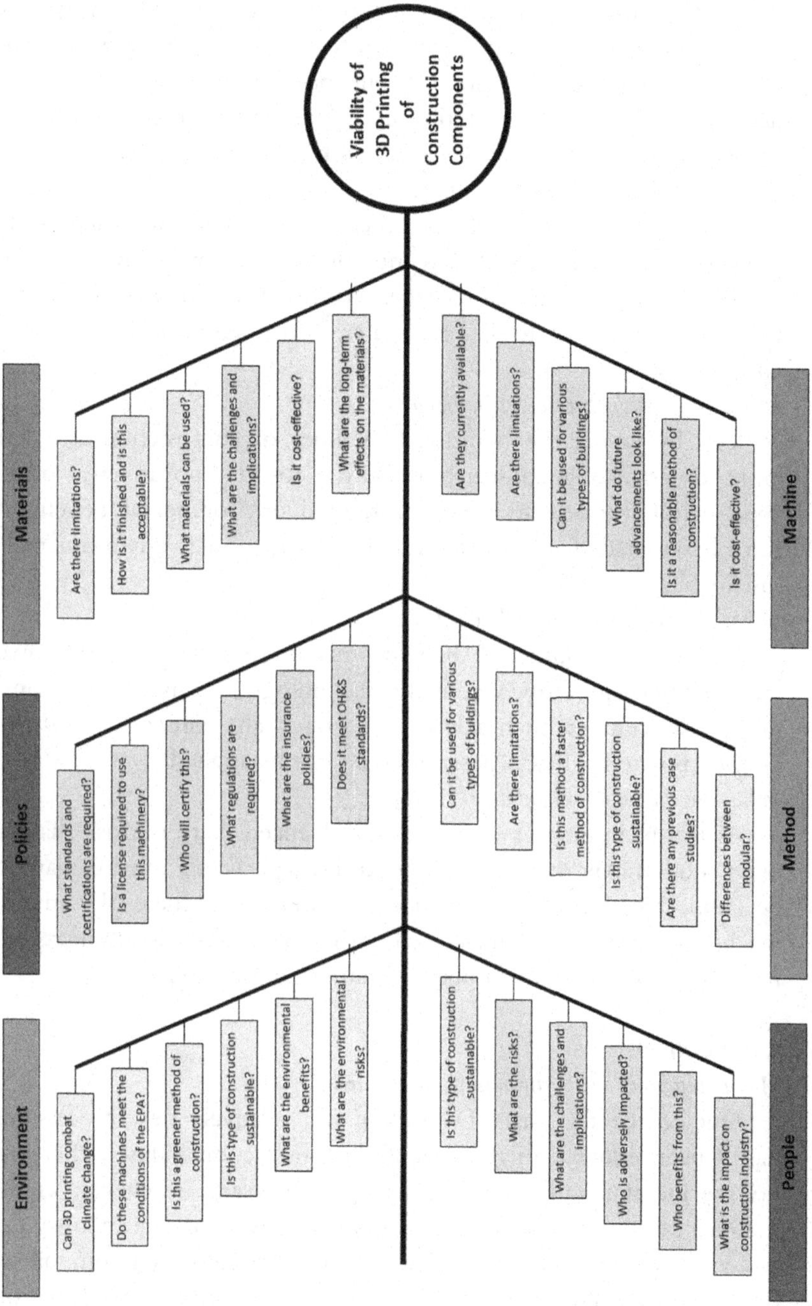

Figure 19.1 Fishbone Diagram and 3D printing of construction components analysis.

segments such as air, climate, water, odor, and waste should have no issues with complying. However due to the restricted tangible examples information on the noise of the printing process is difficult to determine, although due to the innovative nature of this mechanical equipment there should be no issues with noise compliance.

- *Is this a greener method of construction?*

  Yes, although a very similar process to modular construction, 3D printing is a process which has the ability to eliminate the need for transportation and material wastage. As a result, this method of fabrication has the ability to be a greener, more sustainable method of construction.
- *Is this type of construction sustainable?*

  Yes, by taking the traditional methods of construction as a base point for sustainable construction and due to the innovativeness of this construction process and the contributed environmental benefits, it can be assumed that this method of construction is environmentally sustainable.
- *What are the environmental benefits?*

  Current environmental benefits include less wastage and lower embodied energy; however, as this technology becomes more prominent, advancements such as power via renewable energy sources may be incorporated. Thus, a positive rating can be applied.
- *What are the environmental risks?*

  As this technology becomes more advanced there is a risk that the older models may require replacing and that recycling of parts may not be applicable. However, in order to combat this, these 3D printing machines should be manufactured with ecologically friendly parts and recycling in mind. Thus, a negative rating can be applied.

19.4.1.2 Policies

- *What standards and certifications are required?*

  Due to the fact that 3D printing is a relatively new process and hasn't been adopted in the Australian construction industry there are no specific standards or procedures put in place. However, as this method of construction becomes more prominent in the industry this will need to be developed. However, for the fabricated components the current set standards will still need to be applied to ensure that the printed components are structurally sound and industry compliant. Thus, a negative rating can be applied.

- *Is a license required to use this machinery?*

  Although this process of fabrication involves heavy machinery, the need for licensing may not be necessary as the printer is autonomous. Due to minimal human involvement and as long as the machine is set up to safely print components then there is low risk of harm, thus, training and acknowledged competency may be sufficient. Thus, a positive rating can be applied.
- *Who will certify this?*

  The 3D printing process may be different to traditional construction methods however the components of the building will still require the necessary structural capacities to withstand the designed loading. Therefore, similar to traditional circumstances both builders and building surveyors will be liable for construction integrity and building certification. Thus, a positive rating can be applied.
- *What regulations are required?*

  Currently there are no outlined regulations associated with 3D printing of construction components as this innovation has not entered the Australian market. However, in order for fair practice and common ground within the industry rules and regulations must be established around copyrighting digital content and privacy of intellectual property. Thus, a negative rating can be applied.
- *What are the insurance policies?*

  An issue which has surfaced from the manufacturing industry is that although traditionally manufacturers were liable for any product defects, 3D printing makes it difficult to distinguish whether the manufacturer, the proprietor of the printer or the material supplier is at fault. Assuming that similar implications of 3D printing will be experienced in the construction industry clear policies will need to be developed to address this complexity. Thus, a negative rating can be applied.
- *Does it meet OH&S standards?*

  3D printing is seen as a safer method of construction due to its automated fabrication and minimal human involvement. As a result, it can be assumed that if traditional method of construction is occupational health & safety (OH&S) compliant than utilizing 3D printing for component fabrication will also be compliant. However, a set of procedures and processes will need to be developed to provide users with the appropriate knowledge and understanding of what it is required for OH&S compliance. Thus, a positive rating can be applied.

19.4.1.3 Materials

- *Are there limitations?*

  Currently the materials applicable for 3D printing of construction components are limited and require development. Due to the innovative nature of this printing process the existing materials used for construction are not suitable for extrusion, therefore, new material compositions need to be developed. Although there are few available materials they are not without imperfections. Thus, a negative rating can be applied.

- *How is it finished and is this acceptable?*

  As the printing method of 3D components is of a layering nature, the finishes are not smooth but consist of a ribbed pattern which is resultant of the individual filaments. However, similar to the standard method of construction the internal finishes are generally lined with plasterboard which will conceal the ribbed pattern and the external finishes can be clad with any facade treatment or even rendered. Therefore, although the finish of the printed components is not of an appealing texture, it can easily be covered via traditional concealments. Thus, a positive rating can be applied.

- *What materials can be used?*

  Currently the only materials which have been adequately developed for 3D printing and have suitable characteristics for extrusion, stability, solidity, and strength are specific polymers, ceramics and cement. Thus, a negative rating can be applied.

- *What are the challenges and implications?*

  Due to the innovative method of layer printing, current material compositions will not be applicable for this type of fabrication. Thus, optimal material compositions still need to be determined in order to develop components which do not lack in compressive strength and are similar in comparison to traditionally manufactured components. However, in order to derive this optimal material composition this will require further research, development and testing. Thus, a negative rating can be applied.

- *Is it cost-effective?*

  3D printing of construction components offers all the cost benefits which come with modular construction however adds the assurance that each component will be printed with quality and precision. Although the cost of the printer is still arguable, this technology will swiftly pay for itself as it increases product efficiency whilst simultaneously reducing the number of laborers, therefore, making it a cheaper and more reliable method of component fabrication and building construction. Thus, a positive rating can be applied.

- *What are the long-term effects on the materials?*

  Due to the innovativeness of this technology and limited examples of this fabrication process, there are no long-term tangible case studies which can be used as a basis for assessment. However, since the components are being printed with materials which are customary in the industry, there should be no issue with the longevity of these components. Thus, a negative rating can be applied.

19.4.1.4 People

- *What is the impact on construction industry?*

  Since 3D printing is an autonomous process for modular construction, the need for laborers during the component fabrication phase is eliminated. Thus, the amount of laborers required may dramatically drop, forcing builders who decide to utilize this technology to make redundancies. On the other hand, due to the innovativeness and the integration of digital technology, 3D printing also has the capacity to create new jobs in the industry which previously have not been required. Although there are both positives and negatives for employment, a positive rating is being applied as innovation is essential for business growth and progression.

- *Who benefits from this?*

  The innovative nature of this technology will provide the platform for builders to progress in the industry, branch out to sectors which previously were not possible and assist in gaining more contracts and a larger market share. Additionally, it will create new positions within the construction industry and diversify its employment range through 3D printing manufacturers, digital designers and material developers. Ultimately by advancing businesses and increasing stakeholders within the industry this will advantageous for the construction sector and, in turn, the Australian economy. Thus, a positive rating is applied.

- *Who is adversely impacted?*

  Since this technology is autonomous, it eliminates the need for laborers during the component manufacturing period. This reduction in human involvement will force laborers to become redundant and will reduce the employment numbers in this area. Although a reduction of laborers may occur this level of industry employment should stabilize by the addition of new stakeholders. Due to the adverse impact on laborers a negative rating can be applied.

- *What are the challenges and implications?*

  As 3D printing of construction components enters the market it will require a sufficient period of time prior to having an impact on the industry. It will require builders to learn new techniques and develop new skills, which may retract from usual business. However, it is a positive for businesses to innovate and learn new techniques and to be challenged and forced away from their comfort zones. This forces businesses to keep ahead of the market and secure a stable position within the market. Thus, a positive rating can be applied.

- *What are the risks?*

  As this technology has never been used throughout the Australian construction industry there are tangible case studies which can be used a reference. A possible implication of this is that the industry does not respond well to this technology and that businesses and employees suffer. Thus, a negative rating is applied.

- *Is this type of construction sustainable?*

  If this technology does make a positive impact on the construction industry, it has the opportunity to advance businesses to the next level and boost employment, making it sustainable for all stakeholders involved. Thus, a positive rating is applied.

19.4.1.5 Method

- *Differences between modular?*

  Although the process of 3D printing of construction components is different to traditional stick build of components, the actual onsite assembly of these prefabricated sections is very similar. Therefore, the major difference between modular construction and 3D printing of components is that the prefabrication of components is automated, which guarantees quality and precision in conjunction with cost and time efficiencies. Additionally, unlike traditional modular fabrication, 3D printing opens the door to liberated and unrestricted design. Thus, a positive rating is applied.

- *Are there any previous case studies?*

  There are currently no tangible case studies anywhere in the world which have been used for habitability purposes. Although simulated buildings have previously been constructed in China and Amsterdam, which used this 3D printing of components technique, these buildings were assembled to prove that this process is attainable and not for consumer use. Therefore, due to the limited knowledge and no prior reference points, a negative rating is applied.

- *Is this type of construction sustainable?*

  Yes, statistics shows that robotic dependency is becoming increasingly prevalent and that businesses are adopting this autonomous innovation. It proves that technology which simplifies tasks and removes human involvement appears to be the way of the future. Thus, by infiltrating this autonomous technology into the construction industry it has all the necessities to be a sustainable method of construction.
- *Is this method a faster method of construction?*

  Yes, as 3D printing of components is an autonomous adaptation to modular construction it results in the same time savings which are seen through parallel production. Additionally, further time savings could also be obtained by printing the components on-site which removes the need for transportation and any associated delays. Thus, a positive rating is applied.
- *Are there limitations?*

  Currently there are only three typical methods of 3D printing of components. This limitation can be detrimental in the fact that there may be restrictions in what shapes can be printed and at what detail. Additionally, this method is limited by the type of material. Depending on the materials properties this will dictate the shape and size of the component which can be printed. Thus, a negative rating is applied.
- *Can it be used for various types of buildings?*

  Yes, as this technique prints separate components which are assembled onsite, it can theoretically be used to print completely different shapes and sizes, creating flexibility, and freedom for various building construction. Whether for residential or commercial construction this method has the versatility to cater for a broad spectrum of needs and requirements. Thus, a positive rating is applied.

19.4.1.6 Machine

- *Is it cost-effective?*

  Although the 3D printer itself would be a hefty expense, this investment would not come unrewarded. Within a short period of time the machine has the ability to reach break-even point through the cost savings found in the labor reduction, time savings found in the reduction in construction duration and also the ability to expand market share and broaden the targeted construction jobs through the technological advancement in component production. Thus, a positive rating is applied.

- *Is it a reasonable method of construction?*

  Yes, although the machine is limited by the materials which it can extrude, 3D printing is still a reasonable method of component fabrication due to its ability to extrude cement. As cement is one of the most widely used construction materials throughout Australian construction industry, there will be always be a need for these components. Additionally, as this method of fabrication becomes more prominent and reputable in the industry the demand will, in turn, increase making it more than reasonable method of construction. Thus, a positive rating is applied.

- *What do future advancements look like?*

  The innovative nature of this technology is already considered to be futuristic; however, due to it being in its initial stages, alterations and revisions will be made throughout its lifecycle. This development and adaptation will encompass any problems or improvements which may be required to develop the optimum machine possible. Moreover, realistic future advancements which are possible are greater flexibility in component printing whether through new methods, accessibility to more materials and even making it operational via green renewable energy. Thus, from these advancements a positive rating is applied.

- *Can it be used for various types of buildings?*

  Yes, as this machine has the ability to print a wide range of components it is possible that these can be used for various types of buildings. This technology will not only be used for newly constructed buildings but due to the digital modeling which is dictating the printing process, this machine has the ability to analyses and restore deteriorated sections on historical buildings and reprint exact replacements. Thus, a positive rating is applied.

- *Are there limitations?*

  Currently these 3D printing machines are limited by size and two-directional movement. Due to the printable components being restricted by the size of the printer this means that the larger the component requirement the larger the printer. Moreover, due to the limited methods of printing available, the nozzle movement is currently restricted to two-directional freedom, limiting the ability to print certain components. Thus, a negative rating is applied.

- *Are they currently available?*

  The innovativeness of this technology is making purchasing a difficult process. At present there are a limited number of 3D printer manufacturers, restricting production and supply. However, as these machines increase in demand the supply should presumably follow. Thus, due to the limited number of machines available negative rating is applied.

As a result, the Fishbone Diagram analytical tool allowed for an in-depth evaluation to be carried out. From this analysis six critical impact categories were able to be reviewed via another six major segments of each category. The responses to each segment are statistically summarized by the positive or negative impact and tabulated in Table 19.1.

By illustrating this statistical summary via a graph, Fig. 19.2, it is visible as to which category will have a positive impact and which will have a negative impact. Thus, environment, people, method, and machine all have the potential to be positively impacted, policies to be neutrally impacted and materials negatively impacted.

Ultimately, through the Fishbone Diagram analysis of six major segments which are associated with 3D printing of construction components it has been derived that the viability level which may be obtained upon release into the Australian market is 58.33%. Although difficult to precisely determine the effect of such a technology on the local and global segments, this investigation shows that according to the reviewed literature and the rational observations, 3D printing of construction components has the ability to have an overall positive impact on the Australian construction industry and national economy.

Table 19.1 Statistical summary

Category	**Yes/positive**	**No/negative**	**Percentage of category viability (%)**
Environment	4	2	66.67
Policies	3	3	50
Materials	2	4	33.33
People	4	2	66.67
Method	4	2	66.67
Machine	4	2	66.67

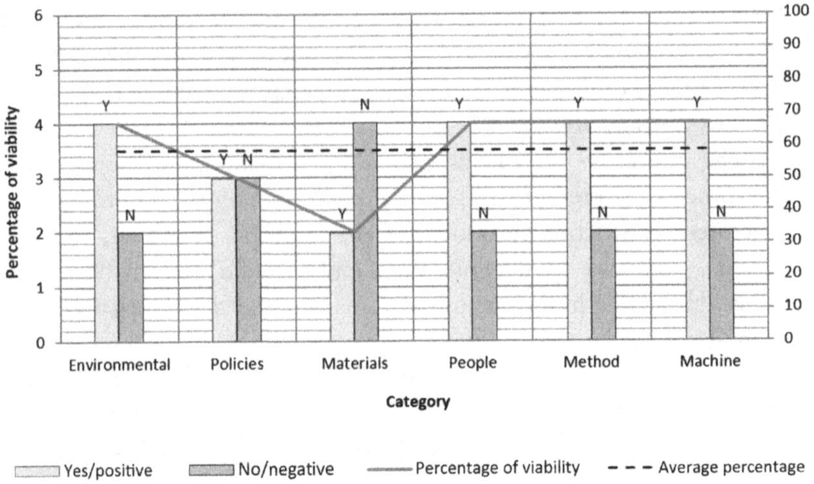

Figure 19.2 Viability of results via statistical format.

19.5 CONCLUSIONS

3D printing is an innovative technological advancement which is quickly overtaking many forms of manufacturing and is now seeping into the industry of construction. The basis of this technology allows for digitally designed 3D objects to be autonomously printed via robotic control, removing the need for the majority of human involvement, thus, capable of changing the entire construction process as currently known.

Since the construction industry is of high importance for the Australian employment segment and its overall economic position, it is essential that this industry is carefully assessed prior to any major changes. Therefore, since 3D printing is in its initial stages in the construction industry, its potential impact on future stability, national employment and economic solidity is unknown and, thus, this chapter reviews various literatures in an attempt to identify all local and global impacts and associated consequences which 3D printing may have on this industry and national economy.

As a result, a review has been undertaken via the Fishbone Diagram analysis tool to determine all the possible causes and effects of this advanced technology. Through the review of six major segments which are associated with 3D printing of construction components it has been derived that the environment, people, method and machine all have the potential to be positively impacted, policies to be neutrally impacted and materials negatively impacted.

Ultimately, the anticipated viability level which may be obtained upon release into the Australian market was calculated to be 58.33%. Although difficult to precisely determine the effect of such a technology on the local and global segments, this investigation shows that according to the reviewed literature and the rational observations, 3D printing of construction components has the ability to have an overall positive impact on the Australian construction industry and national economy.

REFERENCES

[1] B. Khoshnevis, Automated construction by contour crafting-related robotics and information technologies, Autom. Constr. 13 (1) (2004) 5–19.
[2] T. Tampi, D-Shape to 3D print entire luxury estate in New York, <http://3dprintingindustry.com/2015/06/07/d-shape-to-3dprint-entire-luxury-estate-in-new-york/>, 2015.
[3] C. Gosselin, R. Duballet, P. Roux, N. Gaudillière, J. Dirrenberger, P. Morel, Large-scale 3D printing of ultra-high performance concrete–a new processing route for architects and builders, Mater. Des. 100 (2016) 102–109.
[4] I. Perkins, M. Skitmore, Three-dimensional printing in the construction industry: a review, Int. J. Constr. Manage. 15 (1) (2015) 1–9.
[5] R. Rael, V. San Fratello, Developing concrete polymer building components for 3D printing, in: Proceedings of the 31st Annual Conference of the Association for Computer Aided Design in Architecture, October 13–16, 2011, pp. 152–157.
[6] A. Rutkin, Watch as the world's first 3D-printed house goes up, 2014.
[7] I. Hager, A. Golonka, R. Putanowicz, 3D printing of buildings and building components as the future of sustainable construction? Procedia Eng. 151 (2016) 292–299.
[8] E. Krassenstein, D-Shape looks to 3D print bridges, a military bunker, and concrete/metal mixture, <http://3dprint.com/27229/d-shape-3d-printed-military/>, 2014.
[9] K. Ishii, B. Lee, Reverse fishbone diagram: a tool in aid of design for product retirement, Proceedings of the 1996 ASME Design Technical Conference (1996).
[10] G. Ilie, C.N. Ciocoiu, Application of fishbone diagram to determine the risk of an event with multiple causes, Manage. Res. Practice 2 (1) (2010) 1–20.
[11] J. Corr, Cause and effect analysis using the Ishikawa Fishbone & 5 whys, City Process Management, 2008.
[12] A.M. Blayse, K. Manley, Key influences on construction innovation, Constr. Innov. 4 (3) (2004) 143–154.
[13] M. Loosemore, Why the construction industry needs to innovate, <https://sourceable.net/looking-over-the-horizon-why-the-construction-industry-needs-to-innovate/>, 2015 (viewed 08.05.17).
[14] D.M. West, What Happens if Robots Take the Jobs? The Impact of Emerging Technologies on Employment and Public Policy, Centre for Technology Innovation at Brookings, Washington, DC, 2015.
[15] S. Baynes, 3D printing and the construction industry. Housing Observer, Article 3, 2015.
[16] A. Pîrjan, D.M. Petrosanu, The impact of 3D printing technology on the society and economy, J. Inf. Syst. Oper. Manage. 1 (2013).

INDEX

Note: Page numbers followed by "*f*" and "*t*" refer to figures and tables, respectively.

A

G

R

T

U

V

W

X

Y

Z

Printed in the United States
By Bookmasters